Secondary Electron Energy Spectroscopy in the Scanning Electron Microscope

SECONDARY ELECTRON ENERGY SPECTROSCOPY IN THE SCANNING ELECTRON MICROSCOPE

Anjam Khursheed
National University of Singapore

World Scientific

EW JERSEY • LONDON • SINGAPORE • BEIJING • SHANGHAI • HONG KONG • TAIPEI • CHENNAI • TOKYO

Published by

World Scientific Publishing Co. Pte. Ltd.
5 Toh Tuck Link, Singapore 596224
USA office: 27 Warren Street, Suite 401-402, Hackensack, NJ 07601
UK office: 57 Shelton Street, Covent Garden, London WC2H 9HE

British Library Cataloguing-in-Publication Data
A catalogue record for this book is available from the British Library.

SECONDARY ELECTRON ENERGY SPECTROSCOPY IN THE SCANNING ELECTRON MICROSCOPE

ISBN 978-981-122-702-8 (hardcover)
ISBN 978-981-122-703-5 (ebook for institutions)
ISBN 978-981-122-704-2 (ebook for individuals)

For any available supplementary material, please visit
https://www.worldscientific.com/worldscibooks/10.1142/12010#t=suppl

Desk Editor: Rhaimie Wahap

Dedications

My first dedication goes to Alan Dinnis, who supervised my PhD between the years 1979 to 1983. It has now been over 40 years since Alan introduced me to the topic of secondary electron energy spectroscopy inside the scanning electron microscope, and we went on to design several electron energy spectrometers together. Alan's ingenuity, insight, and openness, inspired me throughout my academic career, and he was the reason why I subsequently spent over four decades of my life solving problems related to my original PhD thesis topic. He was the founding inspiration behind much of the work reported in this book.

The second dedication goes to my father, Khursheed Mustafa. His exemplary spirit of hard work, independent mindedness, and confidence in problem solving contributed greatly towards my establishing a career in scientific-engineering research.

The third dedication goes to my two daughters, Sarah and Elettra. You are the new generation, the future lies before you, take your time, and enjoy the journey.

Preface

Electron energy spectroscopic techniques are widely used in materials science research to study electronic structure and atomic/molecular dynamics. They include highly successful analytical methods such as X-ray photoelectron spectroscopy (XPS) and Auger electron spectroscopy (AES). At the same time, imaging instruments like electron microscopes also play a critical role in materials science research and development. Some focused electron beam spectromicroscopy analytical techniques, such as electron energy loss spectroscopy (EELS), straddle the border between both spectroscopy and microscopy, bringing together the two powerful disciplines of transmission electron microscopy (TEM) and atomic energy loss spectroscopy. EELS is now an indispensable materials science research tool for a wide variety of different applications.

This book deals with another focused electron beam spectromicroscopic technique, one which involves the scanning electron microscope (SEM) and the energy spectroscopy of its low energy scattered electrons (0–50 eV), known as secondary electrons (SEs). Unlike EELS, SE energy spectromicroscopy in the SEM is relatively unknown. The main purpose of this book is to demonstrate that it has the potential to become another powerful spectromicroscopic analytical tool for materials science research.

Ironically, SE energy spectroscopy inside SEMs is not a subject that was proposed recently, within say the last decade or so, but was started over 50 years ago. A summary of its circuitous historical development is outlined in chapter 1 of this book, but an important overall observation can be made here. One of the main obstacles preventing the development of SE energy spectroscopy inside the SEM has been the problem of how to integrate an electron energy spectrometer into its objective lens/specimen stage assembly. SEM manufacturers have as yet, not made suitable SE electron energy spectrometer accessories available, and almost all attempts to quantify contrast mechanisms in the SEM so far, have relied upon the use of in-built SE detector systems, as supplied by SEM manufacturers. Very few researchers have attempted to build and test out their own SE energy spectrometer attachments inside SEMs. The important question of what can or cannot be achieved by using a SE energy spectrometer attachment inside a SEM is a question that has therefore, not been answered. One of the main goals of this book is to demonstrate how useful materials science information can be obtained by SE energy spectromicroscopy in the SEM, once the problem of finding a suitable electron energy spectrometer attachment for it is solved.

There are naturally many questions that need to be answered. In what way do the high vacuum (HV) conditions in a SEM specimen chamber limit the kind of material information that can be acquired? For the other focused electron beam energy spectroscopic techniques in materials science, such as AES and XPS, ultra-high vacuum conditions (UHV) are required. Which is the best electron energy spectrometer design for capturing the SE spectrum? Can SE energy spectrometers be designed for different types of SEM objective lenses/detection systems? What kind of applications in materials science are best suited for SE electron energy spectroscopy? How does the presence of surface ad-layers such as native oxide or electron beam induced contamination change the SE energy spectrum? How does SE energy spectroscopy compare with other SEM analytical methods commonly used in the SEM, such as energy dispersive X-ray spectroscopy (EDS)? This book sets out to answer some of these questions.

The following pages are aimed at scientists, engineers, and graduate students that come from a wide range of different disciplines, and who, for one reason or another, need to use the SEM. The book is deliberately written in a non-specialist way where key concepts are mainly conveyed in the context of acquiring experimental SE energy spectra in the SEM for different applications. There are however, places where details about electron energy spectrometer and SEM design are discussed. This occurs in chapter 1, which summarizes past attempts to carry out SE energy spectroscopy in the SEM, and in chapter 6, where new designs for an energy spectrometer attachment are given as part of future work. The purpose of doing this is to provide more details to SEM engineers and applied physicists, so that the task of integrating SE energy spectroscopy into future SEMs is made easier.

Lastly, I would like to acknowledge help from my research group and colleagues. Many of the experimental electron energy spectral results presented in this book were obtained by my former student, Han Weiding, in the course of his PhD work, and then continuing afterwards, as one of my postdoctoral researchers. His PhD work included all the signal-to-noise data and comparison of experimental spectra with theoretical distributions in chapter 2, the SE energy spectral analysis of different metal samples in chapter 3, and the characterization of doped semiconductor test samples and devices in chapter 4. He then went on as a postdoctoral researcher to carry out experiments monitoring the deposited charge on an intrinsic silicon sample by a focused gallium ion beam, reported at the end of chapter 4, and the monitoring of local surface charge distributions on a patterned Metal-Insulator-Semiconductor (MIS) sample, which he designed. This is summarized in chapter 5. More recently, he carried out experiments which were able to acquire bulk density of states information of metal samples, a breakthrough discovery for SE energy spectroscopy inside a SEM, reported at the end of chapter 3. All these results were the fruit of a long period of sustained experimentation, investigation, and collaboration, lasting well over 7 years. Weiding also provided the invaluable help of carefully reading, checking and editing this book. Thank you Weiding for your steadfast cooperation and support. This book would certainly not have been possible without you.

There were two other former PhD students/postdoctoral researchers that greatly contributed to the work reported in this book. The first was Hung Hoang, who produced the first prototype of my wide-angle toroidal energy analyzer design. He obtained the first initial experimental results from it sometime between the years 2008 and 2009. The second was Avinash Srinivasan, who subsequently developed several new applications for the energy analyzer, and found ways to make its experimental spectral signals more reliable. Contributions by both Hung and Avinash are summarized in chapters 2 and 5.

Thanks go to Amit Banerjee, a former postdoctoral fellow in my group, who helped with some electron energy spectral experiments and sample designs, and to one of my research assistants, Matthew Chew, who ran Lorentz-2D software simulations for the redesign of my wide-angle toroidal energy spectrometer, reported at the end of chapter 6.

Beyond my immediate research group, there were significant contributions from other colleagues. In that category, I would first like to thank Filip Mika working in the ISI research Institute, Brno, Czech Republic. He supplied us with very useful semiconductor test samples containing striped doped patterns which we used extensively. Experimental results obtained from them are reported in chapter 4. Filip also came on regular visits to our research lab in order to assist us with our experiments. Thanks go to Milan Matějka, also working in the ISI research institute, who fabricated patterned MIS test samples for mapping primary beam induced charging in the SEM and helped us carry out experiments using them, reported in chapter 5. I would also like to thank Cornelia Rodenburg working in the University of Sheffield, England, for our many fruitful discussions on the topic of SE energy spectroscopy in the SEM, and especially for providing me with relevant reference papers on the topic, too many to acknowledge here one by one. Thanks go to Toshi Agemura working for Hitachi High-Tech in Japan, who supplied me with reference information about his papers on quantitative dopant contrast, and for his willingness to start a collaborative project on SE energy spectroscopy for SEMs between my research group at the National University of Singapore and Hitachi High-Tech in Japan. Finally, I would like to thank Theo Zouros, working in the University of Crete, Greece, and Alberto Tagliaferri working in the Politecnico di Milano, Italy, for our many useful discussions on the subject of electron energy spectroscopy and electron energy spectrometer design. Theo visited Singapore on a couple of occasions to assist us with our energy spectrometer designs.

Anjam Khursheed
National University of Singapore

13 July 2020

Contents

1. SE Energy Spectrometer Design 1
   - 1.1 The first SEM SE energy spectrometer proposal 14
   - 1.2 Voltage contrast spectrometers 19
   - 1.3 In-lens multi-channel position sensitive SE energy analyzer designs 39
   - 1.4 Retarding field energy analyzer attachments 53
   - 1.5 Energy filtered SEM SE detector systems 60
   - 1.6 Band-pass electric toroidal sector spectrometer attachments 72

2. The Wide-angle Toroidal Energy Analyzer Attachment 83
   - 2.1 The experimental setup and modes of operation 83
   - 2.2 Analyzer energy resolution and shot-noise considerations 87
   - 2.3 Comparison of experimental SE energy spectra to theoretical analytical energy distributions 102
   - 2.4 Surface field effects and voltage contrast 111

3. Quantitative Material Contrast 125
   - 3.1 Quantitative material contrast using SE energy spectral peak shape for elemental metal samples 127
   - 3.2 Enhanced material contrast through tracking gradient variations in the spectral signal rising edge 131
   - 3.3 Material contrast through full-range energy spectra 133
   - 3.4 Quantitative material contrast mapping for close atomic numbers 140
   - 3.5 Detection of the Auger hydrocarbon contamination peak in the SEM 154

3.6 Mitigating contamination effects by spectral background normalization 156
3.7 Gallium ion beam generated SE energy spectra on metals 158
3.8 SE spectral fine features and the bulk density of states 164

4. Dopant Profiling and Semiconductor Characterization 179
4.1 Doped stripe test samples 182
4.2 SE energy spectral signal shape dependence on probing position, primary beam voltage, and native oxide thickness 185
4.3 Dopant contrast measurements from p-stripes of different concentrations 187
4.4 *In-situ* estimate of native oxide thickness and surface band-bending energy 202
4.5 Contamination effects 205
4.6 Lateral dopant profiling across doped stripe edges 208
4.7 Dopant concentration measurements of bulk doped silicon wafer samples 214
4.8 Solar cell dopant profiling 221
4.9 Dopant mapping of intrinsic silicon irradiated by a gallium ion beam 229

5. Probing and Mapping Charge Distributions 239
5.1 Monitoring localized charging in metal-insulator-semiconductor (MIS) samples 245
5.2 Probing and analyzing buried charge at multifunctional oxide interfaces 258

6. Conclusions and Future Work 265
6.1 Conclusions 265
6.2 Future SE energy spectral experiments 267
6.3 Future instrument improvements 274

Appendix 1 The Electric Radial Mirror Analyser (RMA) Attachment Design 293

Appendix 2 The Parallel Radial Mirror Analyser (PRMA) Attachment Design 299

Appendix 3 TCAD Simulation of Test Semiconductor Samples with *p*-doped Stripes in *n*-substrate 303

Bibliography 307

Index 325

Chapter 1

SE Energy Spectrometer Design

The Scanning Electron Microscope (SEM) is a high resolution (nm) imaging instrument that is indispensable for research and development in a wide variety of different subjects, such as biology, geology, forensic science, industrial manufacturing, medicine, materials science, and the failure analysis of semiconductor devices (Reimer, 1998a; Goldstein *et al.*, 2017; Ul-Hamid, 2018; and McVicar *et al.*, 1997). It functions by generating and directing a focused energetic primary electron beam (1 to 30 keV) to strike a specimen, raster scanning it, and capturing some of the electrons that scatter back in order to form an image of the specimen's surface topography.

As the primary beam enters the specimen, it loses energy through a complex cascade of collisions occurring inside an "interaction volume", generating electrons and photons, some of which are able to leave the specimen surface, as illustrated in Fig. 1-1a. The low energy scattered electrons, known as secondary electrons (SEs), have a well-defined peak in the energy spectrum, as shown in Fig. 1-1b. Surface topographic imaging, which is the main imaging mode of a SEM, relies on the detection of these low energy electrons, usually defined to lie within the 0 to 50 eV energy range, where the upper limit is arbitrary, taken for convenience.

In the higher portion of the scattered energy spectrum, there is another peak, also shown in Fig. 1-1b, which is formed by primary electrons that scatter more elastically back from the sample, known as Backscattered Electrons (BSEs). The BSE energy range is usually defined by the remaining portion of the scattered electron energy spectrum, from 50 eV to the primary beam energy. Since the number of BSEs scattered from a specimen is highly dependent on its atomic number, BSEs are used in SEMs to provide a material contrast image of the specimen's surface, in which regions of high atomic number appear bright, and are distinguishable from regions of low atomic number, which appear dark.

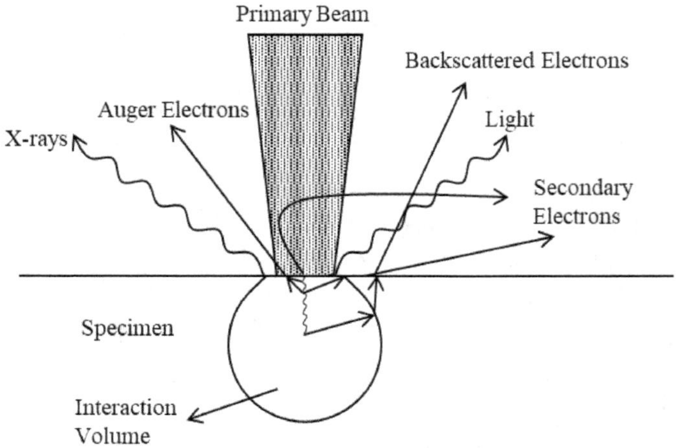

(a)

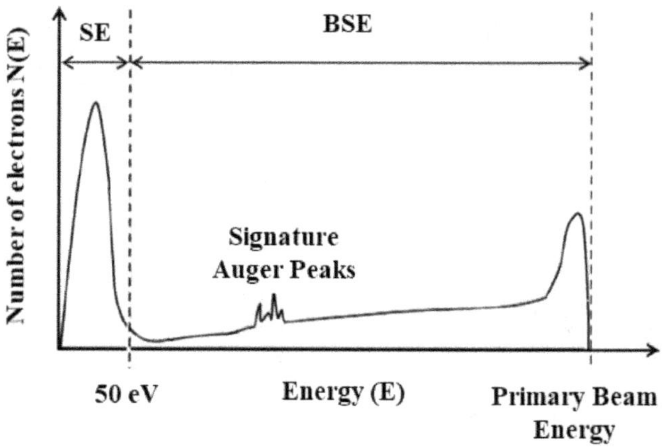

(b)

Fig. 1-1. Primary beam interaction with the sample inside a SEM. (a) Primary beam penetration into the sample. (b) Scattered electron energy distribution.

Auger electron spectral peaks embedded in the scattered electron energy spectrum (see Fig. 1-1b) are also present, but the relatively poor vacuum level of the SEM specimen chamber, (10^{-6} to 10^{-5} Torr) in combination with the primary beam irradiation creates a surface carbon contamination layer that masks their presence. SEMs are therefore usually not able to detect the presence of Auger electrons. Detection of Auger electrons typically requires ultra-high vacuum (UHV) conditions ($< 10^{-10}$ Torr), around five orders of magnitude better than the high vacuum (HV) conditions present in the specimen chamber of an SEM. Operating in HV conditions, and avoiding the need to maintain UHV levels inside the SEM specimen chamber has important technical advantages. It means, for instance, that the down time required for changing the specimen is several minutes to an hour, instead of the tens of hours required for UHV systems, and it gives much greater flexibility on the kinds of materials that can be used inside the SEM chamber, materials that would otherwise be unsuitable in a UHV system due to issues such as out-gassing. On the other hand, it also means that surface contamination is an inherent feature of SEM imaging.

The two main electron detectors inside a conventional SEM, the secondary electron (SE) detector for topographic imaging, and the Backscattered electron (BSE) detector for material contrast imaging, are positioned close to the lower pole-piece of the final lens inside the specimen chamber, as shown in Fig. 1-2a. The BSE detector usually lies above the specimen, while the SE detector is typically located to one side of it. Although the shape of the scattered electron energy spectrum contains valuable information about the sample, that information is not usually captured in a SEM, since the SE detector signal is formed from scattered electrons that have a wide range of different emission energies and angles. Information about the initial energy distribution is therefore averaged out.

There are several problems involved with fitting in an electron energy analyzer attachment into the SEM chamber in order to capture the SE energy spectrum. One of the most important requirements for the energy analyzer attachment is that it must be small. Figure 1-2b is a schematic diagram illustrating the limited space around the specimen. The distance between the final lens lower pole-piece and the specimen, the working distance, usually ranges between 3 to 20 mm, and for high resolution, it should be as small as possible. This means that the analyzer attachment height should typically be less than 20 mm. Another important requirement for the analyzer is that it should have a wide angular range of collection and detection, since the scattered electrons from the specimen travel out in all directions. Ideally, the analyzer should have a 2π radian azimuthal angle collection/detection range, enabling it to accept and detect electrons emitted over a wide range of angles. This is because gains in scattered electron transport efficiency translate directly into gains in output spectral signal-to-noise ratio.

The complications of integrating a SE energy analyzer attachment into the higher spatial resolution through-the-lens SEM objective lens designs that expose the specimen to electric/magnetic fields, are much more challenging than for the conventional objective lens design shown in Fig. 1-2, since the space above or inside their objective lenses is not usually accessible (Khursheed, 2011a). Instrument difficulties of this kind help explain why SE energy spectroscopy has as yet, not been widely incorporated into SEMs.

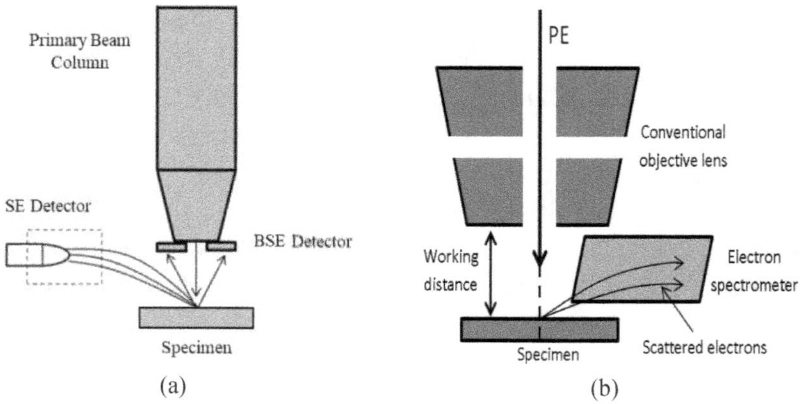

Fig. 1-2. Objective lens and specimen layout in a conventional SEM specimen chamber. (a) Position of electron detectors. (b) Space available for a SE energy analyzer attachment.

SE electron energy emission studies are usually carried out in specialized UHV chamber electron beam energy spectral systems. Most of these instruments do not have any imaging capability, and they typically consist of an electron gun which directs a beam of electrons at a specimen target, and an electron energy analyzer/detection system which captures the energy distribution of scattered electrons that leave the specimen surface. Unlike the primary beam of a SEM, their electron beam irradiation of the specimen is broad (unfocused) and usually not scanned. They also incorporate some form of *in-situ* cleaning of the specimen surface, such as bombarding it with a low energy beam of ions. In order to maintain UHV conditions inside their specimen chambers, additional ion pumps are required, and there are limitations on their overall size, as well as the type of materials that can be used within them. These instruments are therefore not general purpose instruments like the SEM, and are usually designed for a specific purpose. There have been a wide variety of different specialized electron beam energy spectral systems, each designed for a different application. Only a few of them have ever been specifically designed to capture SE energy spectra.

Figure 1-3a shows experimental SE energy spectra for 10 different metals obtained by Kollath (1947), reported around 80 years ago. His studies were made on one of the few UHV energy spectral systems designed specifically to capture SEs. The results shown in Fig. 1-3a indicate that the form of the SE energy distribution is highly asymmetric, where its rising edge is much sharper than its trailing edge. The well-known Chung and Everhart (1974) analytical SE energy distribution was later fitted and compared to this experimental data, *N(E)*, given by the following equation

$$N(E) = \frac{6W^2 E}{(E + W)^4} \tag{1.1}$$

where W is the work function of the sample and E is the SE energy.

Figure 1-3b shows how later experimental data taken for an aluminum sample from another UHV electron beam analyzer instrument closely agrees with that predicted by the Chung and Everhart model. This comparison was made by Cazaux (2011), who used experimental results reported by Bindi *et al.* (1979). Points on the rising edge can differ by up to around 20%, while the overall average deviation is around 5%.

An analytical model for semiconductors was also developed from SE energy spectral data generated by specialized UHV electron beam energy analyzer instruments. Figure 1-3c shows experimental and theoretically predicted SE energy spectra for a silicon sample taken here from a paper presented by Cazaux (2011), using an experimental spectrum reported by Shatalov *et al.* (1988). In this case, the analytical formula for semiconductor samples developed by Henke *et al.* (1979) is used for the comparison

$$N(E) = \frac{F'' E}{[(E + \chi - E_G)^2 (E + \chi)]^{5/2}} \tag{1.2}$$

where F'' is a normalisation factor, χ is the electron affinity and E_G is the bandgap energy of the semiconductor. The agreement between the experimental and theoretical spectra shown in Fig. 1-3c is on average, around 10%, with the error for some points on the rising edge and trailing edge of the spectrum reaching 30%.

Close agreement of SE energy experimental data derived from specialized UHV systems with Monte Carlo simulation methods has also been obtained. This has been reported for metals (Azzolini *et al.*, 2018), and doped semiconductors (Dapor *et al.*, 2008). In the case of doped semiconductors, Dapor *et al.* successively demonstrated that the change in shape of the SE energy spectrum can be understood in terms of an effective change in electron affinity χ.

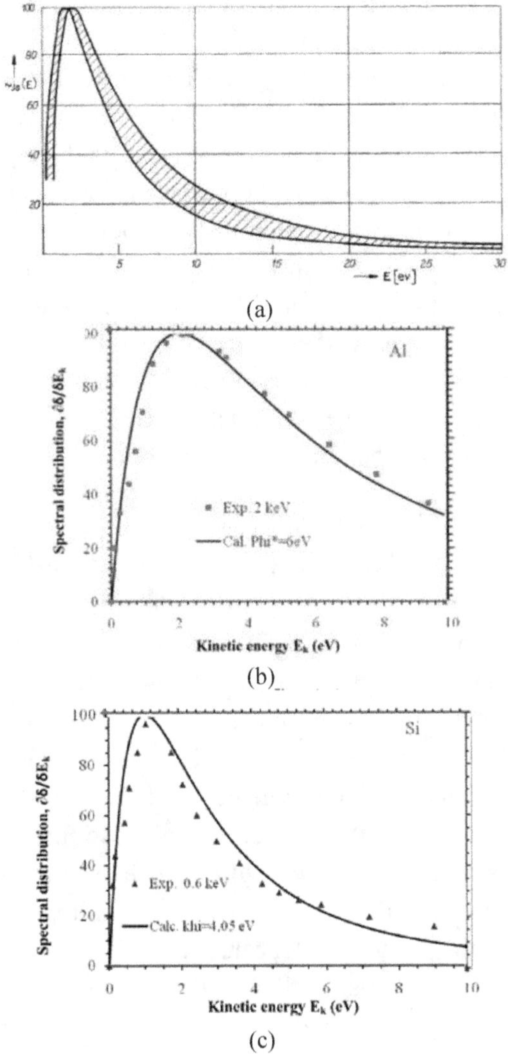

Fig. 1-3. Examples of SE energy spectral data obtained by UHV spectral systems. (a) Range of energy spectra measured for 10 different metals according to Kollath (1947). Copyright 1947 by Wiley VCH. Reprinted with permission. (b) Comparison of theory and experiment for an aluminium sample reported by Cazaux, (2011) where the original results were obtained by Bindi *et al.* (1979). Copyright 2011 by AIP. Reprinted with permission. (c) Comparison of theory and experiment of a silicon sample reported by Cazaux, (2011) where the original experimental results were obtained by Shatalov *et al.* (1988). Copyright 2011 by AIP. Reprinted with permission.

An exact match between the experimental and theoretical shapes of the SE energy spectra shown in Figs. 1-3b and 1-3c is not expected for a variety of different reasons. Firstly, the analytical formulae are based upon theoretical models that make various assumptions about SE generation and transport inside the specimen, and they are therefore inherently approximate. Secondly, the shape of the SE energy spectrum depends not only on specimen material type, but also on various SEM operating conditions, such as the primary beam voltage and vacuum level in the specimen chamber, which differ for each experimental setup. Thirdly, different analyzer systems have their own distinctive types of errors, explaining why only broad agreements between the SE energy spectral shape for the same materials were reported by different researchers (Hachenberg and Brauer, 1959).

In the case of the four-grid hemispherical retarding field spectrometer design, Lanteri *et al.* (1977) demonstrated that spurious peaks were generated by electron scattering on the electrodes inside the spectrometer, and only by reducing the number of grids (to two grids) and using AC modulated signals, could reliable results be obtained. Even for AES spectral systems having the same type of analyzer, inconsistent SE energy spectral results have been obtained. For instance, the SE energy spectral results obtained by a Perkin Elmer 660 scanning Auger microprobe (Joy *et al.*, 1999) were very different to the ones obtained by a PHI 680 AES system (Joy *et al.*, 2004) for the same type of bulk test samples, despite them both operating with a cylindrical mirror analyzer (CMA). The Perkin Elmer 660 scanning Auger microprobe system produced a shoulder in the rising edge of the SE energy spectrum at 2 eV for an aluminum sample, while the corresponding spectral signal of the PHI 680 AES system did not have any such feature. Moreover, the overall shape of the PHI 680 AES system spectral signal was a factor of over two times narrower than the signal predicted by the Chung-Everhart formula, given by equation 1.1. The most likely explanation for this is that different AES systems have their own way of calibrating Auger spectral peaks, which usually involves biasing the specimen, and this in turn generates local electric fields between the specimen and analyzer entrance, causing distortions in the trajectory paths of low energy electrons (SEs) as they travel through the energy analyzer. A special specimen holder in order to screen these fringe electric fields was found to partially reduce this effect in some instances (Mizuhara *et al.*, 2003 and Mizuhara *et al*., 2002). Dapor *et al.* (2008) on the other hand, were able to obtain a relatively good match between Monte Carlo simulations and experiment for the shape of the SE energy spectrum produced from doped silicon samples by using an AES hemispherical deflector analyzer (HDA) system, without any special adaptation.

In general, there are many factors which influence the overall shape of the SE energy spectrum. They not only include the complex cascade of inelastic/elastic collisions occurring inside the specimen, but also involve the primary electron beam's operating conditions, transportation of SEs through the analyzer to the detector, and the generation of background tertiary electrons on surrounding electrode surfaces and specimen chamber walls. All these kinds of effects make it difficult to accurately determine the intrinsic shape of a SE energy distribution for a given material. A breakthrough in the subject of SE energy spectroscopy came with the discovery that electronic energy band information could be extracted from fine structure features present in the SE spectral signal under certain conditions. Willis and Christensen (1978) successively demonstrated the technique of angle resolved secondary electron emission spectroscopy (ARSEES) in a dedicated UHV electron energy spectral system, a schematic drawing of which is shown in Fig. 1-4a. They used an electron gun unit to direct a low energy primary beam (around 100 eV) on to a target specimen, incident at 45°. The specimen was systematically tilted at different angles, and the SE energy spectrum was captured by a 130° band-pass electric field energy analyzer positioned vertically above it. The band-pass energy analyzer was found to be effective for acquiring fine spectral features, while an additional rotatable electric retarding field analyzer was used to provide a cross-check for broad spectral variations. An example of the SE energy spectral results obtained by Willis and Christensen is shown in Fig. 1-4b for a W (100) specimen. They show the presence of small fine spectral peaks which roughly correspond to peaks in a theoretically predicted density of states (DOS) distribution, shown in Fig. 1-4c.

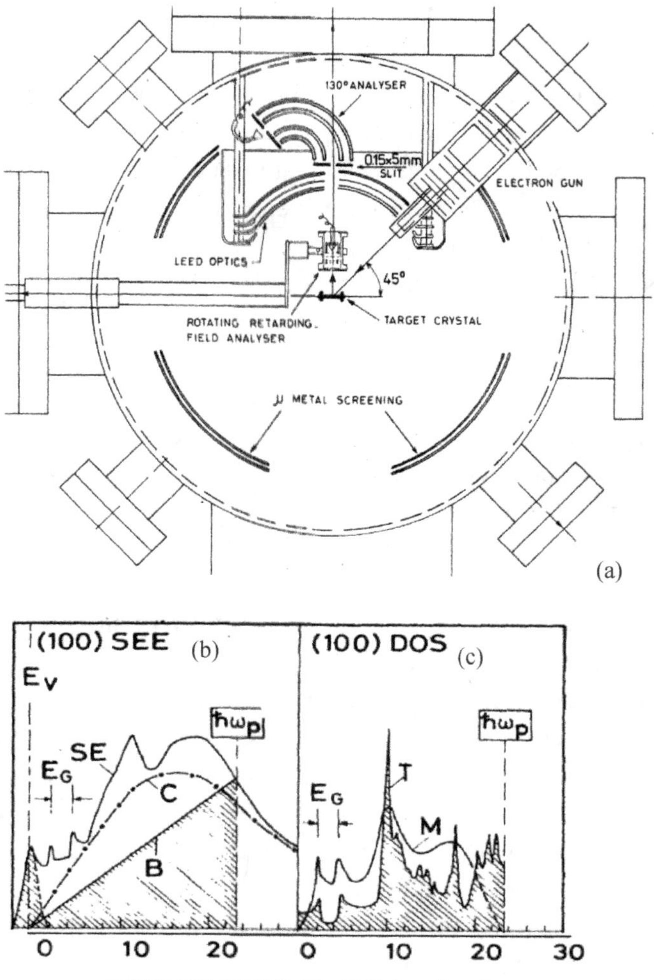

Fig. 1-4. Demonstration of angle-resolved secondary electron emission spectroscopy (ARSEES) for acquiring electronic band-structure information (Willis and Christensen, 1978). A 100 eV primary beam has an incident angle of 45° to the specimen. (a) Schematic drawing of the dedicated UHV electron energy spectral system. (b) Experimental SE energy spectrum for a W (100) test specimen containing fine structure spectral features. C represents a schematic presentation of the SE cascade. (c) Comparison with a density of states (DOS) prediction (T). The experimental signal (M) is obtained by subtracting out the background signal B. Copyright 1978 by American Physical Society Publisher. Reprinted with permission.

Willis and Christensen went on to obtain SE energy spectra for different specimen tilt angles, enabling them to acquire detailed energy-momentum electronic band-structural information. The technique was subsequently implemented by others (Shafer *et al.*, 1981), who, although confirmed some of the fine spectral structures obtained by Willis and Christensen, produced a very different background cascade signal shape. This difference was most likely due to the fact that they had developed their own dedicated UHV electron beam spectral system, using another chamber layout/energy analyzer design to the one used by Willis and Christensen. ARSEES is now a recognized discipline of surface science (Modinos, 1984), and has been used to study highly orientated pyrolytic graphite (HOPG) samples (Hoffman *et al.*, 1990), and graphene on nickel (Riccardi *et al.*, 2010). However, compared to the angle-resolved photon emission spectroscopy (ARPES) method (Suga and Sekiyama, 2014), it is not widely used in surface science studies for capturing electronic band-structure information.

Apart from the fine spectral features associated with acquiring electronic band energy information, some fine spectral features in the SE energy spectrum of aluminum have been identified with surface and bulk plasmon interactions (Henrich, 1973; Pillon *et al.*, 1976; Everhart *et al.*, 1976; Bindi *et al.*, 1979; Kuhr and Fitting, 1979; and Pierron *et al.*, 2017). These plasmon interaction effects are usually correlated to corresponding reflection electron energy loss spectroscopy (REELS) data. Werner *et al.* (2008) demonstrated that there are different component parts to the SE energy spectrum linked to bulk and surface plasmon peaks in the REELS data experimentally by using coincidence spectroscopy. Their results indicate that there is a surface plasmon effect centered around a SE energy of 6.2 eV, and a bulk plasmon effect centered around a SE energy of 10.6 eV. Monte Carlo simulations also predicted the existence of small fine spectral structures due to surface and bulk plasmon interactions (Kuhr and Fitting; 1979; Pierron *et al.*, 2017). The appearance of small humps in the tail end of experimental SE energy spectra for some metals have been correlated to Monte Carlo predictions (Ding *et al.*, 2008). However, overall, the experimental results appear to be inconclusive, a more recent experiment using a hemispherical deflector analyzer (HDA) did not confirm their existence and were significantly different to Monte Carlo predictions (Pierron *et al.*, 2017). One possible explanation for this may come from the different types of energy analyzer systems used.

The early SE energy spectral experiments were carried out using a four-grid hemispherical retarding energy analyzer (Henrich, 1973; Pillon *et al.*, 1976; Everhart *et al.*, 1976). Lanteri *et al.* (1976) later demonstrated the need to minimize artifacts in the SE energy spectrum by using two-grids instead of four grids for this analyzer design. Bindi *et al.* (1979) using the more accurate two-grid version of the analyzer noted that the surface plasmon shoulder located at 5.5 eV was only present for the lower primary beam incident energies of 150 and 300 eV, and was not present at 600 eV.

The HDA, used for the recent SE energy spectral results (Pierron *et al.*, 2017), differs significantly from the earlier energy analyzer systems, it is a band-pass energy analyzer. It has a significantly smaller energy and angular detection window. The HDA has entrance angular spreads of only a few degrees, one to two orders of magnitude smaller than the earlier hemispherical retarding analyzers. The much larger angular range in the case of the hemispherical retarding field energy analyzers may be important for observing plasmon effects. Another difference is that the primary beam incident angle to the specimen surface was 22.5° in the HDA experiment, while it was perpendicular to the sample in the hemispherical retarding field analyzer experiments. These kinds of differences are not taken into account in Monte Carlo simulations, which usually assume that all SEs leaving the sample surface are collected.

More work, both from a theoretical and experimental point of view is required to better understand the presence of plasmon effects in the SE energy spectrum of aluminum and the specific role played by the energy analyzer system.

Da *et al.* (2019) recently demonstrated the appearance of a relatively small shoulder in the SE energy spectrum rising edge at 2.3 eV from a polycrystalline Au film sample having alternating micro/nano-size single-phase grains, which they attributed to surface plasmon energy gain. Da *et al.* correlated this SE spectral fine structure to a peak occurring in the corresponding energy loss spectrum. Their experiments were carried out in UHV conditions inside a scanning Auger electron spectrometer SAM650 system using a primary beam energy of 10 keV, and it is one of the first demonstrations of SE energy fine spectral features at relatively high primary beam energies. Previous work reporting fine structure features in the SE energy spectrum was typically carried out with primary electron beam energies in the hundreds of eV range.

Just how precise and accurate SE energy spectral data in a SEM can be, is not something that has been investigated before. The vacuum conditions inside the specimen chamber of a SEM are more than three orders of magnitude poorer than those in UHV chambers, so significant differences are to be expected. The primary electron beam energy for most commercially available SEMs typically lies in the 0.5 to 30 keV range, much higher than the 100 eV primary beam energy used in some earlier dedicated UHV SE energy spectral systems. Moreover, the SEM is mainly an imaging instrument, not designed to acquire scattered electron energy spectral information.

Experimental results presented in this book will demonstrate that precise SE energy spectroscopy can be carried out in a SEM, and how it is first and foremost a problem of finding the right electron energy spectrometer attachment design. They will show that it is possible to capture both small changes in the SE energy spectrum shape and small fine structural features embedded within the spectrum under normal SEM operating conditions, enabling it to acquire valuable materials science information about the sample in a way that goes far beyond conventional SE imaging.

1.1 The first SEM SE energy spectrometer proposal

The idea of using SE energy spectroscopy in the SEM was first proposed around 50 years ago in a paper by Wells and Bremer (1968), shortly after the pioneering work that led to the SEM becoming a commercial instrument in 1965 (Oatley *et al.*, 1965 and McMullan, 2006). The paper proposed the use of an electron energy spectrometer attachment in the SEM to quantify certain image contrast effects that were not related to specimen surface topography, like the image contrast produced by a reverse-biased *p-n* junction sample. This effect was observed in the late 1950s on several occasions by the University of Cambridge (England) research group led by CW Oatley. An example of this is shown in Fig. 1-5, where a Germanium *p-n* junction device was reverse biased to 1.5 and 3 volts (Everhart *et al.*, 1959).

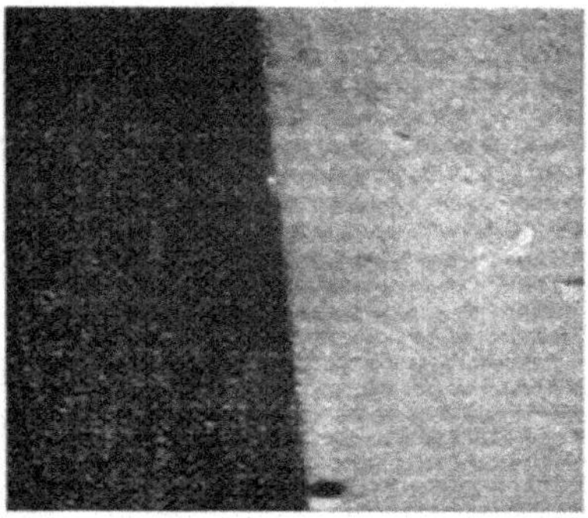

(a) Reverse bias 1.5 V, M = 2300

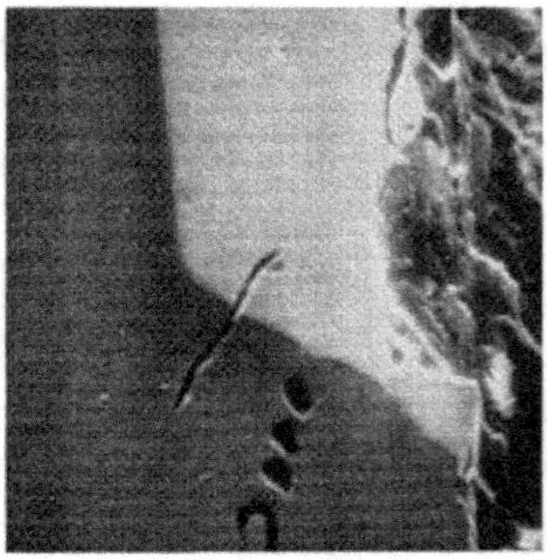

(b) Reverse bias 3 V, M = 1400

Fig. 1-5. Contrast effect observed in a reversed biased Germanium *p-n* junction by Everhart *et al.* (1959) (a) 1.5 V, magnification of 2300 (b) 3 V, magnification of 1400. Copyright 1959 by Taylor and Francis. Reprinted with permission.

It was later realized that the image contrast of a reversed biased *p-n* junction was actually produced by a combination of two separate contrast effects. The first of these effects, known as voltage contrast, relates surface potential to image brightness: regions on the specimen that have a negative potential appear brighter, whilst areas that have a positive potential appear darker. Figure 1-6a shows a lead on an integrated circuit sample brightening up due to it charging up negatively (Thornhill and Mackintosh, 1965). Dopant contrast was later observed separately in a passive (unbiased) integrated circuit planar transistor sample by Chang and Nixon (1967), shown in Fig. 1-6b. A difference in brightness is observed in the *n*-type collector, *p*-type base, and n^+-type emitter regions at different primary beam voltages, indicating that this contrast mechanism is dependent on semiconductor doping.

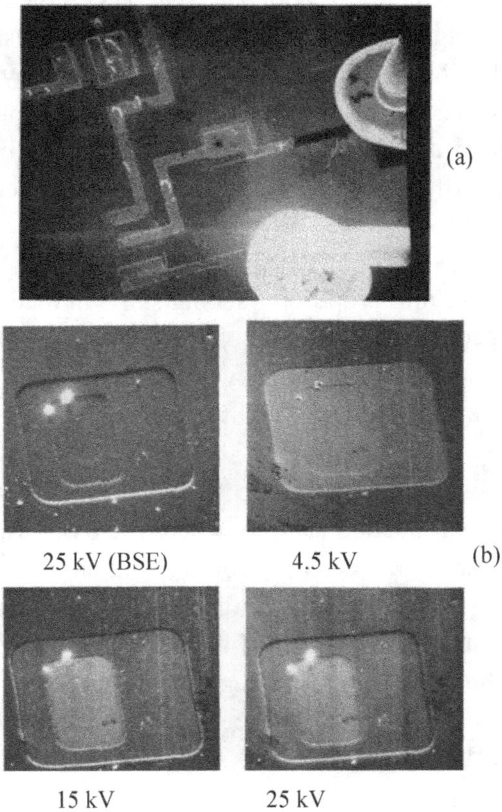

Fig. 1-6. Voltage and dopant contrast effects in the SEM. (a) Voltage contrast reported by Thornhill and Mackintosh (1965) of a lead brightening up by charging negatively on an integrated circuit sample. Copyright 1965 by Pergamon Press. Reprinted with permission. (b) Dopant contrast on a planar transistor by Chang and Nixon (1967) having *n*–type collector, *p*-type base, and n^+-type emitter regions. Copyright 1967 by Pergamon Press. Reprinted with permission.

Wells and Bremer (1968) installed a miniaturized Cylindrical Mirror Analyzer (CMA) design into a SEM which was orientated horizontally, to be perpendicular to the primary beam, as shown in Fig. 1-7a. Leakage electric fields from the analyzer defocused the primary beam, and the final energy resolution of the analyzer was relatively poor (3 to 5 eV), which had the effect of significantly broadening the shape of the SE energy spectrum. Wells and Bremer went on to develop an improved analyzer design which they reported a year later (Wells and Bremer, 1969), shown in Fig. 1-7b. The improved design consisted of a 63° cylindrical deflection analyzer (CDA) followed by an electron collimator. This design is simpler than their previous CMA one, predicted to give a higher transmission factor and produce less defocusing on the SEM primary beam, but no subsequent experiments were reported for it.

Wells and Bremer (1968) proposed tracking shifts in the peak position of the SE energy spectrum in order to quantify voltage changes that occur on the specimen surface, and quantify material contrast by monitoring SE energy spectrum shape and height changes for different materials. Figure 1-8a shows their recording of SE energy spectral data taken from gold and silicon samples. Figure 1-8b shows SE energy spectral signals from a passivated *pn* junction sample, where the oxide is removed from slot regions. The specimen consisted of (1) bare silicon, 2 Ωcm *p*-type, (2) bare silicon with phosphorus diffusion 3 μm deep, (3) as (1) but covered with 670 nm silicon dioxide, (4) as (2) but covered by 310 nm of oxide. Figure 1-8b shows that a change in the SE spectral signal peak height, its position, and overall shape was produced by using different primary beam voltages (5 and 20 kV) at different locations on the specimen. They indicate a consistent shift to the right for the spectral signals from the *n*-type regions compared to the ones from the *p*-type regions. There seems to be a SE peak signal height reversal effect between the passivated spectral signals when the primary beam voltage is changed from 5 to 20 kV: at 5 kV, the height of the SE spectral signal from the passivated *n*-type slot (4) is higher than the one from the passivated *p*-type slot (3), while at 20 kV, the situation is reversed, the height of the SE spectral signal from the passivated *p*-type slot (3) is greater than that of the one from the passivated *n*-type slot (4). Image contrast reversal effects were also obtained by varying the primary beam scan-rate. Wells and Bremer noted that although similar brightness contrast reversal effects were also observed in the normal SE detector image, in that case, the cause of the contrast was difficult to interpret. Their work therefore demonstrated how the extra information contained in the SE energy spectral signal was able to better understand and interpret complex dopant contrast/charging effects inside the SEM.

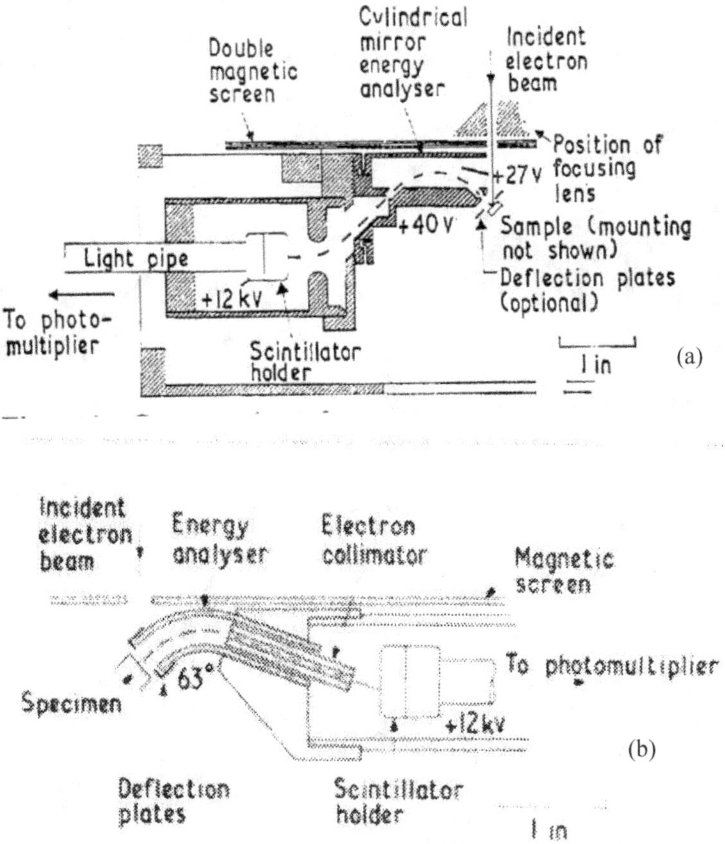

Fig. 1-7. The Wells and Bremer SE energy spectrometer attachment designs. (a) Initial CMA design (Wells and Bremer, 1968). Copyright 1969 by IOP Publishing. Reprinted with permission. (b) Improved 63° CDA design (Wells and Bremer, 1969). Copyright 1969 by IOP Publishing. Reprinted with permission.

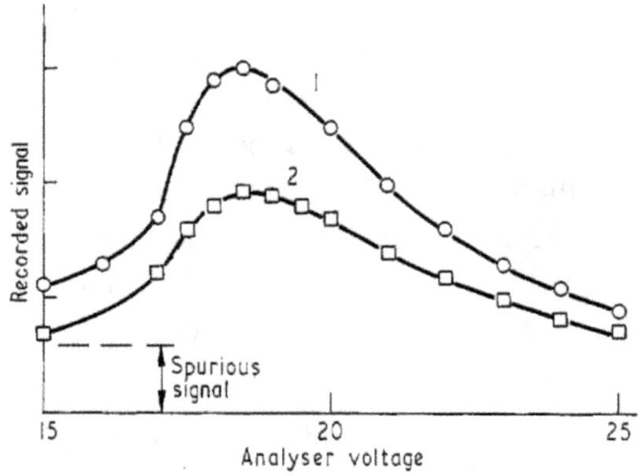

(a)

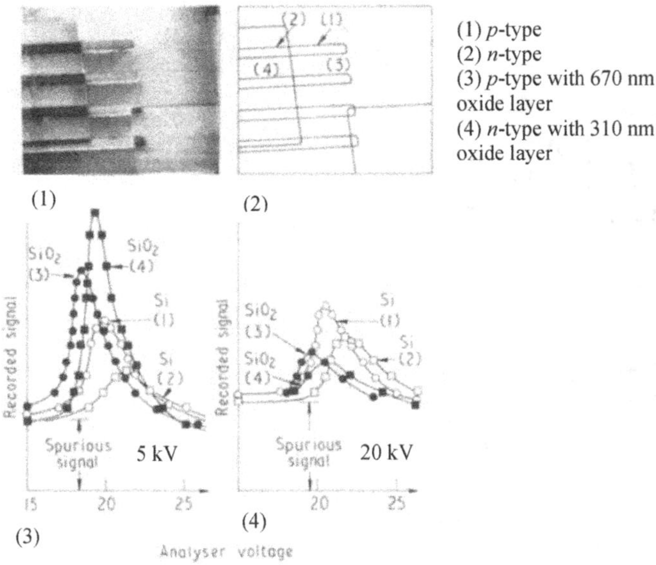

(b)

Fig. 1-8. Experimental SE spectral results reported by Wells and Bremer (1968). Copyright 1968 by IOP Publishing. Reprinted with permission. (a) 1 Gold; 2 Silicon, beam current of 30 nA at 5 kV. (b) Passivated *pn* junction with oxide removed from slot-shaped regions. (1) micrograph (2) identification of areas; (3) and (4) analyzer curves at 5 kV and 20 kV incident beam energy respectively. (Magnification as printed x 150; angle of incidence 45°). (1) Bare *p*-type (2) Bare *n*-type (3) *p*-type covered with a 670 nm thick oxide layer (4) *n*-type covered with a 310 nm oxide layer.

1.2 Voltage contrast spectrometers

It is interesting to note that although the main objective of the 1968 Wells and Bremer paper was to quantify voltage contrast, it did not present any experimental evidence to support it. Instead, they presented experimental results to illustrate the benefits of SE energy spectroscopy for quantifying material and dopant contrast in the SEM. The step of applying their proposal for quantifying voltage contrast however, was subsequently taken up by Hannah (1974), who improved the 63° CDA Wells and Bremer (1969) SE electron spectrometer design (shown in Fig. 1-7b), and used it to monitor shifts in the SE energy spectral peak position caused by voltages changing on the tracks of Integrated Circuits.

A schematic diagram of Hannah's analyzer design, and an example of the experimental SE spectral signals obtained from it, are shown in Fig. 1-9. The SE energy spectral signal was found to shift linearly with specimen voltage variations, as predicted by Wells and Bremer. Hannah re-orientated the spectrometer so the sample could be placed horizontally, and used an accelerating lens field above the specimen to counteract the effect of local transverse surface fields. By optimizing his design through ray tracing simulation, he was able to improve the analyzer voltage resolution to around 0.1 V.

Hannah proposed the use of a peak detection scheme to speed up data-acquisition times. One disadvantage of the Hannah analyzer design was that it had a relatively large working distance (65 mm), over twice the value that is normally used in conventional SEMs today. A large working distance of this kind naturally limits the SEM's ability to acquire high resolution images.

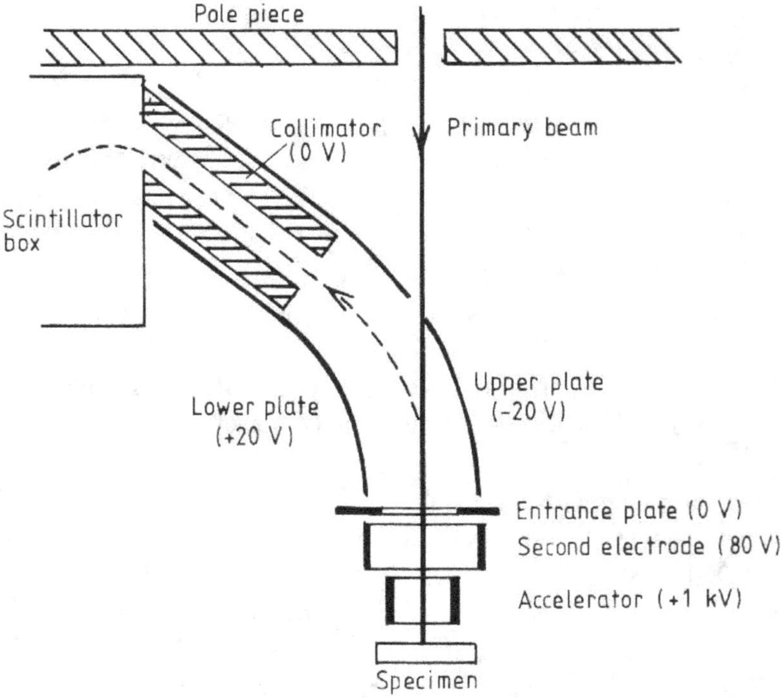

(a)

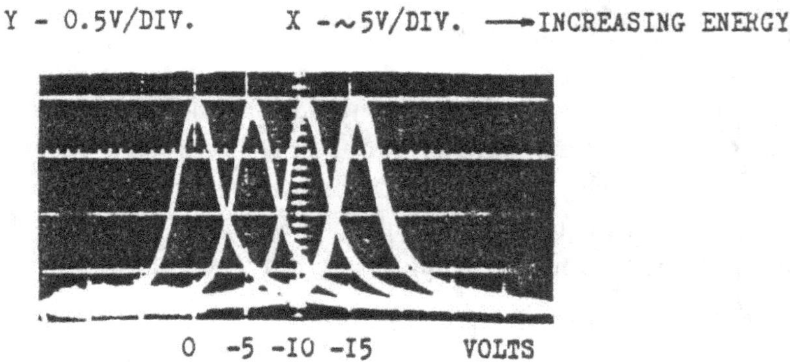

(b)

Fig. 1-9. The 63° CDA Hannah voltage contrast spectrometer (Hannah, 1974). Copyright 1974 by J. M. Hannah (PhD publication from University of Edinburgh). (a) Spectrometer layout. (b) Experimentally acquired SE spectra for different specimen voltages.

The voltage contrast spectrometer developed by Hannah (1974) was one of the very few that involved the generation of a SE energy spectral output signal. Subsequent voltage contrast spectrometer designs were of the retarding field energy analyzer type, whose principle of operation is depicted in Fig. 1-10. These spectrometers function by detecting only those secondary electrons that can surmount a potential barrier created within the analyzer, and are, in effect, high pass energy filters. The potential barrier is created by biasing a retarding electrode or grid to a voltage V_R. The strength of the potential barrier is given by V_R-V_S, where V_S is the specimen voltage. Only electrons having initial energies greater than $e(V_R-V_S)$ will surmount the potential barrier and contribute towards the output signal (represented by the shaded area in Fig. 1-10a). When V_R is varied, an integrated form of the secondary electron energy spectrum is obtained, as illustrated in Fig. 1-10b. This output signal is sometimes referred to as a "S-curve". Shifts in the output signal are therefore obtained when the specimen voltage changes (indicated by V_{S2} and V_{S1} in Fig. 1-10b). In order to help the secondary electrons escape from the specimen, an extraction field is applied by the grid V_E, which is biased positively with respect to the specimen voltage.

For quantitative voltage measurements, retarding field spectrometers are normally operated in the closed feedback loop mode, a method first suggested by Flemming and Ward (1970), where the strength of the potential barrier is adjusted to keep the output current constant, as illustrated in Fig. 1-10a. The output current is continually compared to an off-set current level. Adjustments in the retarding electrode voltage to keep the output current constant are therefore equal to specimen voltage changes, $\Delta V_R = V_{S2} - V_{S1}$. This feedback loop method of surface specimen voltage measurement has the merit of simplicity and is easily automated. It avoids the need to generate the SE spectrum and interpret it, but it has the disadvantage of discarding useful information within the SE spectrum (lower energy part) and suffers from the presence of unwanted background noise (generated from higher energy scattered electrons).

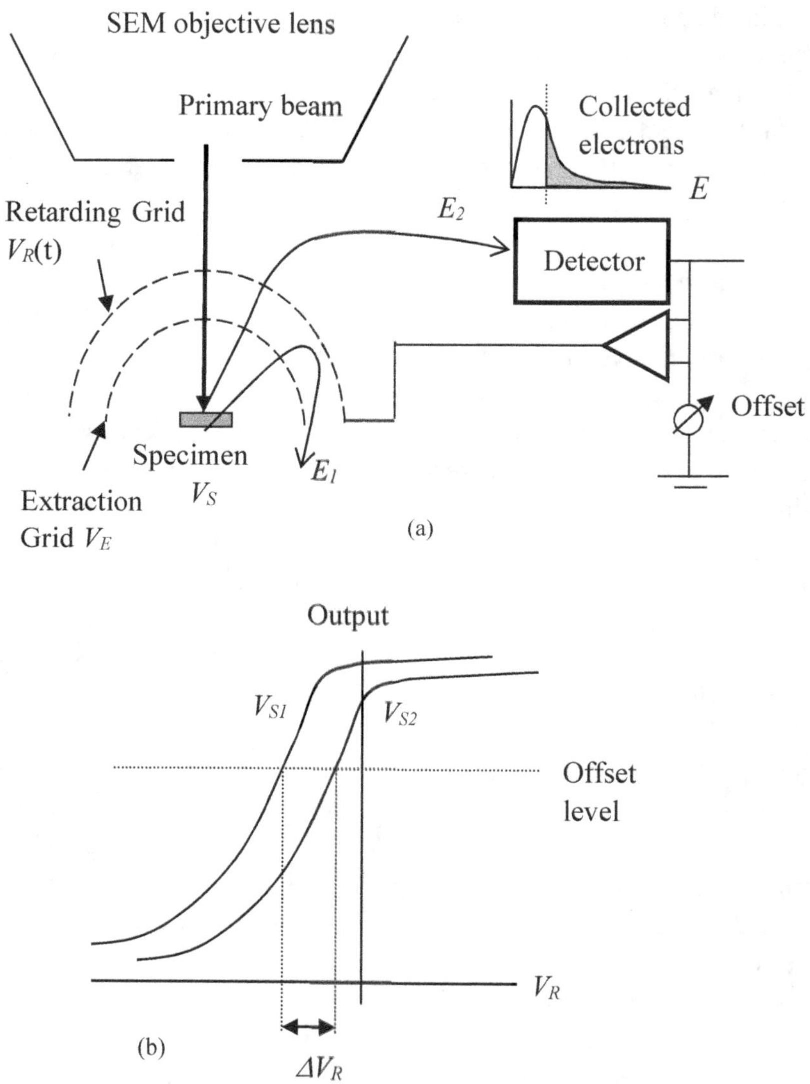

Fig. 1-10. Principle of closed loop retarding field spectrometers for voltage contrast. (a) Spectrometer layout. E_1 lies below the potential barrier and E_2 lies above it. (b) Output S-curve signals.

Two examples of electric field retarding field voltage contrast attachments designed for the SEM are shown in Fig. 1-11. They both function by having planar extraction and retarding field grids, and deflect SEs away from the electron optical axis by asymmetric electric fields on to a conventional SE detector. Both these spectrometers were simulated to have over 90% collection efficiency (Khursheed and Dinnis, 1984), and were designed to minimize interference from backscattered electrons. Examples of the experimental S-curves generated by these spectrometers are also shown in Fig. 1-11. These S-curves appear regular and continuous, and both spectrometers give similar experimental results. Although in principal, experimental S-curves of this kind can be differentiated in order to produce SE energy spectral data, this was rarely done, since it was not required in order to make quantitative voltage contrast measurements. The Feuerbaum spectrometer was simulated by Khursheed and Dinnis (1984) to have a collection efficiency of around 97% for all electrons below 10 eV (neglecting scattering from the grids). This high collection efficiency, along with its relatively short working distance of 10 mm, makes it one of the better performing spectrometers amongst the early retarding field voltage contrast analyzer designs.

For over two decades, from around 1970 to 1990, quantitative voltage contrast in the SEM was developed to measure signal waveforms propagating on the surface of integrated circuits. This world-wide effort eventually culminated in the creation of commercial instruments for the semiconductor failure analysis industry in the late 1980s, known as Electron Beam Testers (Thong, 1993a). These instruments were essentially specialized SEM systems designed to have voltage contrast spectrometers integrated into their final objective lenses. They achieved a small working distance by allowing secondary electrons to go up through the objective lens bore, above which electrodes/grids were installed to energy filter and efficiently deflect secondary electrons on to detectors placed off-axis. Blanker plates were also installed into the column so that the primary beam could be pulsed and synchronized to the signal waveforms propagating on the integrated circuit device under test.

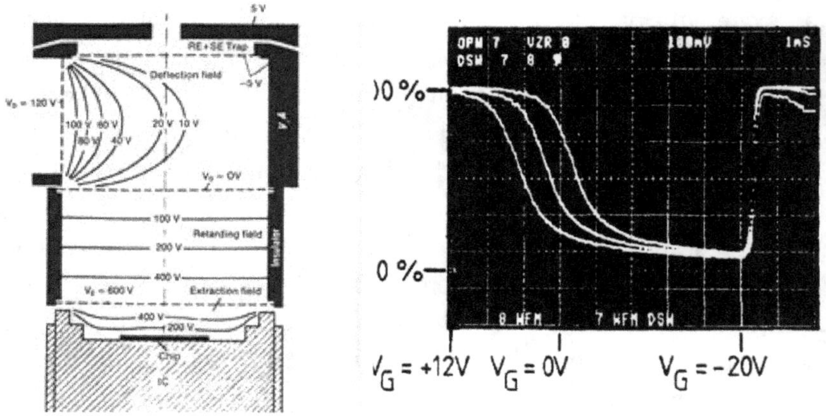

(a)

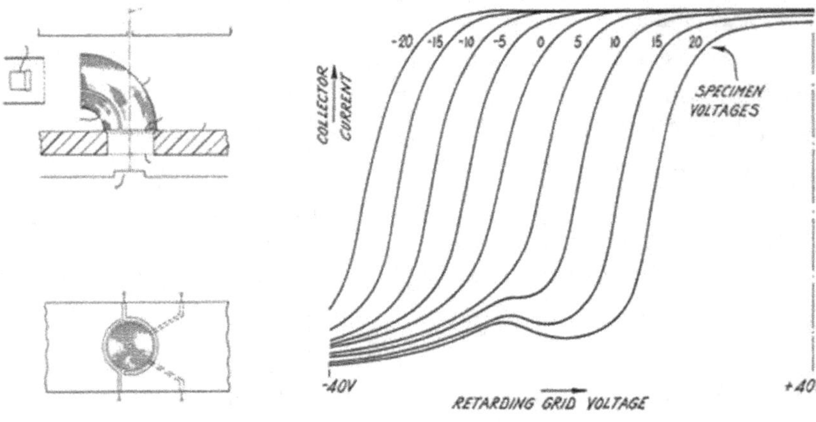

(b)

Fig. 1-11. Two high transport efficiency retarding field voltage contrast spectrometer attachment designs for the SEM and examples of their experimental S-curves. (a) Feuerbaum (1979). Copyright 1979 by IITRI, Chicago. Reprinted with permission. (b) Khursheed (1983a).

Two commercial Electron Beam Tester instrument designs are shown in Fig. 1-12. In the case of the mixed field lens Frosien and Plies (1987) design, shown in Fig. 1-12a, the optics of the secondary electrons ensures that they are accelerated by the extraction field and brought to a focus within the objective lens bore, spreading out radially and subsequently energy filtered by hemispherical shaped grids. The analyzer design combines the application of an extraction field above the specimen, a short working distance preserving high resolution optics for the primary beam, and high transmission of the secondary electrons.

In the case of the wide-bore immersion lens designed by Richardson and Muray (1988), shown in Fig. 1-12b, the specimen is immersed in the magnetic field of the objective lens, and the secondary electrons spiral up vertically through the lens bore after which they are energy filtered by a planar retarding grid and then deflected off-axis on to a SE detector. This voltage contrast spectrometer has better optics for the primary beam, since the specimen is immersed in a magnetic field, similar to the objective lenses of transmission electron microscopes, but this has the disadvantage of not being able to apply a strong extraction field above the specimen. The transmission of secondary electrons through the bore of the specimen is high since they are strongly collimated by the decreasing strength of the immersion objective lens magnetic field. If the SEs were to be accelerated by an extraction field, the collimation effect would be greatly reduced (Kruit and Read, 1983).

The two commercial Electron Beam Tester designs shown in Fig. 1-12 have their relative strengths and weaknesses, but share the common feature of operating in a retarding field energy filtered feedback mode, where a single output SE detector signal is automatically adjusted to be constant by applying voltage variations on a retarding grid. These voltage contrast spectrometers are designed to efficiently extract and capture secondary electrons that leave the specimen while maintaining high resolution primary beam optics in the SEM, but they were not optimized nor designed for the purpose of acquiring the SE electron energy spectrum.

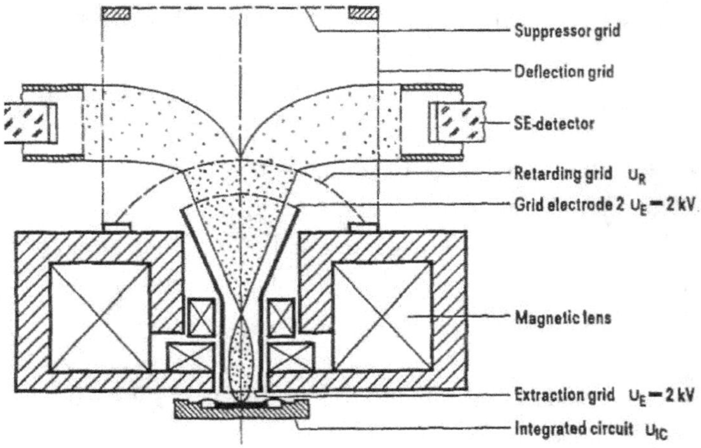

(a)

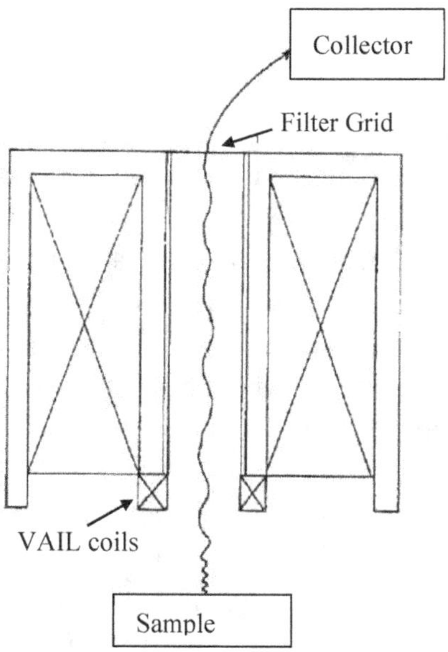

(b)

Fig. 1-12. Commercial Electron Beam Tester designs. (a) Mixed field lens design by Frosien and Plies (1987). Copyright 1987 by Elsevier. Reprinted with permission. (b) Wide-bore immersion lens design by Richardson and Muray (1988). VAIL coils denotes Variable Axis Immersion Lens scan coils, scan coils that can achieve a wide-field of view with magnetic immersion lenses. Copyright 1988 by AIP. Reprinted with permission.

The accuracy to which Electron Beam Testers can measure voltages on an integrated circuit sample is limited by a variety of different effects, but foremost amongst them, is the problem of transverse electric surface fields. Surface fields are typically generated by the presence of neighboring tracks carrying different voltages, as shown by the simulations reported by Khursheed and Dinnis (1990) in Fig. 1-13. In order to counteract this effect, a vertical extraction field is usually applied immediately above the specimen by a grid electrode positioned a few millimeters above the sample. Figure 1-13a shows simulated equipotential lines above a three track conductor layout in the presence of a 1 kV/mm extraction field strength, where the outer two tracks are grounded and the middle track is biased to a positive voltage, a potential barrier region is present above the central track. For this example layout, a potential barrier of 1.4 V is created, in general, the size of the potential barrier depends on the extraction field strength and the distance between conductor tracks. The track widths and separation distances in this case are all set to 1μm. Figure 1-13b shows simulated 1 eV SE ray paths emitted over a wide range of initial angles from the middle track surface, indicating that they are pulled back to the specimen surface. Figure 1-13c shows how the strength of the potential barrier varies with the applied extraction field strength, the extraction field strength needs to be significantly greater than 2 kV/mm in order for low energy SEs to be collected. This requirement of course, becomes more stringent for smaller IC structures, where track widths/gaps are in the sub-micron range.

Since electrical breakdown considerations and disturbance of high electric fields on the primary beam make it impractical to apply more than say 5 kV on the spectrometer extraction grid, errors caused by transverse surface field effects impose fundamental limits on the accuracy that can be attained by retarding field voltage contrast spectrometers.

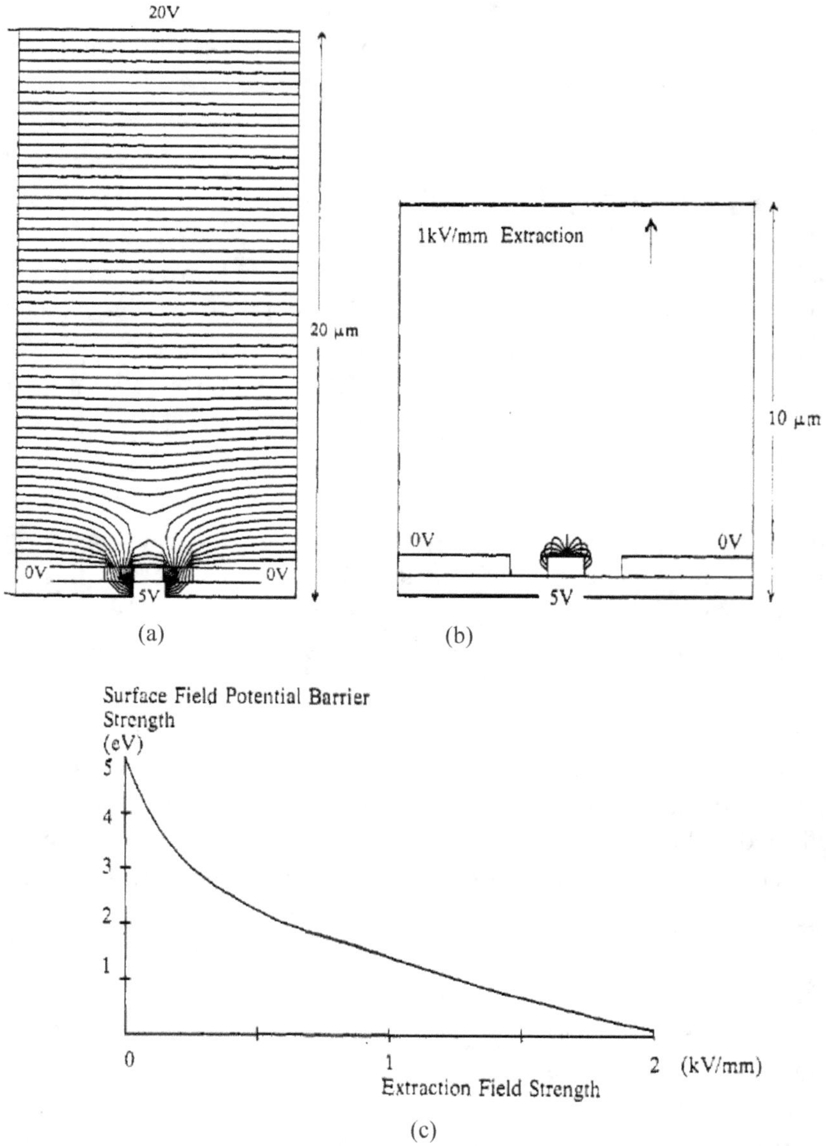

Fig. 1-13. Simulated influence of surface fields by Khursheed and Dinnis (1990). (a) Equipotential distribution. (b) Ray tracing of 1 eV SEs emitted over a wide range of angles. (c) Dependence of potential barrier strength with extraction field strength.

How transverse surface fields distort the output S-curve signal has been investigated both theoretically and experimentally. Simulations showed that the surface field potential barrier above a central track surrounded by a 0 volt track is predicted to reduce the S-curve output signal height, imposing large errors for automated feedback voltage measurement systems that function by keeping the output current constant (Khursheed, 1983b, and Khursheed and Dinnis, 1984). These simulations were in good agreement with experimental S-curve results that had been reported by Nakamae *et al.* (1981), shown in Fig. 1-14, obtained from a planar grid retarding field analyzer. The drop in S-curve height is caused by the local potential barrier effect (known as type I local field effect), while the broadening of the signal is caused by an asymmetric transverse field effect (known as type II local field effect). A specimen having 8 μm width tracks separated by a distance of 12 μm was used, and the extraction grid-specimen distance was set to be around 1.6 mm. Although the extraction field strength is increased by an order of magnitude, there is still a significant distortion of the output S-curve signal (Fig. 1-14b) produced by the surface field effect, which will obviously become more pronounced as the size and separation of the sample track dimensions decrease.

The surface field effect was an unresolved problem for retarding field voltage contrast spectrometer systems. In the case of low strength extraction field voltage contrast spectrometer designs, such as the Richardson and Muray (1988) wide-bore immersion lens analyzer shown in Fig. 1-12b, the operating point of the analyzer was deliberately set to a higher energy point on the S-curve (-5 V retarding grid), increasing the strength of the potential barrier, and filtering out low energy SEs so that local surface fields have less of an effect. However, this comes at the price of incurring much higher levels of shot noise in the output signal.

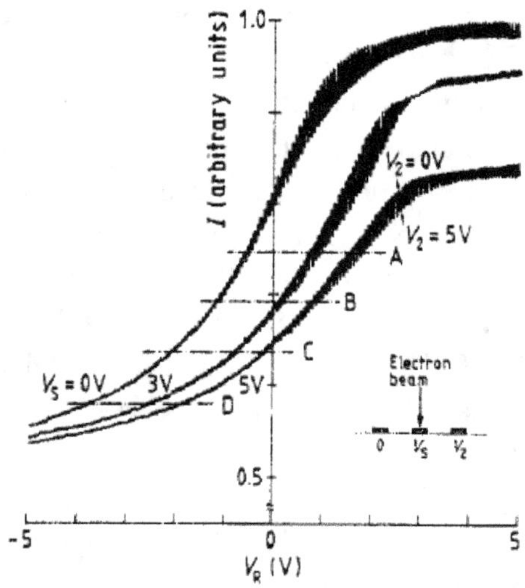

(a)

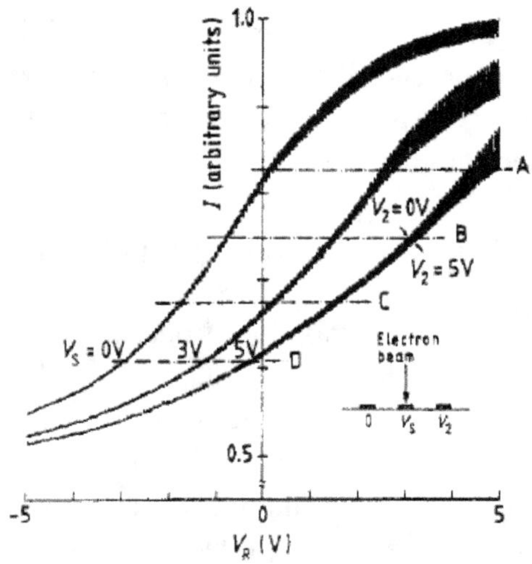

(b)

Fig. 1-14. S-curve distortion by surface fields measured by Nakamae *et al.* (1981) from a planar grid retarding field analyzer. The sample has 8 μm width tracks separated by a distance of 12 μm and the extraction grid-specimen distance is around 1.6 mm. (a) Extraction voltage of 100 V. (b) Extraction voltage of 1000 V. Copyright 1981 by IOP Publishing. Reprinted with permission.

Apart from transverse surface field effects, the main source of errors for quantitative voltage contrast measurement systems comes from primary beam shot noise. The shot noise limit is of course, a well-known feature of SEM imaging, which is usually reduced to an acceptable level by averaging over multiple scan frames. In the subject of voltage contrast, it sets a limit on the precision to which changes in the sample voltage, ΔV_S, can be monitored. Gopinath (1977) proposed a useful guide to characterize the voltage resolution of retarding field analyzers, he defined the voltage resolution ΔV_S in terms of the detector current, I_D, bandwidth Δf, and a detector constant C, as

$$\Delta V_S^2 = C^2(\Delta f / I_D) \tag{1.3}$$

In the derivation of this formula, a unity signal-to-noise ratio in the output voltage measurement is assumed. The formula helped establish a practical figure of merit for retarding field analyzers and is useful for estimating the data-acquisition time required to attain a given voltage resolution. It was widely used to compare the performance of different retarding field spectrometer designs (Dubbledam, 1993).

The above formula can be recast into a more general expression by linking the voltage resolution to the number of electrons in the primary beam that strike the specimen, N. If shot noise is the dominant form of noise limiting the measurement, the *rms* value of the noise is approximately given by $\sqrt{N}$ and the voltage resolution ΔV_S can then be related to the number of electrons in the primary beam by the following simple formula (Khursheed and Dinnis, 1990),

$$\Delta V_S = \frac{\chi}{\sqrt{N}} \tag{1.4}$$

where the constant χ in this expression is obviously dependent on many different parameters, such as the secondary electron emission coefficient of the sample being probed, the transport efficiency of secondary electrons to the detector and the detector quantum efficiency. It also depends on the type of voltage contrast spectrometer being used.

The work of Dubbledam and Kruit (1987a) in the late 1980s highlighted the signal to noise advantage of detecting shifts in the SE energy spectrum, as opposed to monitoring shifts in the S-curve (integrated form of the SE energy spectrum). They showed for instance, that if variations in the SE signal could be tracked as a function of SE energy, then the rising edge of the SE spectrum is much more sensitive to specimen voltage variations than its trailing edge. This information is averaged out in the S-curve signal, and as a consequence, was predicted to have much lower levels of output signal-to-noise. Dubbledam and Kruit used the Chung-Everhart SE distribution for their theoretical derivation with a work function value of 4 eV, and plot the results shown in Fig. 1-15, in terms of the minimum number of electrons needed to produce a voltage shift of 1 mV (with a signal-to-noise ratio equal to 1) as a function of the working point of a conventional retarding grid analyzer (A) and a multichannel analyzer placed behind its retarding grid (B). The working point here refers to the strength of the retarding field analyzer potential barrier (normalized to the work function). For zero potential barrier, all energy channels in the multi-channel analyzer are used, and as the potential barrier strength grows, the energy channels below it are not used in the multi-channel analyzer estimate B. Note that Fig. 1-15 is useful mainly as a graph predicting how much better the voltage resolution is expected to be for a multi-channel analyzer compared to a retarding field analyzer, particularly when the full range of secondary electrons are collected (zero potential barrier). For detection of the full secondary electron energy range, the multi-channel analyzer is predicted to have over two orders of magnitude better signal-to-noise characteristics than retarding field analyzers. The actual numbers do not take into account important parameters such as the specimen secondary electron coefficient, or analyzer secondary electron transport efficiency. Another important feature of the theoretical predictions shown in Fig. 1-15 is that the retarding field analyzer curve A has a minimum optimum value around a working point of 0.4 (1.6 eV). The signal-to-noise characteristics of retarding field analyzers are predicted to degrade significantly on either side of the optimal value, particularly on the low potential barrier energy side (top of the S-curve). These theoretical predictions provided the motivation for acquiring the SE energy spectrum directly, through the use of multi-channel SE energy spectrometers.

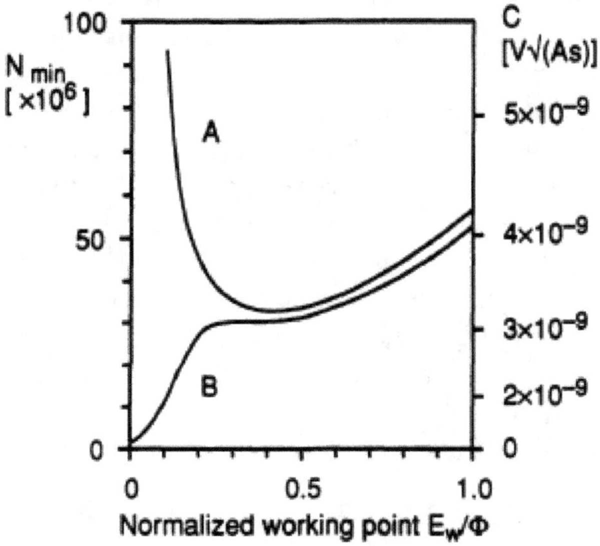

Fig. 1-15. Minimum number of electrons needed to detect a voltage shift of 1 mV with a signal-to-noise ratio equal to 1, as a function of the working point for a conventional retarding grid analyzer (A) and a multichannel analyzer behind the retarding grid (B) by Dubbledam and Kruit (1987a). First published in Scanning Microscopy. Copyright 1987 by Scanning Electron Microscopy, Inc. of Chicago IL.

Dubbledam and Kruit (1987b) proposed a dual channel analyzer in combination with a magnetic immersion lens, which used the trochoidal deflection field principle, in which crossed electric and magnetic fields are applied and electrons travel in the direction of the magnetic field (Granneman and Van Der Wiel, 1983). This analyzer was in effect, a hybrid multi-channel-retarding field analyzer, since a retarding grid was placed in front of each detector and the output signal of each channel was an integrated signal. One disadvantage of this voltage contrast spectrometer design was that the trochoidal analyzer could only be optimized for a given secondary electron energy, limiting its voltage resolution.

The first fully multi-channel voltage contrast spectrometer was designed by Khursheed and Dinnis (1990). Their proposal was based upon using the time-of-flight of secondary electrons. The primary beam is pulsed, as is normally required for time resolved voltage contrast measurements on integrated circuits. The specimen is placed in a magnetic immersion lens, and the secondary electrons are collimated as they travel up the objective lens bore, as shown in Fig. 1-16. A high bandwidth detector system collects the secondary electrons at the top of the objective lens. The lens bore acts as a drift-tube region for the secondary electrons. The sharply decreasing axial field strength through which the secondary electrons initially travel will not only collimate the secondary trajectory paths, but make their time-of-flight relatively independent of their initial emission angle. Under these conditions, the output signal will be directly related to the form of the initial secondary electron energy spectrum (Kruit and Read, 1983).

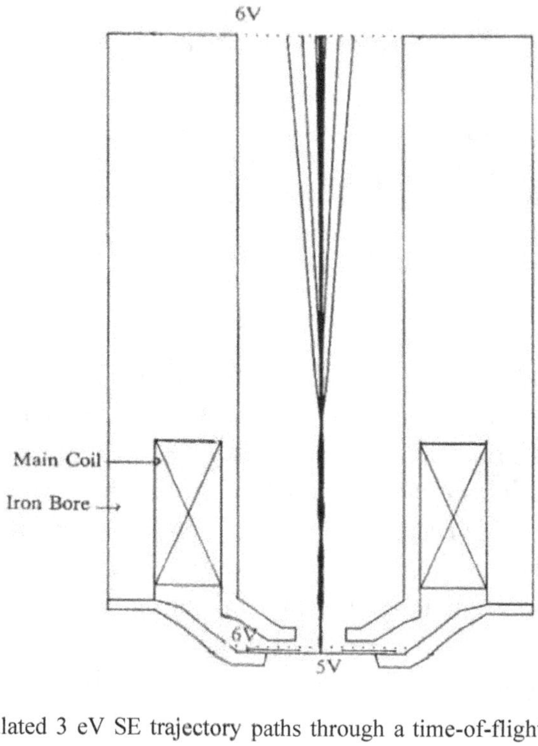

Fig. 1-16. Simulated 3 eV SE trajectory paths through a time-of-flight voltage contrast analyzer (Khursheed and Dinnis, 1990) for a wide-variety of different emission angles.

Neglecting the effect of initial emission angle, a secondary electron having an initial energy E (in eV), has a time-of-flight T given approximately by

$$T = \frac{d}{v} = \frac{c}{\sqrt{E + e(V_d - V_s)}} \tag{1.5}$$

where c is a constant directly related to the drift tube distance d, V_d is the drift tube voltage, and V_s is the specimen voltage. If the energy spectrum of secondary electrons at the specimen is $Q(E)$, then the time-probability density function for electrons arriving at the electron detector is given by

$$P(T) = \frac{dN}{dT} = \frac{dN}{dE}\frac{dE}{dT} = 2\frac{c^2}{T^3}Q(E) \tag{1.6}$$

The limited bandwidth of the detection system will tend to distort the output signal, and its influence can be approximated by a low pass filter response with a single time constant τ, in which case, the output signal $S(T)$ is given by

$$S(T) = (1/\tau)\exp(-T/\tau)\int_0^T P(x)\exp(x/\tau)dx \tag{1.7}$$

The Chung and Everhart (1974) energy distribution can be used for the secondary electron energy distribution $Q(E)$. By using the above equations, the output signal of the time of flight spectrometer can be predicted theoretically.

Dinnis *et al.* (1992) used a modified Cambridge S100 SEM to test the time-of-flight spectrometer concept. A schematic drawing of the experimental layout they used is shown in Fig. 1-17. They pushed a small copper stub specimen into the middle of the S100 objective lens magnetic gap, and used a multi-channel plate detector with a high bandwidth trans-impedance head amplifier to obtain the output signal.

The final signal was displayed on a 400 MSa/s digitizing oscilloscope that was also linked to a personal computer for data capture.

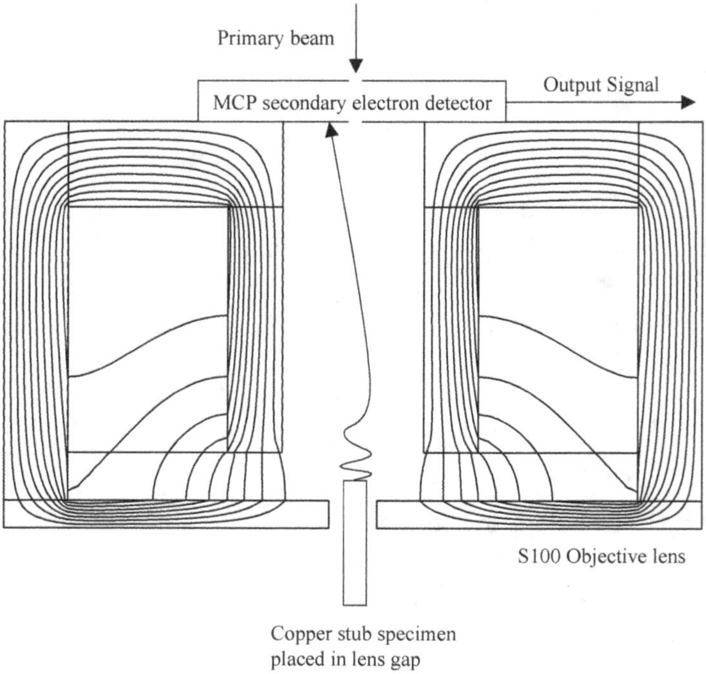

Fig. 1-17. Experimental setup to test the SE energy time-of-flight spectrometer concept using a Cambridge S100 SEM objective lens.

Originally, it was thought that the detection system bandwidth was around 140 MHz, but subsequent analysis of the output signals revealed that it was in fact around 10.6 MHz (τ = 15 ns).

Experimental results, for the specimen voltage changing in 1 volt steps from –1 to –5 volts and a drift tube voltage of 0 volts are shown in Fig. 1-18a (Khursheed, 1992), 500 time channels were used and the effective drift distance was 11.9 cm. Figure 1-18b shows that theoretical predictions based upon using the above equations match quite well with experiment for a work function of 4 eV in the Chung-Everhart SE energy distribution.

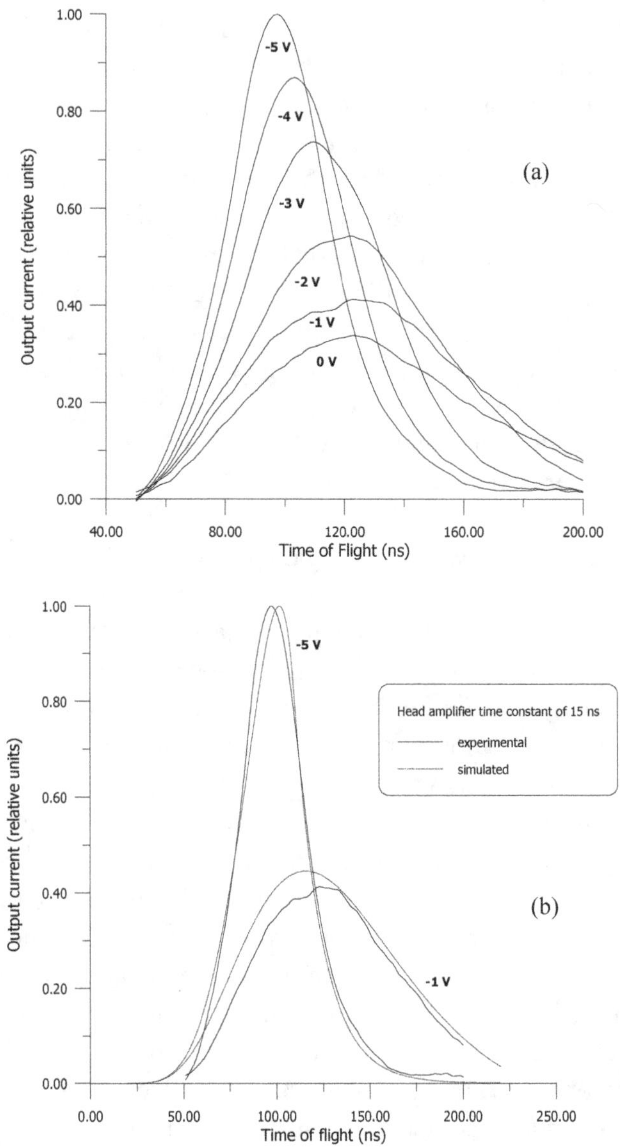

Fig. 1-18. Experimental results from a time of flight voltage contrast analyzer (Khursheed 1992). (a) Specimen voltage change from 0 to -5 V in steps of 1V. (b) Comparison of experimental signals and theoretical predictions for specimen voltages of -1 and -5 volts, a work function of 4 eV is used in the Chung-Everhart distribution.

Khursheed (1992) demonstrated that the time-of-flight voltage contrast analyzer gave around 30 times better signal-to-noise characteristics than retarding field contrast spectrometers for the same voltage resolution (a 30 times lower spectrometer constant). The time-of-flight voltage contrast analyzer appeared towards the late 1980s, at a time when interest in Electron Beam Testers began to diminish. It was from this time that integrated circuit manufacturers started to fabricate circuits covered by a top ground plane, making it difficult for Electron Beam Testers to directly probe voltage signals on their internal conductors. Since the main driving force spurring developments in voltage contrast spectrometer designs up to that point had come from the testing of integrated circuits, from around 1990 onwards, very few new voltage contrast analyzer designs were reported.

Although the SE energy time-of-flight spectrometer has not so far been used for materials science applications, it is in many ways, ideally suited for them. Its high signal to noise performance has been demonstrated experimentally, and small changes in the spectral signal shape can be tracked directly. Its magnetic immersion objective lens design is able to provide high image resolution. Moreover, it is intrinsically a full range SE energy spectrometer, which also captures BSE energy information. The SE energy spectra shown in Fig. 1-18 were subtracted out from full-range energy signals containing the BSE peak (Dinnis *et al.*, 1992). The main disadvantage of the SEM time-of-flight energy spectrometer is that it needs the primary beam to be pulsed to nanosecond widths, and this requires the use of a specialized beam blanker or a photocathode (Thong, 1993b). In addition, the pulsing of the primary beam needs to be synchronized to the image scan rate and the spectrometer requires a drift-tube region to be integrated into the magnetic objective lens bore. Most SEMs are not usually equipped to function in this way, nor is it easy to build such features into an add-on attachment that can fit inside the SEM specimen chamber. At present, SEM columns need to be completely redesigned in order to implement the multi-channel time-of-flight SE energy spectrometer concept.

1.3 In-lens multi-channel position sensitive SE energy analyzer designs

The search for SEM multi-channel SE energy spectrometers, started first within the subject of quantitative voltage contrast, later led on to the proposal of several other in-lens multi-channel designs, developed specifically for materials science applications.

Khursheed and Karuppiah (2001) proposed a multi-channel/deflection secondary electron analyzer in the context of developing permanent magnet high resolution add-on lens attachments. Figure 1-19 shows a schematic diagram of the experimental setup they used in a conventional SEM chamber. This diagram also indicates simulated flux lines of the permanent magnet lens designed to focus a 1 keV primary beam on to the specimen. The distance from the specimen to the top plate is 0.8 mm, and the top plate hole has a diameter of 4 mm.

Although the add-on lens attachment is essentially an immersion magnetic lens, it differs in several ways from previous voltage contrast analyzers that made use of magnetic immersion lenses. In this case, the lens field is confined to a much smaller region than in any of the previous voltage contrast immersion lens spectrometer designs, and the lens field falls by over an order of magnitude within the space of only a few millimeters, so that there is a much higher degree of collimation for scattered low energy secondary electrons. Secondly, since the specimen is placed very close to the lens top plate (0.8 mm), this helps in filtering out wide-angle higher energy electrons (those that travel more than 2 mm off-axis through the top plate hole), reducing the presence of background noise. There is no extraction field, so that when the specimen is biased with positive voltages, a potential barrier between the specimen and top plate (0 V) causes lower energy secondary electrons to return back to the specimen.

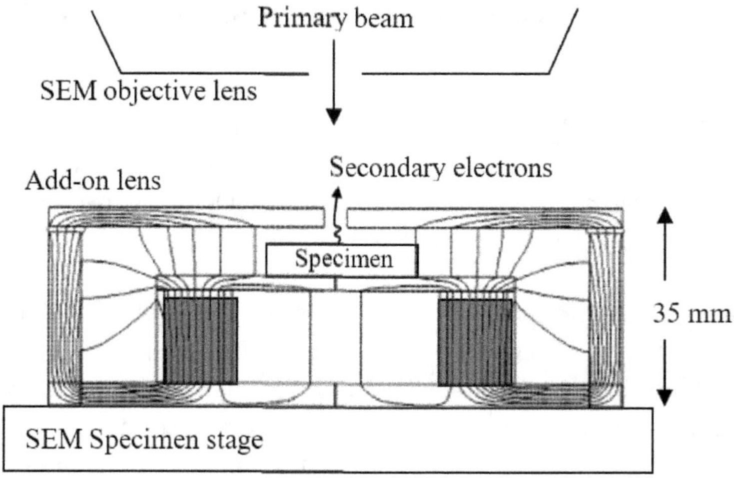

Fig. 1-19. Schematic diagram of an add-on immersion lens experimental setup within the specimen chamber of a conventional SEM. Simulated flux lines for the add-on lens are indicated for the focusing of a 1 keV primary beam. The ray path of the SE trajectory is deliberately exaggerated for reasons of scale.

Figure 1-20 shows simulated trajectory paths for secondary electrons having initial energies of 1, 3 and 5 eV that are emitted over a wide-range of emission angles at the specimen, and subsequently deflected by a Wien Filter located above the lens top-plate. The Wien Filter is designed to deflect secondary electrons while leaving the primary beam unaffected.

The trajectory paths shown in Fig. 1-20 indicate that since low energy electrons experience a high degree of collimation, dispersion due to initial emission angle is expected to be small, and this gives rise to the possibility of detecting the energy spectrum in parallel via the use of a multi-channel detector. On the other hand, high energy secondary electrons or backscattered electrons will not be collimated, and not be able to exit from the lens, nor will they be deflected on to the detector by the Wien Filter. The background noise level on the output signal generated from higher energy scattered electrons is therefore expected to be relatively small.

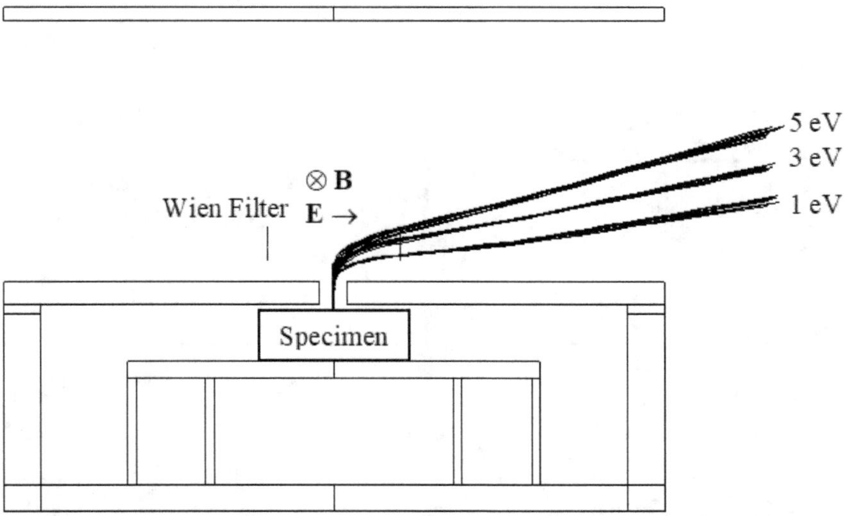

Fig. 1-20. Direct ray tracing of low energy secondary electrons through an immersion add-on lens with a Wien Filter placed above it. The secondary electrons are emitted over a wide-range of emission angles at the specimen. The Wien filter has a radius of 1cm, with 1 AT magnetic excitation.

Khursheed and Karuppiah (2001) generated secondary electron spectra from the add-on immersion lens spectrometer by using it in the band-pass energy mode, where an effective slit is cut into a foil that surrounds the scintillator cage, as shown in Fig. 1-21. Instead of a Wien Filter, an electrostatic deflector was found to be sufficient. Since only low energy secondary electrons are collected, relatively small deflection voltages on electrostatic deflectors are required (less than 12 V), which typically shift the primary beam, but do not degrade the image resolution. These shifts can be minimized by designing the deflector to have a compensating electric field in its upper part, as shown in Fig. 1-21. The deflector consisted of a single cylindrical wall plate having a diameter of 30 mm, which had a rectangular opening in its lower portion to allow secondary electrons to exit and be attracted to the scintillator cage. An opening on the top portion compensates for deflection on the primary beam, see Khursheed and Karuppiah (2001) for more details on the deflector design.

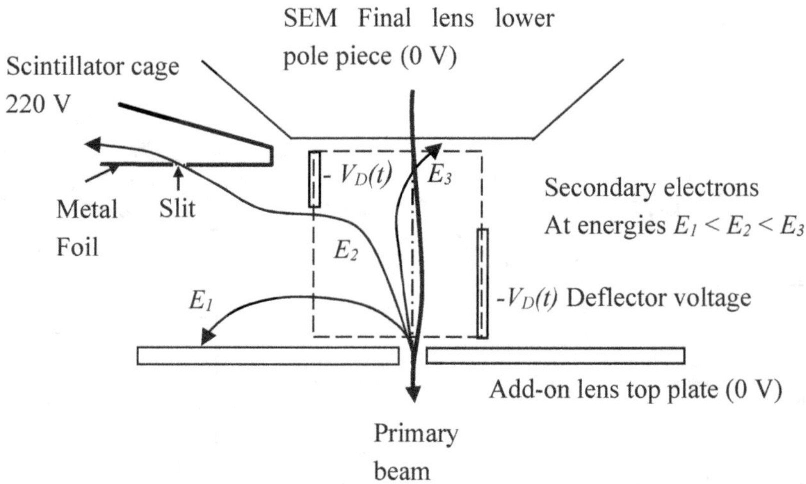

Fig. 1-21. Schematic layout of an experiment operating the immersion add-on lens spectrometer in band-pass mode (Khursheed and Karuppiah, 2001). The central cross-section of the deflector is shown.

Experimental secondary electron spectra obtained by varying the deflector from 0 to −13 volts for specimen voltages of –5, 0 and 5 volts are shown in Fig. 1-22a. They show that a truncated secondary electron spectrum is obtained, where high energy electrons are filtered. Not only is there considerable voltage contrast for the left-hand edge of the spectrum when the specimen changes from 0 to −5 volts, but there is also significant changes to the right-hand side edge, indicating that the total number of secondary electrons reaching the scintillator as the specimen is biased more negatively is higher. This mostly likely comes from the fact that the pass energy is higher for negative specimen potentials, effectively allowing a wider band of energies to be transmitted through the slit. To obtain an estimate of the minimum detectable specimen voltage difference associated with the output signals shown in Fig. 1-22a, 16 signals were recorded for a specimen voltage of –5 volts. The mean and variance of each signal were calculated, and an *rms* noise value for the mean was found (Khursheed and Karuppiah, 2001). The *rms* noise value for the signal mean, 0.01485 volts, was then compared with the change in mean value for the specimen voltage switching from –5 to –4 volts. The output signals corresponding to the specimen voltages of –5 and –4 volts are shown in Fig. 1-22b.

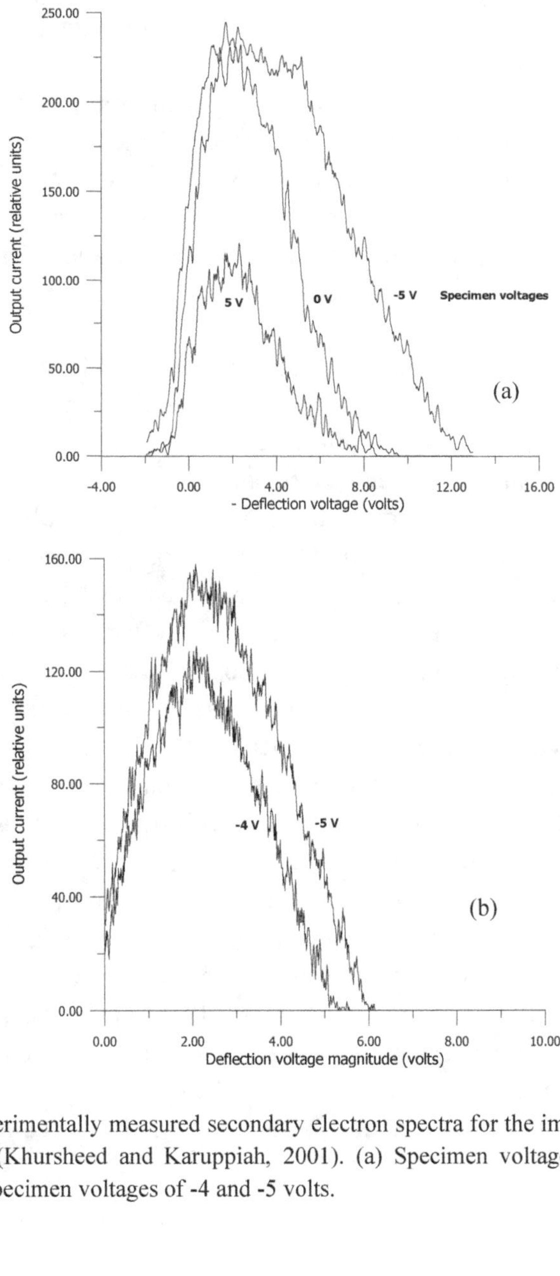

Fig. 1-22. Experimentally measured secondary electron spectra for the immersion add-on lens analyzer (Khursheed and Karuppiah, 2001). (a) Specimen voltages of -5, 0 and +5 volts. (b) Specimen voltages of -4 and -5 volts.

The change in the mean value for the specimen changing from −5 to −4 volts was found to be 0.2751 volts. This provides a signal to noise ratio of 18.52, giving a minimum detectable specimen voltage difference of 54 mV. The noise level is of course, dependent on data acquisition time, which in this case was 0.16 seconds for each signal. Since the spectrometer's signal to noise ratio depends on many operating factors such as the primary beam current and detector efficiency, it is difficult to compare it to other voltage contrast systems. However, these results obtained from the add-on magnet immersion lens spectrometer show that it can potentially provide high voltage resolution and operate as a SE energy spectrometer.

Figure 1-23 shows that the secondary electron immersion add-on lens spectrometer is able to provide quantitative material contrast. Experimental secondary electron spectra were taken from a specimen having copper and brass regions separated by glue. The specimen is coated with a 25 nm layer of carbon. In this case, the deflection voltage is synchronized with the vertical slow scan, and varies from 2 to −18 volts. The secondary electron image, shown in Fig. 1-23a, indicates that the maximum intensity is obtained for a deflector voltage of around −2.5 volts. By examining vertical line scans taken on either side of the glue junction, shown in Fig. 1-23b, considerable contrast is obtained between brass and copper regions, particularly with respect to the change of the output signal width. This contrast between spectra was found to be invariant to the sample's orientation, confirming that it is a genuine material contrast effect, and not an artifact of the spectrometer. The contrast was quantified by recording 16 such lines scans from the secondary electron image, and calculating the sample mean and standard deviation. The standard deviation was found to be 0.0256 V for copper line scans, while the change in standard deviation between the copper and brass signals was calculated to be 0.3, giving an approximate signal to noise ratio of 11.72. This result is significant, since in the standard SE detector image, the copper and brass parts of the specimen were not distinguishable.

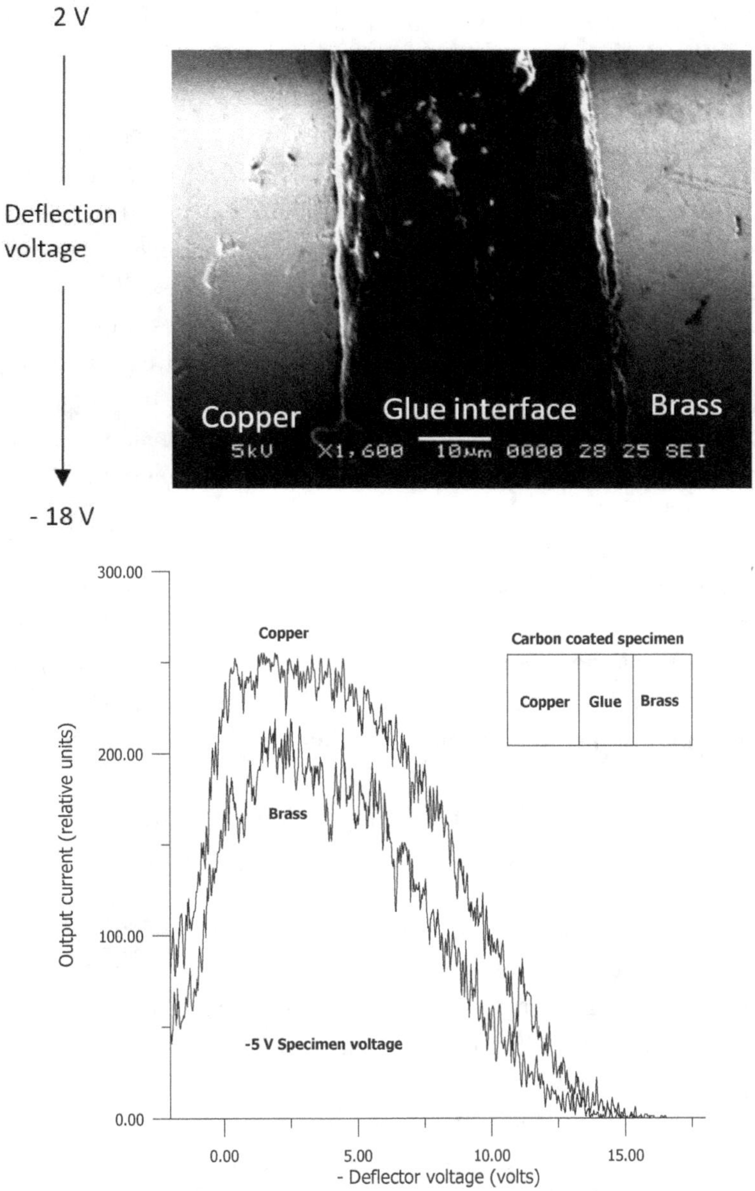

Fig. 1-23. Experimental material contrast results with the immersion add-on lens secondary electron spectrometer on a Copper-Brass specimen (Khursheed and Karuppiah, 2001). (a) Secondary electron image. (b) Vertical line scans.

Although the initial experimental SE spectral results from the add-on lens spectrometer were obtained by it operating in the band-pass mode using an exit slit aperture, it can in principle easily be upgraded to a multi-channel detector by using a position sensitive detector to one side, capturing secondary electrons after they are deflected from the central axis. This is the next stage in its development. Despite the add-on lens multi-channel spectrometer having several limitations, such as only being able to capture SEs (and not BSEs), or not being able to retard the primary beam to low landing energies, it has the desirable feature of being a self-contained immersion lens attachment that can be placed on the specimen stage of conventional SEMs, in the field free region below the objective lens lower pole-piece. It not only makes SE energy spectroscopy possible, but it also improves the image resolution of conventional SEMs.

Kienle and Plies (2004) presented a secondary electron analyzer design for a mixed field objective lens, where the primary beam is retarded just before it reaches the specimen, and additional focusing is provided by a single-pole/wide bore immersion lens. This type of mixed field lens gives the highest image resolution of all in-lens designs and provides the possibility of operating the SEM in the low landing energy mode (LVSEM). In the analyzer design reported by Kienle and Plies, the primary beam travels through most the column at around 9 kV, and is decelerated down to a landing voltage of 1 kV at the specimen. Secondary electrons are typically accelerated to 8 kV as they travel back up the column, and are deflected off-axis by a Wien Filter. A schematic diagram of their SE energy spectrometer is shown in Fig. 1-24. Another Wien filter is placed further up the column in order to cancel out adverse energy dispersion and second-order geometrical aberration effects on the primary beam, while the remaining astigmatism is corrected using a conventional stigmator. The two Wien filters are excited in opposing directions with respect to one another. The secondary electrons after having been deflected by the lower Wien Filter are further deflected and dispersed by an electric spherical deflector analyzer (SDA), after which their image is magnified and focused on to a YAG scintillator. The photon image from the YAG scintillator is further magnified by a light optical objective lens before it is detected on a CCD camera. The CCD camera acts as a position sensitive detector which simultaneously captures the secondary electron spectrum from 0 to 20 eV over 1280 channels. Several stages of magnifying the energy dispersion are required in this analyzer, since the lower Wien Filter only deflects the secondary electron beam by 15°, requiring the SDA to have a deflection angle of 75°.

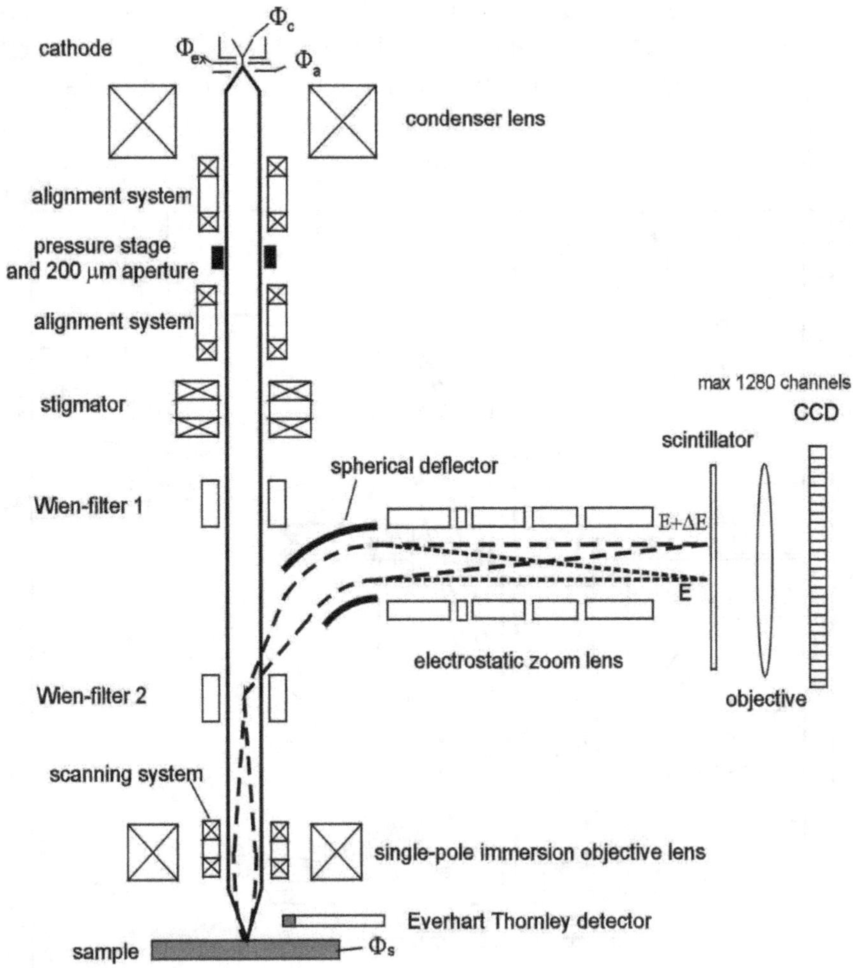

Fig.1-24. Schematic layout for the multi-channel secondary electron off-axis analyzer reported by Kienle and Plies (2004). Copyright 2004 by Elsevier. Reprinted with permission.

Simulated trajectory paths for secondary electrons through the analyzer reported by Kienle and Plies (2004) are shown in Fig. 1-25, while experimental secondary electron spectra for the specimen voltage changing between 0 to −5 volts in −1 volt steps are shown in Fig. 1-26. The experimental voltage resolution lies around 50 mV with an average data acquisition time of 0.5 s per spectrum. The analyzer is able to completely separate secondary electron and backscattered electron signals from one another. This analyzer, like the secondary electron time-of-flight analyzer, is fully multi-channel and is able to capture spectral energy information in parallel. It is not a full range multi-channel energy spectrometer, since it cannot acquire the SE and BSE energy spectra at the same time, but it can acquire them separately.

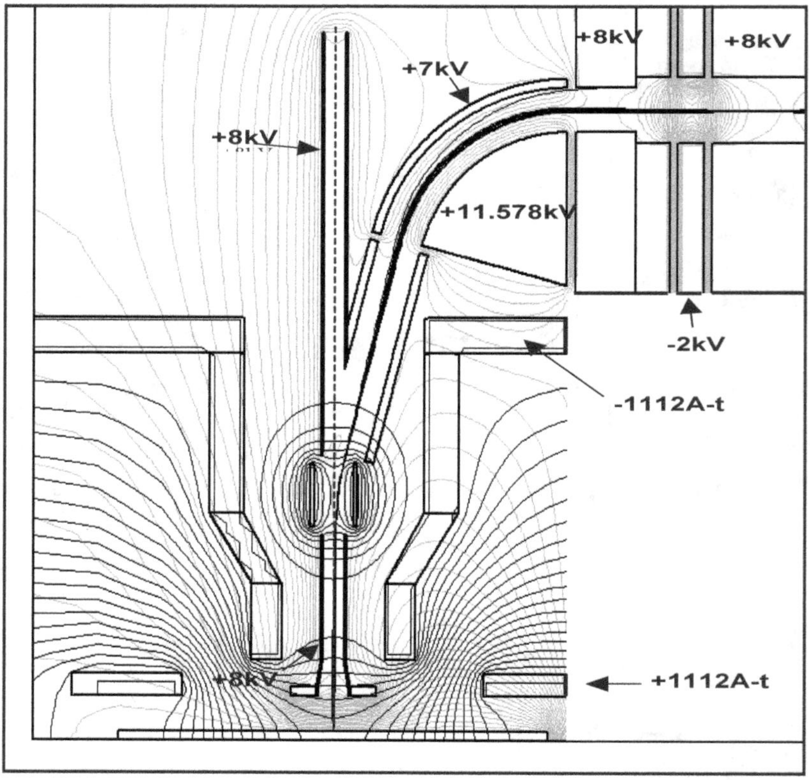

Fig. 1-25. Simulated trajectories through a multi-channel off-axis analyzer, as reported by Kienle and Plies (2004). Copyright 2004 by Elsevier. Reprinted with permission.

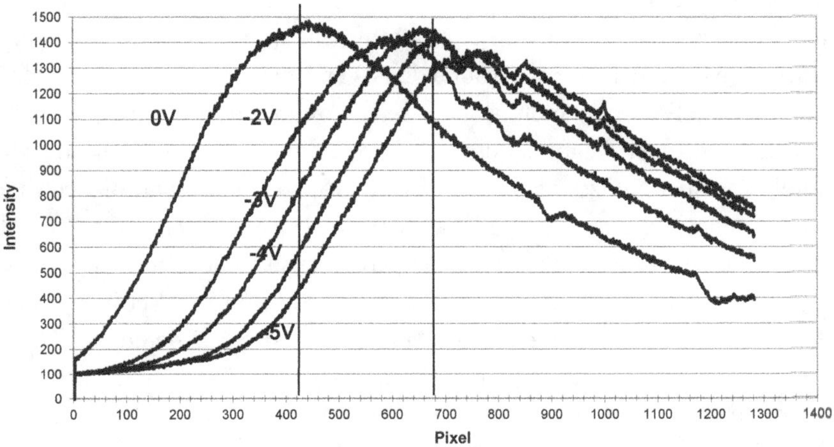

Fig. 1-26. Experimental secondary electron spectra for the multi-channel off-axis analyzer as reported by Kienle and Plies (2004). Copyright 2004 by Elsevier. Reprinted with permission.

The SE energy spectrometer designed by Kleine and Plies (2004) requires a custom made SEM column. However, it may also be possible, using the same principles, to design it to be an add-on energy spectrometer attachment for conventional SEMs.

Khursheed and Osterberg (2006) proposed another multi-channel energy spectrometer design for magnetic immersion/electric retarding field objective lenses, suitable for low voltage SEM. Their design is based upon using a round magnetic sector beam separator placed before the objective lens, both to bend the primary beam on to the objective lens optical axis, and to disperse the scattered electrons in energy, using it as the first stage of a full range multi-channel energy spectrometer. A schematic diagram of their proposal is shown in Fig. 1-27.

The SEM column, not shown here, directs and focuses a primary beam into the center of a circular magnetic sector deflector. The SEM column has the usual gun, condenser lens, stigmator, aperture, and scan coil components, and employs an auxiliary lens to focus the primary beam into the magnetic sector unit. The magnetic sector unit deflects the primary beam by 90 degrees, operating in the stigmatic focusing condition, where it acts as a round lens. The primary beam then goes through a transfer lens which is set to have a relatively long focal length, so that the primary beam enters the bore of the objective lens nearly parallel. The specimen is biased (typically around −5 kV) and placed in a strong magnetic field (between say 0.3 to 1 Tesla). The subsequent scattered electrons from the specimen are focused back into the sector unit by the transfer lens, and deflected away from the SEM column. The wider-angle backscattered electrons are collected by a detector placed below the upper pole-piece of the transfer lens. The scattered electrons transmitted through the transfer lens and separator unit are then further deflected by a magnetic post-deflector unit, which directs and focuses them on to a multi-channel detector. An aperture placed below the beam separator helps to restrict the angular dispersion of scattered electrons.

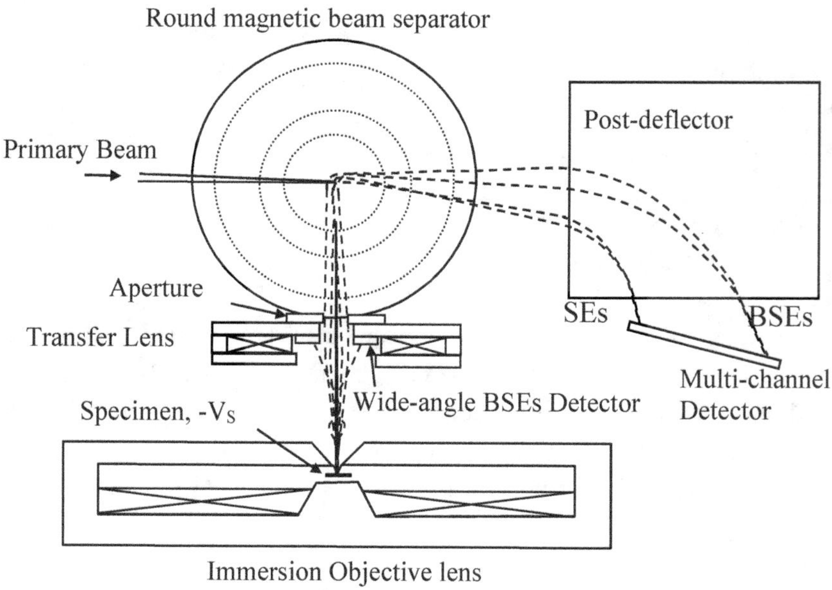

Fig. 1-27. Schematic diagram of the magnetic beam separator spectrometer layout (Khursheed and Osterberg, 2006).

The primary beam aberrations of the magnetic sector unit, and the trajectory paths of the scattered electron paths back through it, were analyzed by ray tracing simulation (Khursheed and Hoang, 2008a), and the spectrometer was predicted to be capable of combining high spatial resolution at low landing energies (0.5 to 1 keV) while at the same time, being able to capture the full range energy spectrum of the scattered electrons in parallel.

Preliminary experiments with test permanent magnet lenses, an electromagnetic sector unit, and a scintillator-PMT electron detector in the SEM confirmed some of these predictions for the primary beam aberrations and SE energy spectrum (Osterberg *et al.*, 2009). The layout schematic diagram for this and preliminary experimental SE energy spectra are shown in Fig. 1-28. The primary beam image aberrations of the round magnetic sector were deduced to be less than 2 nm, confirming simulation predictions (see Osterberg *et al.* (2009) for more details).

SE energy spectral results were generated indirectly, without using a position sensitive detector, by varying the strength of a mirror electrode causing secondary electrons to go through the beam separator for a second pass. The SE signal as a function of the mirror voltage was then differentiated in order to obtain the spectra shown in Fig. 1-28b. This design has the advantage of the BSEs exiting from a different side of the beam separator to the SEs, so that their respective energy spectra can be separated and simultaneously analyzed. This preliminary experiment only establishes the fact that some form of SE energy spectrum can be captured, but it needs further refinement to operate in a multi-channel way by using a position sensitive detector in place of the scintillator PMT detector.

The magnetic beam separator energy spectrometer design for SEMs is easier to implement as a customized SEM design, rather than an add-on attachment, and more experiments are required to confirm that it can operate as a multi-channel spectrometer. There is also the need to design a scanning unit for it, one that is optimized to work with the magnetic beam separator.

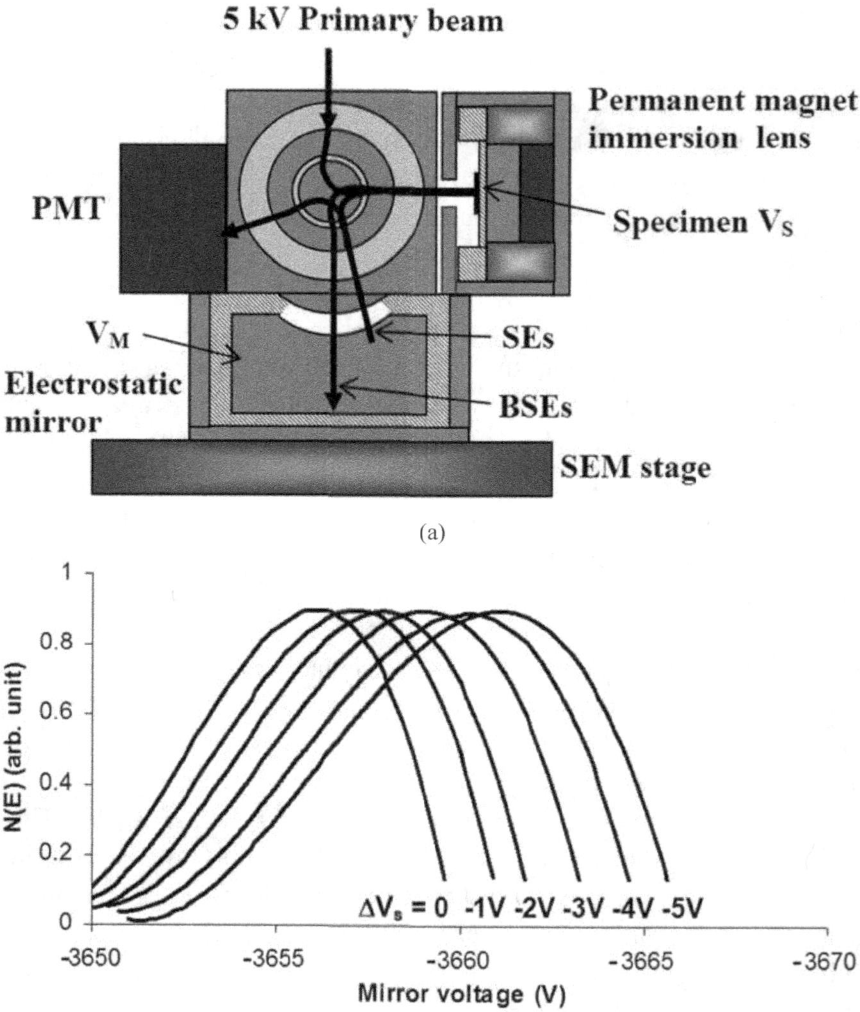

(b)

Fig. 1-28. Preliminary experimental SE energy spectral results for the round magnetic sector multi-channel spectrometer design (Osterberg *et al.*, 2009). (a) Schematic layout. (b) Experimental SE energy spectra obtained by differentiating the SE signal as a function of mirror voltage.

1.4 Retarding field energy analyzer attachments

There have been several attempts at generating the SE energy spectrum inside a SEM for materials science applications using retarding field energy analyzer designs. Most of them made use of preexisting in-built SEM SE detector systems or were adaptations of prior voltage contrast spectrometer systems. Yong and Thong (2000) presented one of the few retarding field energy spectrometer attachments designed specifically to acquire the SE energy spectrum inside a SEM. Their main aim was to obtain SE energy spectra of insulator samples in order to help model and better interpret signal profiles obtained by low voltage SEM (LVSEM) metrology on semiconductor samples. A schematic drawing of their attachment design is shown in Fig. 1-29.

The Yong and Thong spectrometer attachment consisted of blanking plates to pulse the primary beam, a planar retarding grid analyzer, and a pico-ammeter to measure specimen substrate current. The retarding grid was biased to a range of negative values, causing low energy electrons to be mirrored back on to the specimen surface, which in turn, induced a change of charge in the specimen substrate.

Some experimental S-curves together with corresponding SE energy spectra obtained by Yong and Thong are shown in Fig. 1-30. This retarding field energy spectrometer design has the advantage of not distorting the SE energy spectrum through using electric fields to deflect SEs on to a detector positioned off-axis. For the results taken on an Al stub specimen, shown in Fig. 1-30a, there is generally good agreement between experiment and the theoretical Chung-Everhart distribution. Figure 1-30b shows the experimental S-curve and corresponding SE energy spectrum of a Si_3N_4 sample. In the case of insulator specimens, the primary beam voltage was first adjusted in order to reach a charge balance condition, where the net charge going into the sample was balanced with the net charge leaving it (total scattered yield).

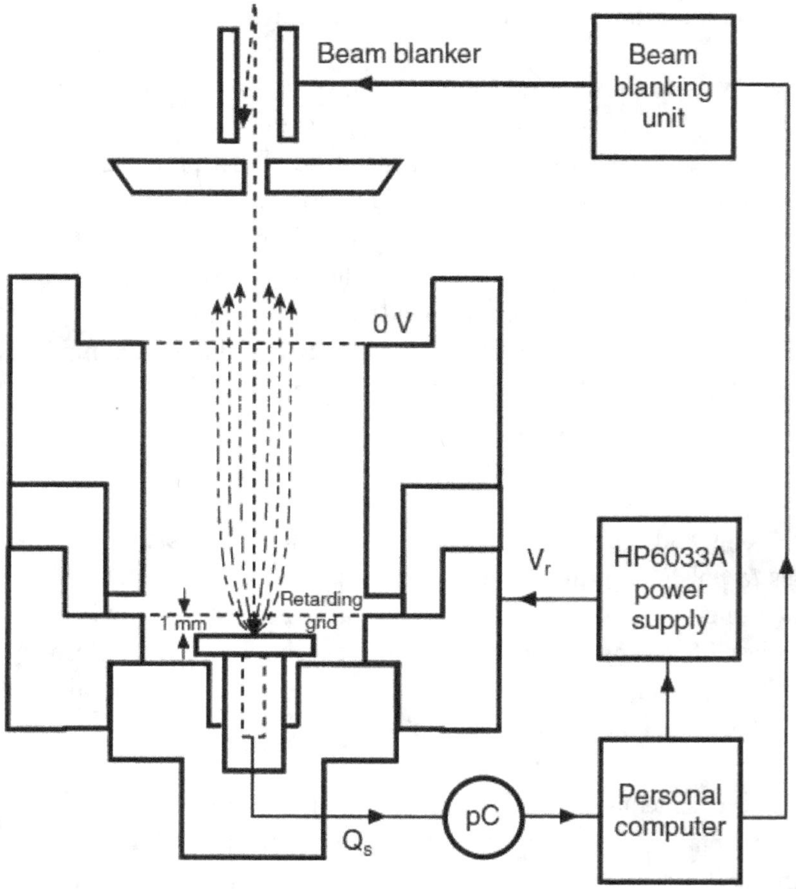

Fig. 1-29. Schematic diagram of the retarding field energy spectrometer attachment design used by Yong and Thong (2000). Copyright 2000 by Wiley Periodicals, Inc. Reprinted with permission.

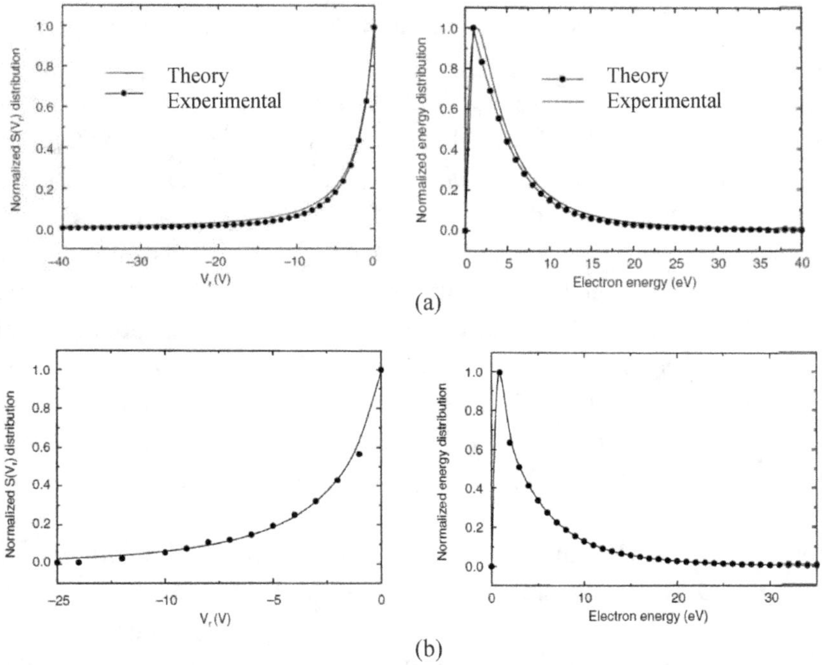

Fig. 1-30. Experimental spectral measurements by the planar grid specimen current retarding field spectrometer attachment (Yong and Thong, 2000). (a) S-curve and SE energy spectrum of an Al sample. (b) S-curve and SE energy spectrum of a Si_3N_4 sample. Copyright 2000 by Wiley Periodicals, Inc. Reprinted with permission.

Yong and Thong (2000) formed an output S-curve signal by subtracting the measured specimen substrate charge away from the net SE charge leaving the sample (with zero retarding field), effectively obtaining a high pass energy filter signal, which was then differentiated to obtain the SE energy spectrum. They used a mathematical expression to compensate for the distortion of the SE energy spectrum created by the planar retarding grid. Pulsing of the primary beam was not only needed for capacitive charge coupling measurements across insulator specimens, but it also provided a way of minimizing charging and contamination effects. The Yong and Thong approach was mainly a spectral energy approach, confined to irradiating large areas on bulk specimens. It was not designed to combine SE spectral information with the SE image.

Another SE energy retarding field energy spectrometer attachment designed specifically for the SEM is shown in Fig. 1-31. This design was first reported by Sekiguchi *et al.* (2016) as way of low pass energy filtering SEs to enhance dopant contrast. The analyzer and detector in this case have a circular shape centered on the primary beam axis. The spectrometer has a retarding planar grid geometry, and as the potential barrier strength is increased, low energy electrons are mirrored back on to a ring shaped Microchannel Plate (MCP) detector positioned below the specimen plane. This SE detector energy filter was referred to as a Fountain Secondary Electron Detector (FSED), since the trajectory paths from the specimen to the MCP detector resemble a fountain shape.

Agemura *et al.* (2017) later used both planar and spherical retarding grids in the FSED attachment to produce experimental SE spectra of an Al stub sample with a working distance of 15 mm. The S-curve and SE energy spectra derived from them for both planar and spherical retarding grids are shown in Fig. 1-32. The results for the spherical grids (solid line) are clearly more distorted and attenuated compared to the planar retarding grid (dotted line). By carrying out more experiments, Agemura *et al.* (2017) concluded that in both cases, significant distortions of the output signal were created by the generation of tertiary electrons scattered from the retarding field grid electrodes, which they referred to as SE3s.

For a later improvement of the FSED attachment, where the retarding field grid was modulated by a small AC voltage and a lock-in amplifier at the MCP, SE3s were still found to significantly distort the output signal (Agemura *et al.*, 2018).

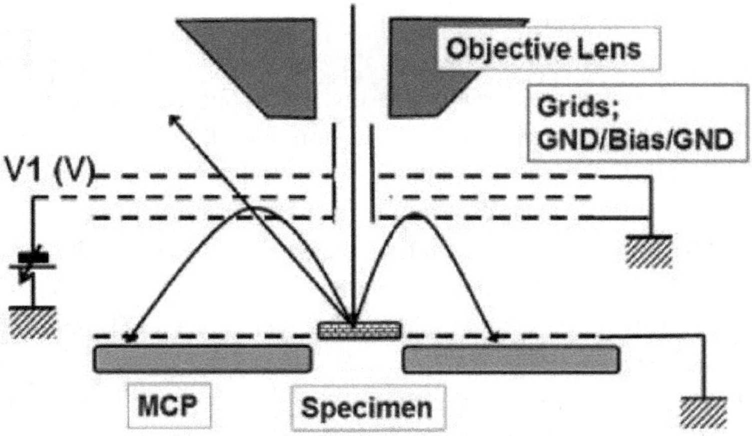

Fig. 1-31. Fountain SE detector (FSED) for low pass energy filtering (Sekiguchi *et al.*, 2016). Copyright 2016 by Elsevier. Reprinted with permission.

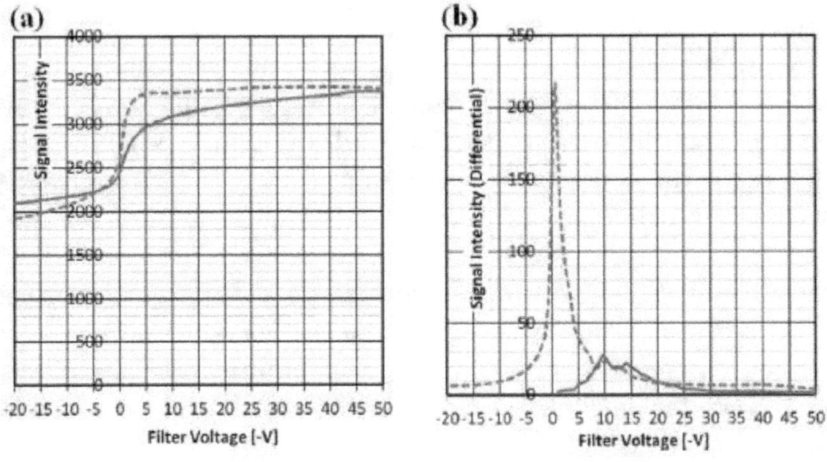

Fig. 1-32. Experimental SE energy spectra from the FSED with planar and spherical retarding grids obtained by Agemura *et al.* (2017). Planar grid represented by dotted line and spherical grid represented by solid line. (a) S-curve obtained by varying filter grid voltage. (b) SE energy spectra obtained by differentiating the S-curve. Copyright 2017 by Wiley. Reprinted with permission.

The Yong and Thong (2000) spectrometer was designed specifically to obtain the SE energy spectra of insulators, something that would be difficult for any other retarding field spectrometer to do, since charging on the specimen surface greatly distorts the output S-curve signal shape. In terms of its geometric layout, it is similar to the FSED proposed by Sekiguchi *et al.* (2016) and Agemura *et al.* (2018). Both of them mirror low energy electrons back. Why the Yong and Thong setup was more successful for generating SE energy spectra than the FSED design is not clear. The fact that Yong and Thong did not form a low pass energy filter signal directly, but subtracted it from the overall scattered yield (at zero potential barrier) may be one factor. Another factor might be due to the fact that Yong and Thong only used a single retarding field grid electrode, instead of the three retarding grid arrangement used by Agemura *et al.* (2018), thereby creating less background grid scattering effects.

More investigation into improving the FSED approach is certainly worth pursuing. This is because, compared to the Yong and Thong spectrometer design, it is better adapted for combining high resolution imaging with SE energy spectroscopy, and can deal with more general specimen types (beyond bulk specimens). It can achieve small working distances (for the planar retarding grid arrangement), and has a relatively fast output detector response (MCP).

Apart from the two SE retarding field energy spectrometer attachments just described, a voltage contrast spectrometer attachment was used to determine charge conditions while observing insulator specimens inside a SEM. Girard *et al.* (1992) used the ABT voltage contrast spectrometer designed by Feuerbaum (1979), shown in Fig. 1-11a, to track shifts in the SE energy spectrum of various types of photoresist while varying the primary beam energy. Their experimental results for KTI photoresist at the primary beam energies of 0.93, 0.94 and 0.95 keV are shown in Fig. 1-33. Figure 1-33a shows an example of S-curves, while Figs. 1-33b–d indicate how the SE energy spectrum shifts in response to the specimen charging: charging positively for a primary beam energy of 0.93 keV, negatively for 0.95 keV, and reaching a charge balance condition for 0.94 keV. These results suggest that a well-designed retarding field voltage contrast spectrometer, such as the one designed by Feuerbaum (1979), may also have some materials science SE energy spectral applications.

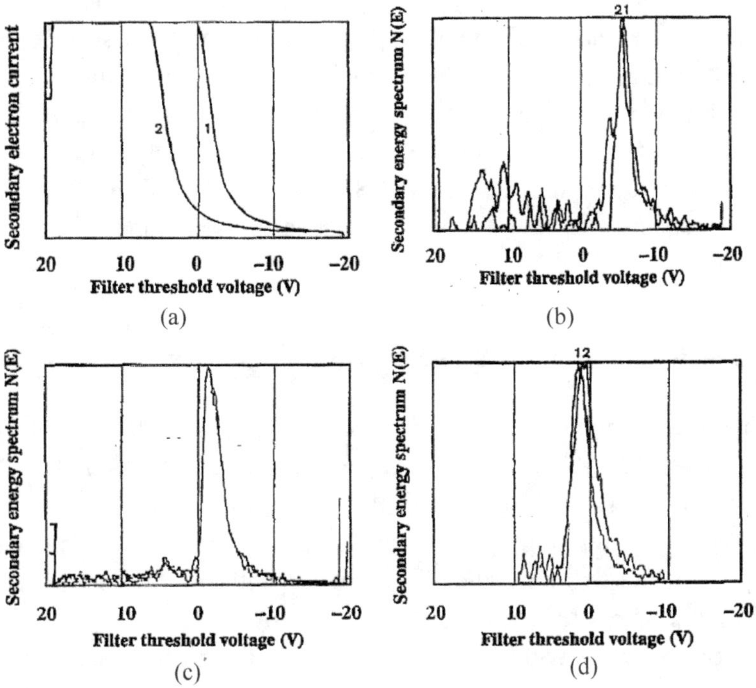

Fig. 1-33. ABT standard voltage contrast electron spectrometer used to monitor charging of KTI photoresist by Girard *et al.* (1992). (a) Example S-curves, (b) 0.93 keV, positive charging, (c) 0.94 keV zero charging, (d) 0.95 keV negative charging. Copyright 1992 by AIP Publishing. Reprinted with permission.

The experimental SE energy spectra shown in Figs. 1-33b–d appear to be truncated, indicating the occurrence of some form of additional high pass energy filtering action after SEs pass through the retarding field grid, most probably coming from the fact that they are deflected off-axis by an asymmetric three dimensional electric field. Not only does this truncate the detected SE energy spectrum, but it inevitably distorts it. Whether this type of SE energy spectrum is acceptable or not depends very much on the application of interest. If only relative changes in the shape are required, this type of spectrum may be acceptable, depending on the output signal-to-noise requirements. Retarding field spectrometers have much higher levels of shot noise than multi-channel spectrometers (Fig. 1-15) and the S-curve shape of retarding field voltage contrast spectrometers is easily distorted by the presence of local surface fields and grid scattering effects. It will not be suitable for applications that require knowledge of the emitted SE energy spectrum shape at the specimen surface. Just how effective the ABT voltage contrast spectrometer attachment can be for SE energy spectral applications in materials science is therefore not clear at present, and needs more investigation.

1.5 Energy filtered SEM SE detector systems

With the development of in-lens SEM columns in the 1990s (such as the Hitachi S-4500 FE-SEM), SE detectors were positioned both above and below the objective lens bore, and this led to a variety of different SE image energy filtered proposals for the SEM, which in some cases, attempted also to capture the SE energy spectrum. The advantage of using such systems for SE energy spectroscopy is that they make use of in-built SE detectors within the SEM, avoiding the need to use an add-on spectrometer attachment.

One of the first in-built SEM SE detector systems proposed for SE energy spectroscopy was reported by Shachal *et al.* (2000) in a patent application entitled, "Secondary electron spectroscopy and system". They proposed the use of SE energy spectroscopy for rapid material identification on wafer specimens as part of the inspection and review process of semiconductor fabrication. The aim was to identify materials such as Si, SiO_x, SiN, Ti, W, and Al, as part of an in-line inspection process. Shachal *et al.* proposed this for SEMs such as the SEMVision™. Figure 1-34 shows a schematic diagram of their proposal, which consists of suitably varying the voltages of the specimen on its holder (35 and 30), V_{wf}, a grid voltage (36) on a SE detector below the lens (34), V_{cl}, a column decelerating electrode voltage (33), V_{dec}, and the bias voltage of a SE detector placed above the objective lens bore (37), V_{id}. These voltages are adjusted so that at any point in time, SEs are detected within an energy band, and subsequently systematically changed to acquire the SE energy spectrum, operating in a band-pass energy mode of spectral acquisition. The proposal does not give any experimental details, such as the width of the detected energy band, but it does give an example of experimentally generated SE energy spectra, such as the one shown in Fig. 1-34b.

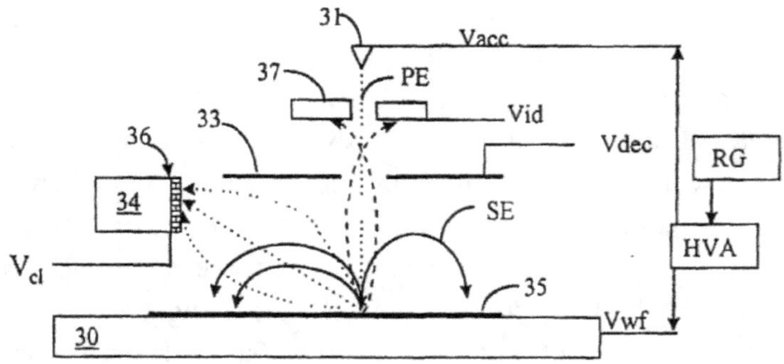

(a)

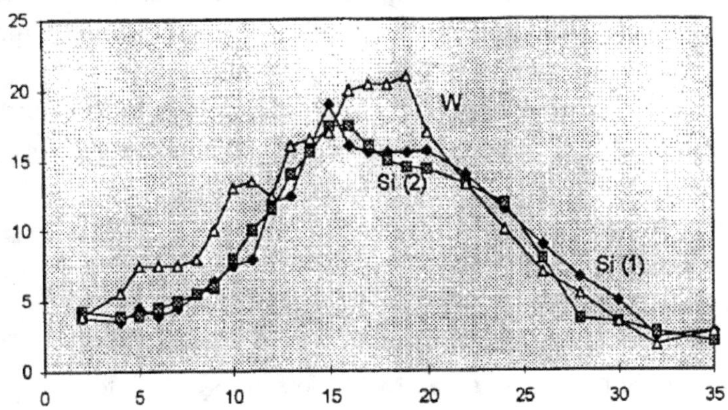

(b)

Fig. 1-34. A band-pass SE energy spectroscopy proposal with an in-built SE detector inside a SEM (Shachal *et al.*, 2000). (a) Schematic diagram (objective lens not shown). (b) Experimental SE energy spectra. Copyright 2000 by Shachal, D. and Bernadicou, M.

Figure 1-34b shows that there is some difference in the shape and peak positions of the SE energy spectral signal from W, compared to the two spectral signals from Si. The Si spectral peak is centered at 15eV, while for W, it is located at 17eV. Also, the area under the W curve is larger than that under both the ones for Si. Shachal *et al.* (2000) claimed that both curves of Si track each other with sufficient repeatability to provide some form of material identification. Obviously, the SE energy spectra shown in Fig. 1-34b are very different to the highly asymmetric ones predicted by the Chung and Everhart analytical model, or those obtained in UHV systems (see Figs. 1-3a–c). They are much broader, symmetric, and have fictitious bumps in them.

One obvious problem with the Shachal *et al.* (2000) ad-hoc SE detector spectrometer design is that it inherently captures a mix of both SE emission angle and SE emission energy information, and is therefore not in principle capable of detecting a narrow band of SE energies. This is very different to the band-pass energy spectrometers used in surface science, such as the CMA or HDA, which are designed to ensure that electrons having different angles focus on to the analyzer exit aperture, while at the same time, dispersing different energies. The relatively poor quality of SE energy spectra obtained by the add-hoc SEM in-built SE detector system shown in Fig. 1-34b highlights the need and importance of using a properly designed SE energy spectrometer, the kind that is routinely used for surface science applications. It is interesting to note that the Shachal *et al.* (2000) original patent application did not proceed as a method to carry out SE energy spectroscopy inside SEMs. A related patent which was subsequently granted (US 6,201,240 B1), was based upon providing a rudimentary form of energy filtering to improve SE image contrast.

The proposal of using in-built SEM SE detector systems to provide some form of SE image filtering/spectroscopy was later repeated on several separate occasions. Kazemian *et al.* (2006), used SE energy filtering in a commercially available FEI XL30 S–FEG SEM. They were able to obtain a low pass energy filtering action of SEs travelling up through the bore of a semi-in-lens magnetic objective lens, by varying the voltages of an extraction tube and electric deflector, as shown in Fig. 1-35a.

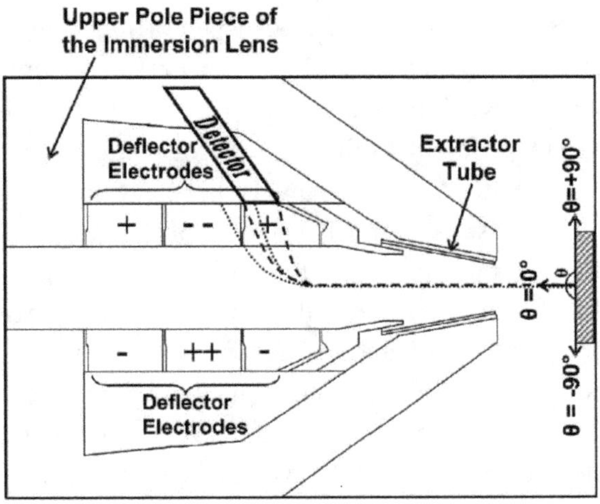

(a)

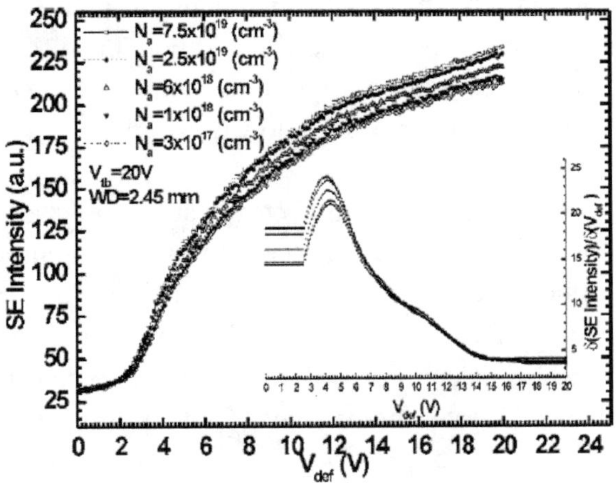

(b)

Fig. 1-35. SE energy filtered detection in the through-the-lens FEI XL30 S-FEG SEM (Kazemian *et al.*, 2006). (a) Schematic drawing of energy filtering in the through-the-lens detector. (b) S-curve output signals and the SE energy spectra data derived from them for a Si test sample consisting of six *p*-doped layers (B doped) on a *n*-doped (Sb-doped) substrate. Copyright 2006 by AIP Publishing. Reprinted with permission.

The Kazemian *et al.* (2006) proposal was made in the context of quantifying dopant contrast, and came from an earlier proposal to detect energy shifts in the SE energy spectrum (Schönjahn *al.*, 2002). An example of the S-curves and SE energy spectra reported by Kazemian *et al.* (2006) are shown in Fig. 1-35b for a test semiconductor doped specimen. The S-curves/SE energy spectra are clearly distorted and truncated, and very different to the kind of SE energy spectra predicted by theory for Si or those obtained by UHV electron energy analyzers systems (shown in Figs. 1-3b and 1-3c). One of the main reasons for this is that SEs leaving the specimen tend to disperse and are not well collimated. This is due to the fact that the magnetic semi-in lens field peak position typically lies above the specimen plane. This is quite different to the situation of the specimen being immersed fully inside immersion objective lenses, where the SEs experience a sharply decreasing field strength as they leave the specimen and are strongly collimated (see Fig. 1-20). Both the lack of collimation and the subsequent non-uniform deflection electric field action mean that electrons contributing to the output signal come from a complex mix of different emission angles and energies. In addition, the electric deflection field above the lens bore is not optimized to capture a wide range of SEs, and in practice, truncates them to be below 6 to 7 eV.

Kazemian *et al.* (2007) simulated trajectory paths in their energy filtered SE detector proposal, and found that the experimental S-curve distortions most likely came from the generation of tertiary electrons (SE3s) scattered from the deflection field electrodes. In practice, the Kazemian *et al.* (2007) SEM SE detector proposal was limited to detecting shifts on linear parts of the S-curve output signal, and was not well adapted to carrying out SE energy spectroscopy. An example of this is shown in Fig. 1-36, where experimental S-curves and their corresponding differentiated SE energy spectral signals were obtained on a test *pn* junction specimen. A voltage shift in the linear part of the S-curve was related to the in-built potential of the *pn* junction, but the SE energy spectral signals appear jagged, truncated and distorted, and have limited use for SE energy spectral applications.

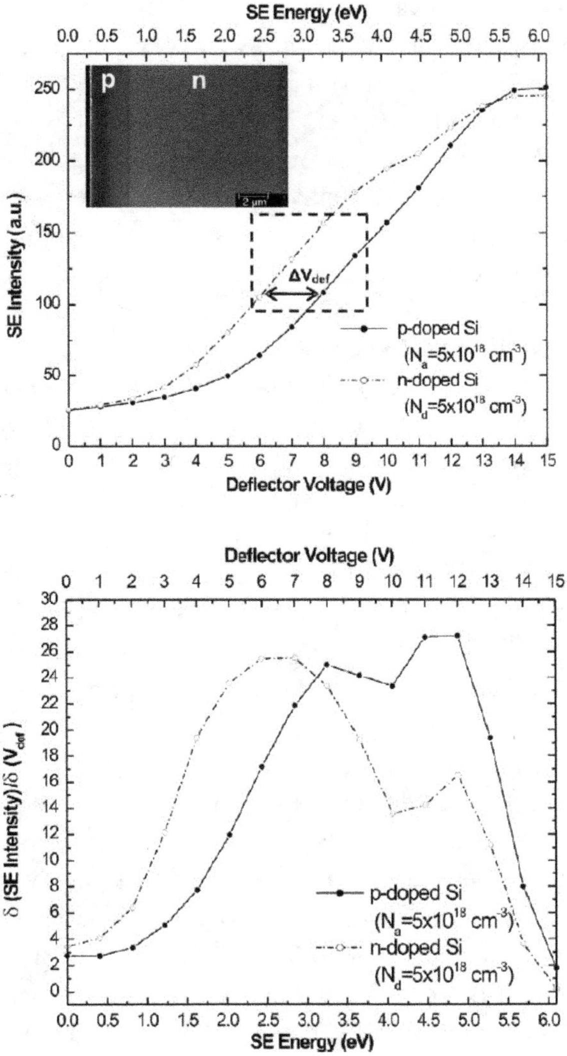

(a)

(b)

Fig. 1-36. Experimental SE energy filtering signals inside the FEI XL30 S-FEG SEM for a *pn* junction test specimen (Kazemian *et al.*, 2007). (a) S-curves from the *p*-doped and *n*-doped regions. (b) SE energy spectral signals derived by differentiating S-curves. Copyright 2007 by Elsevier. Reprinted with permission.

Apart from the initial Kazemian *et al.* (2016, 2017) attempts at monitoring energy shifts in the output S-curve signal, energy filtering inside the FEI XL30 S-FEG SEM has mainly been used for SE detector image contrast enhancement (Chee, 2016). This work has made little or no reference to the errors coming from the ad-hoc energy filtering action of the SE detection system. Recently, S-curves/SE energy spectra from the FEI XL30 S–FEG SEM SE detection system were presented in the context of examining the effect of NH_4F pre-cleaning treatment on dopant contrast (Chee, 2018). They were however, very similar in quality to the truncated/distorted spectral signals originally obtained by Kazemian *et al.* (2007) shown in Fig. 1-36b. Inconsistencies in SE energy spectral shift for the case without pre-cleaning were mainly attributed to the presence of surface ad-layers, such as contamination and native oxide, while the much larger errors in the SE energy spectral shape caused by the ad-hoc energy filtering action of the SE detection system were not mentioned or taken into account. Although effects of surface ad-layers on the SE energy spectral signal were observed, the relatively poor quality of these spectral signals limited the degree to which useful quantitative information could be extracted from them.

Another proposal for a through-the-lens SEM SE detection system being used for SE energy filtered dopant contrast applications was presented by Tsurumi and Hamada (2013). They operated a Hitachi S-4800 SEM through-the-lens SE detector in an ad-hoc high-pass energy filter mode, as shown in Fig. 1-37a. A control electrode at potential V_C acts as a retarding field electrode, and a $\mathbf{E}\times\mathbf{B}$ Wien Filter deflects SEs away from the optical axis to a detector located above the objective lens. This SE detector captures high energy SEs, the ones that have energies above the potential barrier set by V_C, and tertiary electrons coming from BSE scattering on the control electrode. Tsurumi and Hamada demonstrated how ad-hoc high-pass energy filtering in the SEM could be used to counteract the degrading effect of beam induced contamination on dopant contrast, and they produced SE energy spectral signals as part of their analysis obtained from differentiating output S-curves, shown here in Fig. 1-37b.

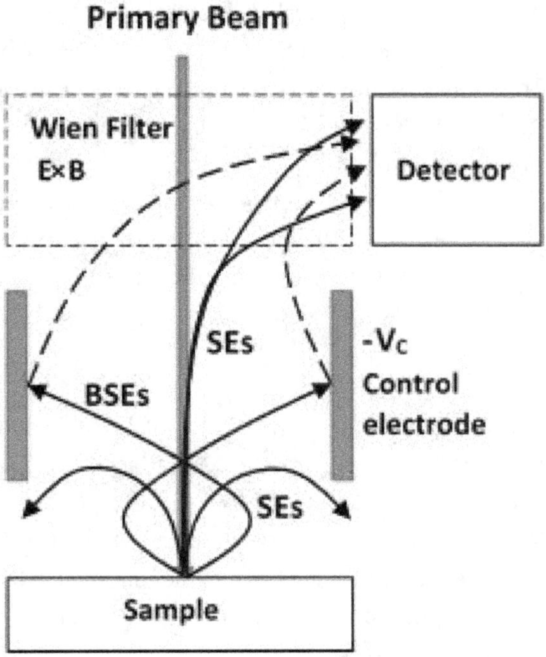

(a)

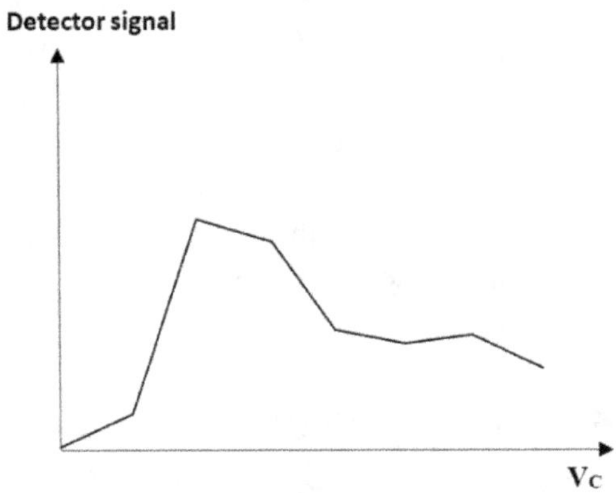

(b)

Fig. 1-37. SE high-pass energy filtering in a Hitachi S-4800 microscope with a through-the-lens detector (Tsurumi and Hamada, 2013). (a) Schematic drawing of SEM column. (b) Experimental SE energy distribution obtained by differentiating the total signal intensity with respect to Vc. Adapted from Tsurumi and Hamada. Copyright 2013 by Sumitomo Electric Industries Ltd.

The spectral signals shown in Fig. 1-37b clearly have a truncated energy range, and show evidence of there being a high energy background signal, which most likely comes from the tertiary electrons created by BSE scattering. As with other ad-hoc energy filtered SE detection system proposals, errors are produced by the detected SE signal being formed from SEs having a wide mix of emission angles and energies. Tsurumi and Hamada mainly restricted their work to showing how contamination effects on dopant contrast could be reduced by high pass energy filtering of SE detector images, they did not propose it for SE energy spectroscopy.

Other ad-hoc energy filtered through-the-lens SEM SE detection system proposals include the one made for the Magellan 400 FEG SEM (Mika *et al.*, 2016, and Konvalina *et al.*, 2019) and the Helios Nanolab SEM (Rodenburg *et al.*, 2018). The through-the-lens detector system in this case comprises of a suction tube which accelerates SEs up into the column, deflects them off-axis by a push/mirror electrode arrangement, after which they are attracted towards the high voltage of a scintillator/SE detector, as shown in Fig. 1-38a. Typical voltage values for these electrodes are, 150 V for the suction tube, -140 V for the push electrode, 20 V for the mirror, and 10 kV for the scintillator. Konvalina *et al.* (2019) simulated secondary electron trajectory paths for this column for a primary beam energy of 1 keV and working distance of 1 mm, helping to understand how it functions in an ad-hoc band-pass energy mode when the specimen voltage is varied positively, in the 0 to 16 V range. Their simulations predict that it has a detection energy bandwidth of around 3 eV.

Mika *et al.* (2016) succeeded in operating the Magellan 400 FEG SEM SE detection system as an ad-hoc band-pass energy filter, capable of providing a SE spectral signal when the specimen voltage was biased positively, as shown in Fig. 1-38b. They used a specimen consisting of Ag/Cr stripes in silicon substrate and biased it in order to collect SE energies varying up to around 15 eV. The spectral signal obtained by Mika *et al.* shown in Fig. 1-38b, are similar to the distorted/truncated spectral signals obtained by the other earlier through-the-lens SE detection systems.

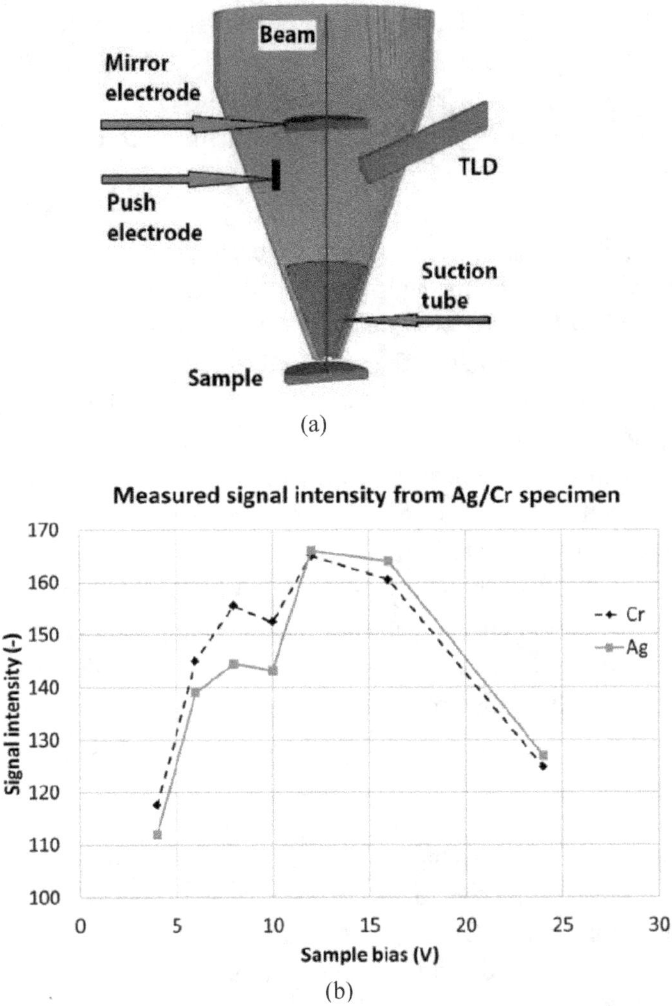

Fig. 1-38. SE energy filtering with the Magellan 400 FEG SEM through-the-lens SE detection system. (a) Schematic drawing of the column layout (Konvalina *et al.,* 2019). First appeared in Materials. Open Access. (b) SE energy spectral signals generated by operating in an ad-hoc band-pass energy mode (Mika *et al.*, 2016). First appeared in Proceedings of the 15th International Seminar on Recent Trends in Charged Particle Optics and Surface Physics Instrumentation. Open Access.

Rodenburg *et al.* (2018) obtained several puzzling peaks in their ad-hoc energy filtered output SE energy spectral signal from a Ag/Cr stripe specimen using the through-the-lens SE detection system of the Helios Nanolab SEM, which has a column design similar to the Magellan 400 FEG SEM (Fig. 1-38a). Their spectral signals differed significantly from corresponding Monte-Carlo predictions, and it was not clear whether these peaks were created by spectrometer artifacts or by genuine material properties.

Recently, SE energy spectral signals were reported by Hashimoto *et al.* (2020), who used the through-the-lens SE detection system of a Hitachi Regulus 8220 SEM to provide ad-hoc band-pass energy filtering. Their column layout is similar to the one used by Tsurumi and Hamada (2013) inside a Hitachi S-4800 SEM, but the difference is that Hashimoto *et al.* (2020) deliberately switched off the **E**×**B** Wien Filter, and relied on the electric field from the nearby scintillator to punch through into the Wien Filter assembly to deflect SEs off-axis, as shown in Fig. 1-39a. They were then able to create a SE energy band-pass filter action by applying positive voltages to the specimen in the 0 to + 5 V range, producing an effective detection energy bandwidth of around 0.9 eV. Electrons having energies above the pass energy travel too fast to be deflected on to the scintillator, while electrons having energies below the pass energy (below the potential barrier energy), are attracted back on to the specimen.

The output spectral signals obtained by Hashimoto *et al.* (2020) are much more continuous and cover a larger energy range than the distorted/truncated SE energy spectral signal produced by previous through-the-lens SE detection systems, as shown in Fig. 1-39b. However, they have a relatively high background signal, reaching around 60% of the SE peak height at 0 eV, and 15% of the SE peak height at 20 eV. The background signal distorts the output signal shape throughout the detected energy range, causing it to differ significantly from Monte Carlo predictions. Most the experimental results presented by Hashimoto *et al.* (2020) take the form of energy filtered SE images. The presence of the relatively large background signal makes it difficult to extract meaningful quantitative information by tracking changes in the output SE energy spectral signal shape.

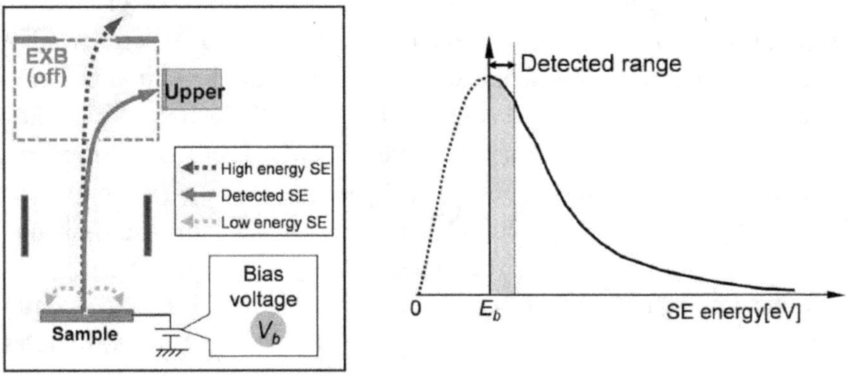

(a)

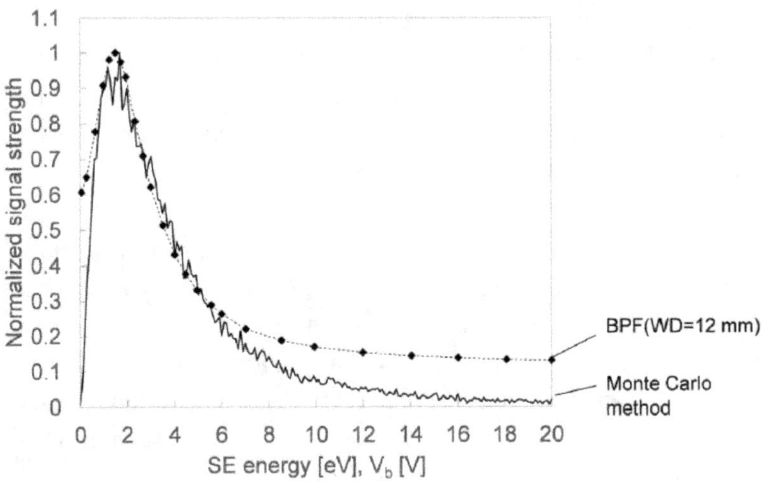

(b)

Fig. 1-39. Band-pass energy spectral signals obtained by a Hitachi Regulus 8220 through-the-lens SE detection system (Hashimoto *et al.*, 2020). (a) Schematic drawing of the band-pass energy mode of operation. (b) Experimental spectral signal obtained for a copper polycrystalline plate sample and the predicted SE energy spectrum predicted by Monte Carlo simulations. Copyright 2020 by Elsevier. Reprinted with permission.

1.6 Band-pass electric toroidal sector spectrometer attachments

The work on in-lens multi-channel SE energy spectrometers for the SEM has mainly been of the instrument development kind, where proof of concept experiments have been carried out to in order to demonstrate that SE energy spectra can be obtained. Their initial spectral results show promise, but they require more development. There is a need for instance, to develop a suitable primary beam scanning system for the in-lens analyzer, and synchronize it to an array of multi-channel detectors in order to provide real-time imaging. It will also be useful to build in some sort of angle filtering action, so that the SE energy spectral distribution can be acquired as a function of different emission angles, providing more data about the nature of SE emission. The Yong and Thong (2000) retarding field analyzer attachment has also generated SE energy spectra in the SEM, but it was designed mainly for the spectral analysis of insulator bulk specimens, and did not combine its spectral information with SE detector imaging. More development is also needed before it can become a general purpose SE energy spectroscopy attachment for the SEM.

Most of the energy filtered SE energy analyzer proposals, including the in-built SEM SE detector ones, have not proved to be successful for acquiring the SE energy spectrum inside a SEM, or precise enough for say tracking small changes in its shape. However, there are two other SEM energy spectrometer attachment designs that have already been successively used for a variety of different materials science applications. Both of these energy analyzers are of the electric sector toroidal band-pass type and represent a move back towards the simpler designs that were originally made by Wells and Bremer (1968) and Hannah (1974), where an electric band-pass energy analyzer attachment is placed in the field-free region above the specimen in a conventional SEM. Although this approach has the disadvantage of increasing the working distance compared to through-the-lens in-lens SE detection systems, it gives more scope for implementing the same kind of angular focusing/energy dispersion principles that are widely used in other band-pass electron energy analyzers, such as the CMA or HDA, which have well established applications in surface science. An introduction to the basic design principles of such electron energy analyzers can be obtained from a variety of different sources (Sar-el, 1967; Zouros and Benis, 2002, and Khursheed, 2011b), and will not be discussed here in detail.

The return to a band-pass energy spectrometer design for the SEM first came with the work of Rau and Robinson (1996) who presented an electric sector toroidal analyzer attachment in order to capture BSE energy spectral information. Their design consisted of a small sector deflection angle analyzer (90° or less), a schematic diagram of which is shown in Fig. 1-40. The version of the electric toroidal energy analyzer shown in this figure is the one that was later adapted by Gostev *et al.* (2013) to capture both the energy spectra of SEs and BSEs. The primary beam travels down the analyzer rotational axis of symmetry and generates scattered electrons that travel out in radial directions. The toroidal geometry is able in principle to transport scattered electrons over a full 2π radian azimuthal angular range, from the field-free region above the specimen on to an exit ring aperture where they are focused and band-pass energy filtered. A toroidal analyzer geometry is therefore well suited to capturing scattered electrons that leave the specimen in a conventional SEM. The relative energy resolution of toroidal analyzers (% detected narrow band energy width divided by pass energy), typically lies in the 0.2 to 3% range, similar to the relative energy resolution range of other band-pass electron energy spectrometers such as the CMA or HDA.

In terms of carrying out SE energy spectroscopy in the SEM, it is important to emphasis that toroidal band-pass energy analyzers are very different to any of the retarding field analyzer designs discussed so far. They reconstruct the SE energy spectrum from a much narrower set of SE emission angles and energies. A 1% relative energy resolution for instance, gives a 10 meV detection energy width for 1 eV electrons, or a 50 meV energy width for 5 eV electrons. Band-pass electron energy analyzers typically restrict their entrance polar angular spreads to within a ± 3° to ± 6° range. This means that their output SE spectral signals will provide a much purer and accurate representation of the SE energy spectrum at the specimen, minimizing the mixing of angular and energy emission information together. Another important aspect of the toroidal band-pass analyzer geometry is that it has an intrinsically high collection range in the azimuthal direction, up to 2π radians, and this partially compensates for its narrow energy detection band and relatively small entrance polar angular range.

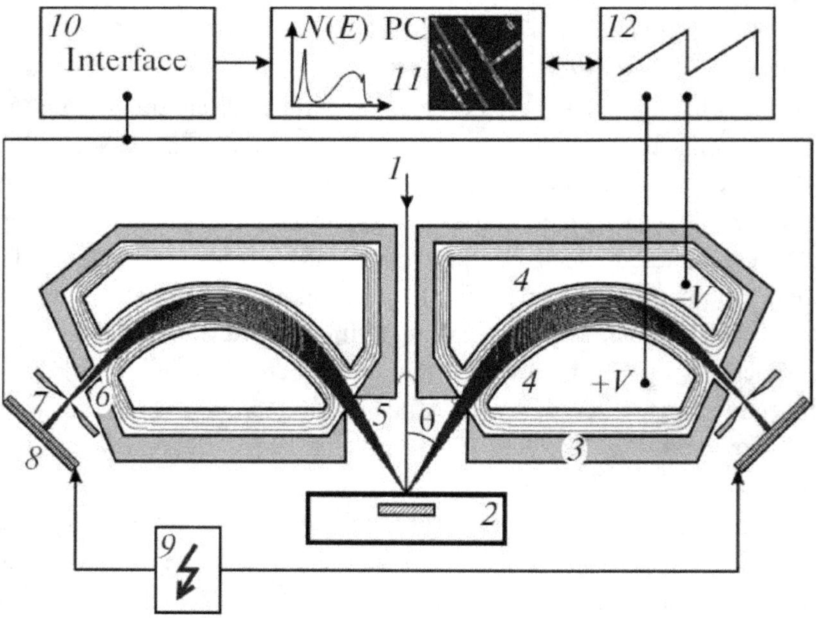

Fig. 1-40. Small deflection angle electric toroidal spectrometer attachment design by Gostev *et al.* (2013). (1) electron probe of SEM, (2) sample, (3) body of spectrometer, (4) radial field toroidal deflector, (5, 6) entrance and exit ring-shaped Herzog compensating electrodes, (7) ring slot diaphragm, (8) MCP detector, (9) power supply and high-voltage isolation of MCP, (10) interface unit, (11) SEM monitor and PC, and (12) constant and saw-tooth feed voltage of spectrometer. Copyright 2013 by Maik Nauka/Interperiodica. Reprinted with permission.

In the first version of the Rau and Robinson (1996) toroidal spectrometer attachment design, the relative energy resolution was estimated to be around 1%, for a polar angle entry of 20° ± 4° and sector deflection angle of around 90°. It was used to obtain BSE energy spectra for primary beam energies in the 15 to 30 keV range and acquire energy filtered BSE images for multi-layered samples. Rau *et al.* (2002) later improved their design through direct ray tracing simulation for 10 keV BSEs. The predicted energy resolution for the optimized analyzer design was improved to around 0.25% for an entrance polar angular spread of ± 3°.

An example of experimental SE energy spectra obtained from the small-angle toroidal energy analyzer attachment are shown in Fig. 1-41. They were obtained from different points on a doped test semiconductor sample (Gostev *et al*., 2013). This was one of the few instances that the small-angle toroidal energy analyzer attachment was applied to SE energy spectroscopy. The quality of these SE energy spectra are in a different category to the jagged/truncated ones produced by most previous energy filtered SE detection systems (Figs. 1-34 to 38), and are more in line with the kind of results produced by UHV energy spectral systems (shown in Fig. 1-3).

The small angle toroidal analyzer attachment SE energy spectral results shown in Fig. 1-41 demonstrate that dopant contrast causes an overall change in the SE spectral signal shape, similar to the original spectral results reported by Wells and Bremer (1968), but it does so with a much better energy resolution (tens of meV compared to several eV), and unlike the CMA attachment designed by Wells and Bremer, it does not deflect or defocus the primary beam.

The spectral results shown in Fig. 1-41 have a similar form to the SE energy spectral signals taken from a *pn* junction sample by Dapor *et al.* (2009) using an Auger electron energy spectrometer in a UHV chamber. Dapor *et al.* were able to obtain a good fit between the shape of their experimental SE energy spectral signals and those predicted by Monte Carlo simulations. The SE energy spectral results obtained by Gostev *et al.* (2013) demonstrate that similar results can be obtained within a SEM.

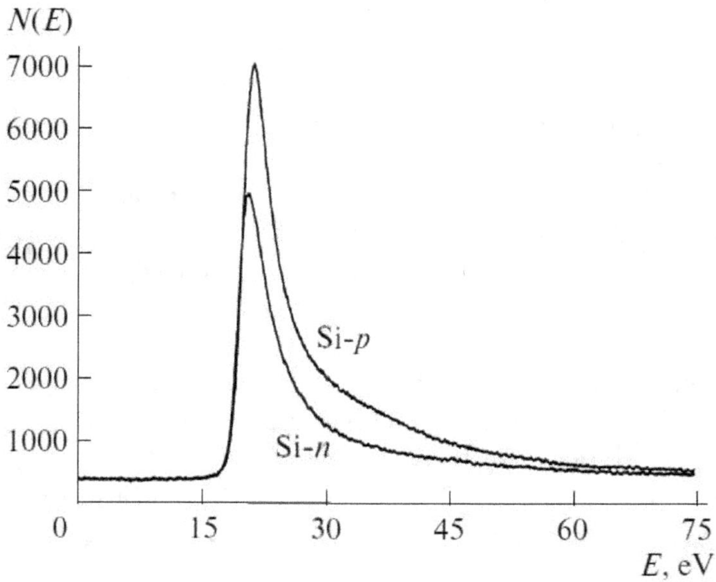

Fig. 1-41. Experimental SE energy spectra reported by Gostev *et al.* (2013) obtained by the small-angle toroidal analyzer attachment taken on a Si wafer sample having 10^{19} cm^{-3} *p*-type regions in a 10^{16} cm^{-3} *n*-type substrate. Copyright 2013 by Maik Nauka/Interperiodica. Reprinted with permission.

Khursheed and Hoang (2008b) developed a wide-angle sector electric toroidal energy attachment design for the SEM. Their design has a sector deflection angle of around 215° and a central polar entrance angle of 45°. Figure 1-42 shows trajectory paths passing through the analyzer with an entrance angular spread of ±6°. The analyzer functions by forming an intermediate focus point in the r-z plane, and having an exit slit at a second focal point. The spectrometer's plates are centered at point 0. This design allows for second-order geometrical aberration contributions accumulated before and after the intermediate focus to cancel, leaving third-order geometrical order aberrations at the exit focal plane.

In the first version of the spectrometer (Khursheed and Hoang, 2008b), the analyzer was simulated to have a relative energy resolution of 0.146% (around 0.3% base resolution) for an input angular spread of ±6°, which is about 7 times better than that of first-order focusing electrostatic toroidal spectrometers (for the same input angular spread), such as the small sector angle one designed by Rau and Robinson (1996) or Gostev *et al.* (2013). This entrance polar angular spread corresponds to a transmission of 20% for a single energy, assuming a cosine polar angular emission distribution (Khursheed, 2011c). These energy resolution–transmission characteristics are comparable to those predicted for the second-order focusing CMA (Sar-el, 1967).

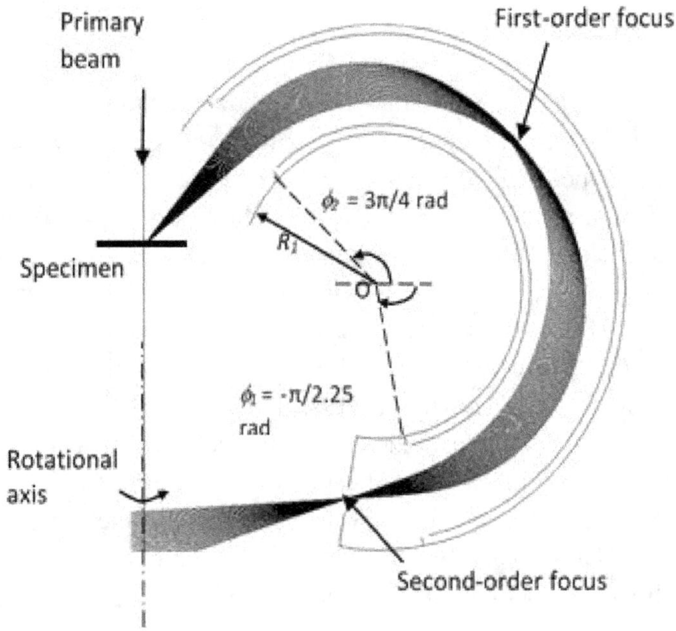

Fig. 1-42. Direct ray tracing in the wide deflection angle second-order focusing toroidal analyzer design by Khursheed and Hoang (2008b) for SE and BSE energy spectroscopy. Inner radius R_1 of 14 mm and outer radius of 22 mm (central radius of 18 mm). The entry polar angular range is 45°±6°, and the sector deflection angle is 215°.

Apart from its better focusing properties, the Khursheed and Hoang wide-angle sector toroidal analyzer design has several other advantages over the small angle sector toroidal analyzer design. It requires, for instance, only one detector to capture all electrons that emerge from its ring slit output aperture, since they exit moving towards the rotational axis of symmetry. For the small angle sector toroidal analyzer, electrons travel out in different radial directions, requiring the use of an array of detectors in the azimuthal direction. Another important advantage is its compact size, which allows for smaller working distances, 15 mm compared to the 30 mm required for the original Rau and Robinson (1996) design. These advantages give it over an order of magnitude better signal-to-noise characteristics (for the same energy resolution) and a higher SEM image resolution than the Rau and Robinson small-angle electric toroidal energy analyzer attachment design.

The most important point about the wide-angle toroidal analyzer attachment in the present context is that it has already been used to demonstrate the benefits of SE energy spectroscopy for a variety of different applications. Experimental results show that it can overcome local surface field effects (Srinivasan and Khursheed, 2015), and has high signal-to-noise characteristics for voltage contrast measurements (Hoang *et al.*, 2011) and quantitative dopant contrast (Srinivasan and Khursheed, 2015), typically over an order of magnitude better than what was achievable previously by energy filtered SEM SE detectors (Kazemian *et al.*, 2006). The wide-angle electric toroidal analyzer was also used to report some new applications, such as monitoring buried electric charge trapped at the interfaces of multifunctional oxides (Srinivasan and Khursheed, 2014), or mapping the build-up of charge in metal-insulator-semiconductor (MIS) samples (Srinivasan *et al.*, 2018).

Over the past decade or so, both the small-angle and wide-angle toroidal energy spectrometer attachments have presented experimental results to demonstrate the potential benefit of using SE energy spectroscopy inside the SEM to quantify contrast mechanisms such as dopant contrast. However, most the papers published on this subject over the same period do not seem to make any reference to them (Agemura *et al.*, 2018; Agemura *et al.*, 2017; Chee 2016; Chee 2018; and Sekiguchi, 2016). Taking a broader view, it would seem that the whole subject of SE energy spectroscopy in the SEM is relatively unknown, despite it having been originally proposed over 50 years ago by Wells and Bremer (1968). Instead, research work directed towards quantifying different contrast mechanisms in the SEM has, for the most part, relied upon using in-built SE detectors rather developing separate SE energy spectrometer attachments. Take for instance, the semi-quantitative dopant contrast research work carried out using the in-built SE detectors of field emission SEMs from around the mid-1990s to the present day. Very little reference to the possibility of using SE energy spectroscopy has been explicitly made, even after it had encountered scattering problems caused by the presence of a native oxide layer (Elliot *et al.*, 2002, and Kazemian *et al.*, 2004) or after it had reported complex image contrast reversal effects created by the primary beam voltage or scan rate changing (El-Gomati *et al.*, 2005). Ironically, Wells and Bremer (1968) had specifically presented experimental results to show how SE energy spectroscopy could help understand the complex image contrast reversal effects caused by operational parameters like the primary beam voltage and scan rates changing (Fig. 1-8).

A similar omission is also evident in the proposals which advocated some sort of energy filtering by through-the-lens SE detector systems for quantitative dopant contrast. Not only has there been no reference to the results reported by the small-angle or wide-angle toroidal energy analyzer attachments over the last decade or so, but the whole subject of energy analyzer design has been conspicuously absent. The relatively poor SE energy spectra obtained by these in-built SE detection systems functioning as ad-hoc retarding field analyzers (discontinuous and truncated SE spectra), does not appear to have initiated any wider discussion about the need to design SE energy analyzer attachments specifically for the SEM (Chee, 2016, 2018).

In some instances, proposals in support of using SE energy spectroscopy in the SEM were actually based upon experiments carried out in scanning Auger electron spectrometer systems operating in relatively poor vacuum conditions (HV). Joy *et al.* (1999) advocated the use of SE energy spectroscopy in the SEM for performing microanalysis and defect review, and the SE energy spectral results that they provided in support of it were obtained in a Perkin Elmer 660 scanning Auger microprobe using a Cylindrical Mirror Analyzer (CMA) operating in a vacuum level of 10^{-7} Torr. They later went on to publish more experimental SE energy spectra data from a variety of different metal specimens, for the purpose of better understanding secondary electron imaging in the SEM (Joy *et al.*, 2004), but the instrument they used to do it was a PHI 680 Auger Electron Spectroscopy system (operating in a poorer vacuum level than its standard UHV operating condition). Joy *et al.* did not seem to consider the possibility of actually using a SEM, despite the fact that several previous SE energy spectrometer attachments for the SEM had already been reported (Wells and Bremer, 1968; Khursheed and Karuppiah, 2001; Khursheed, 1992; and Rau and Robinson, 1996). Their attempts at understanding SE energy spectroscopy in the SEM by performing experiments in scanning Auger electron instruments highlights the urgent need for SEM manufacturers to make SE energy spectrometer attachments available as add-on accessories.

This chapter has presented at least three different add-on attachment designs which have already proved capable of generating experimental SE energy spectra in the SEM (Gostev *et al.*, 2013; Khursheed and Hoang, 2008b; and Khursheed and Karuppiah, 2001), and four different proposals for redesigning in-lens SEM columns to incorporate SE energy spectroscopy. All of these four proposals have provided promising proof-of-concept experimental SE energy spectral results. Two of them, are multi-channel fully magnetic immersion objective lens spectrometers (Khursheed, 1992; and Khursheed and Karuppiah, 2001), while the other two, are multi-channel spectrometers integrated into magnetic immersion combined retarding field objective lenses, the kind that are widely used for low voltage SEM (LVSEM) applications (Kleine and Plies, 2004; and Khursheed and Osterberg, 2006). These add-on spectrometer attachments and in-lens SEM column multi-channel spectrometer designs already provide examples of how SE energy spectroscopy might be incorporated into SEMs.

The general lack of reference to the possibility of using SE energy spectrometers in the SEM for materials science applications is the main motivation for writing this book. Beyond SEM instrument redesign challenges, there is a need to highlight what kind of benefits SE energy spectroscopy in the SEM might bring, if the problem of finding the right electron energy spectrometer is solved. The following pages aim to demonstrate how the use of a SE energy spectrometer attachment specifically designed for the SEM can better quantify material and dopant contrast, and in doing so, provides information that goes far beyond the semi-quantitative information captured by conventional SE detectors, effectively transforming the SEM into a powerful analytical tool that can map useful materials science properties on the nano-scale.

The experimental results presented in the following pages were obtained through the use of the Khursheed and Hoang (2008b) wide-angle electric toroidal band-pass electron energy attachment. A summary of the earlier experimental results obtained from this spectrometer will be presented, as well as more recent findings obtained in the last one to two years. Among the more recent findings in the subject of quantitative material contrast is the discovery that fine structure features present in the SE energy spectrum at low primary beam voltages can be used to acquire bulk density of states information. Another discovery is that several parameters of the SE energy spectrum depend linearly on sample atomic number, such as SE peak height, and this opens up the possibility of carrying out rapid high resolution elemental identification using secondary electrons, information that in some instances can obviate the need to use the more conventional Energy-dispersive X-ray Spectroscopy (EDS) technique, or at least complement it. SE energy spectroscopy is also used to map metal elemental regions having close atomic numbers at low primary beam voltages (1 kV and less), demonstrating its potential to play a useful role in the subject of low voltage SEM (LVSEM).

In the subject of quantitative dopant contrast, the recent findings include making *in-situ* native oxide thickness measurements, mapping space-charge regions, characterizing the dopant concentration of bulk silicon wafers, and obtaining reliable dopant profiling results, ones that agree well with results obtained by more conventional industrially accepted dopant profiling methods. In addition, SE energy spectroscopy is used to provide two-dimensional dopant profiling of Gallium ion beam irradiated areas on intrinsic silicon specimens, and map surface charge distributions that are present when inspecting non-conductive specimens under charge-balance LVSEM conditions.

Although there have been a variety of different electron energy analyzer attachment designs made specifically for the SEM since the development of the wide-angle toroidal analyzer, no reliable SE energy spectral results have as yet been produced from them. Some, such as the radial mirror analyzer (RMA) design (Hoang and Khursheed, 2011) were proposed specifically for the purpose of SE energy spectroscopy, while others, were proposed for the purpose of carrying out AES in the SEM (Cubric *et al.*, 2011; and Khursheed *et al.*, 2012). The RMA attachment in particular, has some notable design advantages over the wide-angle toroidal energy analyzer, but it also brings additional practical challenges. The RMA's potential benefits and practical difficulties are summarized in Appendix 1. Similarly, the parallel radial mirror analyzer (PRMA) attachment design (Khursheed *et al.*, 2012), has many potential advantages for AES applications in the SEM, where it is used together with a low energy Argon ion gun inside the specimen chamber. In principle, the PRMA design can also be used for parallel SE energy spectroscopy applications. However, it also involves additional practical difficulties. Its potential benefits and additional practical difficulties are summarized in Appendix 2.

Chapter 2

The Wide-angle Toroidal Energy Analyzer Attachment

The wide-angle toroidal band-pass electron energy spectrometer attachment was briefly introduced in the last chapter and has been described in detail elsewhere (Hoang *et al.*, 2011a). Most of the SE energy spectral experimental data presented in the remaining chapters of this book were obtained by using a prototype version of the wide-angle toroidal band-pass add-on energy analyzer attachment. This chapter describes how it was installed into a SEM, its basic modes of operation, its energy resolution/noise characteristics, a comparison of its SE experimental spectral data to analytical energy distributions, and its response to specimen fringe/surface electric fields.

2.1 The experimental setup and modes of operation

Figure 2-1 illustrates how the wide-angle band-pass toroidal energy spectrometer attachment (Hoang *et al.*, 2011a; Srinivasan and Khursheed, 2015b) is installed into a SEM specimen chamber. The attachment is mounted on the SEM stage and aligned to the primary beam axis *in-situ* by using the secondary electron (SE) detector image. The height of the stage is also adjusted such that the working distance to the sample surface is 15 mm. A polar angular limiting aperture of ± 8°, together with its 100° azimuthal angular collection range gives a transport efficiency of around 7.65% for any given pass energy, assuming that the SE emission has a cosine polar angle dependence (Khursheed, 2011c).

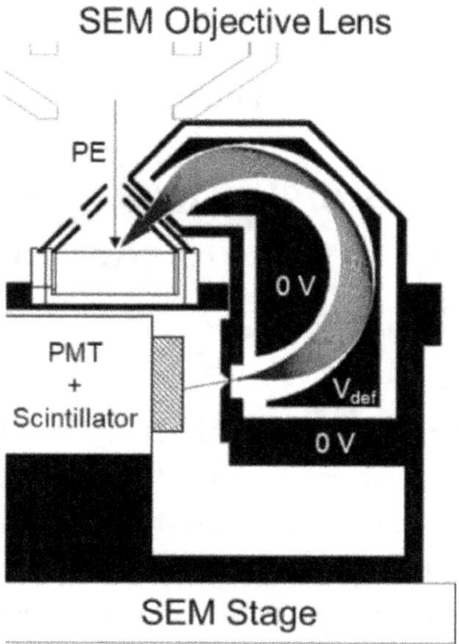

Fig. 2-1. Installation of the wide-angle toroidal energy analyzer attachment on to a SEM specimen stage. Direct ray tracing of electron trajectory paths is shown for a polar angular spread of 45 ± 15° with a ± 8° limiting entrance aperture.

The purpose of the conical cap electrodes around the specimen, shown in Fig. 2-1, is to give the possibility of varying the pass energy of SEs as they go through the analyzer. The inner cap voltage is typically set to the same voltage that is used to bias the specimen, while the outside cap voltage is set to ground potential. This provides a field free region directly above the specimen, while making it possible to slow down or speed up SEs as they travel through the analyzer. It should be noted that the analyzer exit aperture is placed vertically, slightly different to the conical focal exit plane in the original analyzer design (Khursheed and Hoang, 2008). This was done for ease of manufacture.

Figure 2-2a shows a schematic diagram of how the spectral data acquisition system functions. Two methods of data acquisition are used: a spectral acquisition mode and a spectroscopic imaging mode. The first method is a direct SE energy spectral acquisition mode, where a point or a small area on the sample is continuously raster-scanned via the SEM in-built scan coils, while a PC-DAQ system ramps through a series of analyzer deflector voltages V_D (typically from 0 to 25 V). Each value of V_D corresponds to a given analyzer pass energy E_P, which in turn, is directly related to a given SE emission energy. In this way, the SE energy spectrum of the point or small area being irradiated on the sample is directly acquired. The dwell time can be adjusted through the software that controls the PC-DAQ system. A scintillator-PMT detector is positioned beyond the analyzer exit slit, and a current-to-voltage amplifier then outputs the result back to the PC-DAQ system where it is stored for each analyzer deflection voltage (analyzer pass energy value E_P) in order to subsequently form the output SE spectral signal. The deflection voltage step size was selected to typically vary E_P in 0.1 to 0.25 eV step changes within a 0 to 15 eV range.

The second method of data acquisition is a spectroscopic imaging mode, where the PC-DAQ system acquires a series of images for different analyzer deflector plate voltages V_D (pass energies E_P). In this case, the SEM scan system is controlled externally via software control, through the PC-DAQ system, as illustrated in Fig. 2-2b. The SE energy spectrum for every pixel in the image is then reconstructed from a series of stored images, each corresponding to a different value of V_D. The spectroscopic imaging mode is used to acquire 2D mapping or line profile data, while the direct spectral method is used to acquire the SE energy spectrum for a few selected points/areas on the specimen.

In the original analyzer design (Khursheed and Hoang, 2008), the outer and inner deflection plates were both biased, at $-V_D$ and $+V_D$ respectively, and the fixed ratio between electron pass energy and deflection voltage E_p/V_D was found to be 2.293. However, in the current set-up, only the outer deflection plate was biased at $-V_D$, while the inner deflection plate was grounded. This was done in order to eliminate deflection voltage synchronization errors, and it naturally leads to a change in the E_p/V_D ratio. In order to determine the new relationship, electric field distribution and ray-tracing simulations were performed using Lorentz-2D (Integrated Engineering Software). SEs having a certain pass energy E_P were plot along the analyzer central entrance polar angle of 45°, and the voltage V_D was varied until the exit ray passed through the center of the exit aperture. These simulations were then repeated for a range of different central electron energy values, and a constant E_p/V_D ratio of 1.67 was obtained.

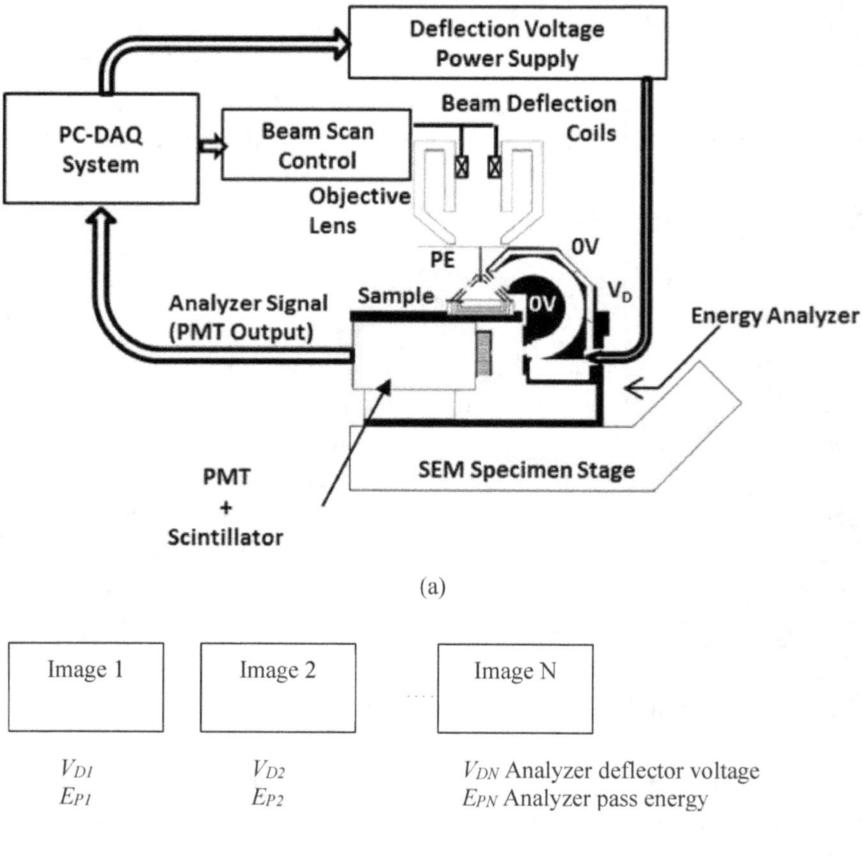

Fig. 2-2. Data acquisition setup for the wide-angle toroidal energy analyzer attachment. (a) Schematic diagram of the spectral data acquisition system. (b) Block diagram of the spectroscopic imaging mode.

2.2 Analyzer energy resolution and shot-noise considerations

Most electron energy spectrometer systems need to be carefully designed in order to reach a viable compromise between energy resolution and output signal-to-noise ratio. In the case of SE energy electron energy spectroscopy however, energy resolution is not the main parameter of interest, and it can be sacrificed in order to maximize the output signal-to-noise ratio. This is because the SE energy spectrum has relatively broad variations across a low energy range (0 to tens of eV). The relative energy resolution of band-pass energy analyzers is in general, dependent on both the analyzer geometric aberrations and the exit slit width at its output focal plane. Previous experiments carried out by Hoang *et al.* (2011b) on the wide-angle toroidal energy analyzer for an entrance ± 8° polar angular spread had measured its relative energy resolution to be 1% for an exit slit width of 200 μm, and 0.6% for an exit slit width of 100 μm, giving an estimated geometric aberration limited energy resolution of 0.38%. This energy resolution corresponds to the Full-Width-Half-Maximum (FWHM) of a Gaussian distribution response at a single energy. In the present experimental setup, the width of the exit slit was deliberately increased to 500 μm in order to give better output signal-to-noise characteristics. Since the spectrometer is effectively operating in an exit aperture limited mode, its relative energy resolution has therefore been scaled up to be approximately 2.5%.

Since the SE energy spectrum is reconstructed by varying the analyzer pass energy, the accuracy of the captured SE energy spectrum naturally depends on how rapidly the spectral signal varies in comparison to the analyzer detection energy (width) ΔE. Energy dispersion at the analyzer output focal plane varies linearly with the pass energy, E_P, so that the detected energy width is given by,

$$\Delta E = k\, E_P \tag{2.1}$$

where k is a constant. The analyzer signal is therefore inherently different in form to the SE emission energy distribution. At any given SE energy, the analyzer signal is integrated over a finite bandwidth ΔE around the pass energy E_P, and since ΔE increases linearly with E_P, the analyzer spectral signal $S(E)$ expressed in terms of an SE emission distribution $N(E)$ is given approximately by,

$$S(E) \approx N(E).\Delta E = k.N(E).E \tag{2.2}$$

This formula assumes that ΔE is relatively small. Figure 2-3 shows both the normalized theoretical SE emission distributions for gold and silicon and their corresponding theoretical SE analyzer signals calculated by using equation 2.2, where the Chung and Everhart (1974) distribution is used for gold, and the Henke *et al.* (1979) distribution is used for silicon (see equations 1.1 and 1.2 in chap. 1). The work function (W) for gold is taken to be 5.1 eV (Lide, 2009) and the electron affinity χ for silicon is 4.05 eV (Dapor *et al.*, 2008). The theoretical analyzer signals are much broader than the SE emission distribution, caused by the fact that the analyzer signal is proportional to $N(E).E$, not $N(E)$. For metal, the SE peak position for $N(E)$ should occur at $W/3$, while for $S(E)$, it lies at W.

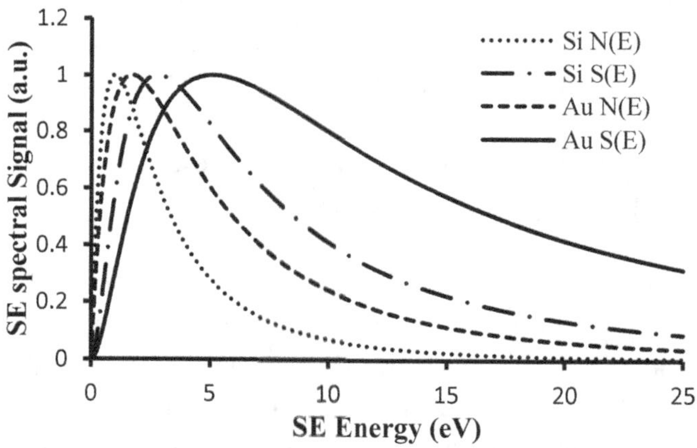

Fig. 2-3. Normalized theoretical SE emission distributions N(E) for Si and Au and their corresponding SE analyzer signals S(E).

From an energy resolution point of view, the energy analyzer has two main requirements. The first, is to reproduce the overall SE energy distribution shape, while the second, is to capture fine spectral features.

The output signal, S(E), is produced by the convolution of the initial SE energy distribution and the analyzer transfer response, which in the case of a band-pass analyzer, is a detection impulse response centered at the pass energy E_p. The detection response is usually approximated by a Gaussian distribution whose standard deviation, σ, is related to its FWHM by FWHM/2.355. The analyzer detection response can then be estimated for each SE energy. Figure 2-4a superimposes the Gaussian distribution detection response calculated in this way at the selected SE energies of 0.25, 0.5, 1, 2, 3, 4, 5, 6, 7, and 8 eV on to the Chung-Everhart energy distribution (work function of 4 eV) for an analyzer relative energy resolution of 2.5%. It indicates that the analyzer response is extremely narrow for low SE energies (< 3 eV). In general, the SE energy distribution has relatively broad variations with respect to the FWHM of each Gaussian distribution, which means that the sampling error due to the finite energy resolution of the analyzer will be relatively small. Although the width of the Gaussian shaped detection response grows with SE energy, the sampling error is partially compensated by the SE energy spectrum gradient being comparatively smaller in the tail end of the distribution.

Figure 2-4b shows the result of calculating the convolution integral formed between the analyzer Gaussian detection response and the SE energy distribution (Chung-Everhart distribution at a work function of 4 eV) for the relative energy resolutions of 2.5 and 10%. There are no visible differences in the shape of the analyzer output signal from its ideal response given by equation 2.2, which is why percentage deviations from S(E) are plot. The maximum overall deviations are 0.0178 and 0.286% at 15 eV for the relative energy resolutions of 2.5 and 10% respectively. At the relative energy resolution of 2.5%, the percentage deviations in spectral signal peak height, most probable energy, and standard deviation were calculated to be, 0.0033%, 0.024% and 0.00017% respectively. In general, these levels of error were found to be relatively small in comparison to those incurred by shot noise.

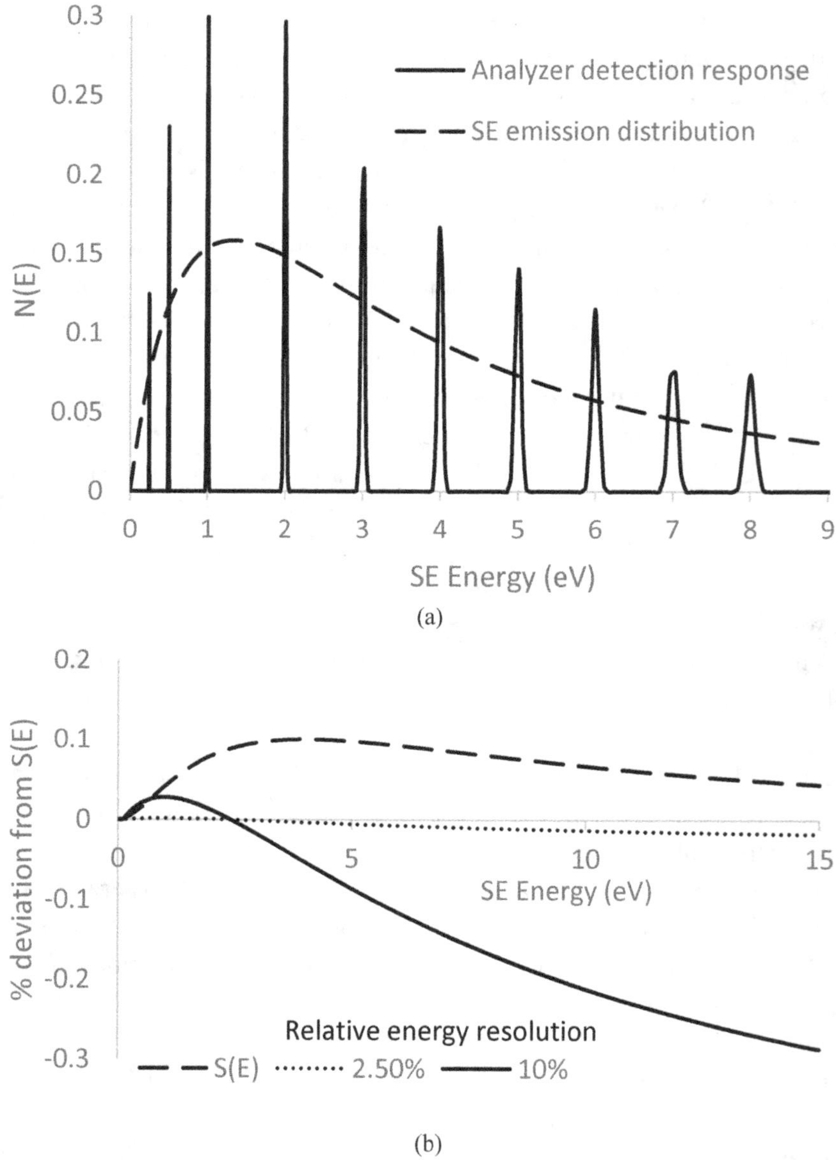

Fig. 2-4. Analyzer output signal dependence on relative energy resolution. (a) Selected energies of 0.25, 0.5, 1, 2, 3, 4, 5, 6, 7 and 8 eV on the Chung-Everhart SE energy spectrum (W = 4 eV) for a relative energy resolution of 2.5%. (b) Ideal analyzer output signal S(E) and percentage deviations from it for relative analyzer energy resolutions of 2.5, and 10%.

Figure 2-4a gives an approximate indication of what can be expected in terms of resolving fine structural spectral features on a slowly varying SE background signal shape. A relative resolution of 2.5% should be able to resolve signal changes that occur over fractions of an eV: 0.025 eV at a SE energy of 1 eV, 0.1 eV at a SE energy of 4 eV, and 0.25 eV at a SE energy of 10 eV. The smallest valence band fine structure spectral features reported by Willis and Christensen (1978) typically had a FWHM of around 0.75 eV at 3 eV. The present analyzer, operating with a relative energy resolution of 2.5% for the same energy, should be capable of capturing fine spectral features having FWHMs down to 0.075 eV.

SE energy spectroscopy energy resolution requirements are much less demanding than those for AES or XPS, where relative energy resolutions of better than 1% are required in order to resolve the presence of sharp narrow peaks (FWHM of 1 eV) in the 50 eV to 2 keV energy range (Prutton and El Gomati, 2006). Capturing a spectral peak having a FWHM of 1 eV at 1 keV for instance, requires a relative energy resolution of 0.1 %. The much lower relative energy resolution requirement in SE energy spectroscopy means that it can have better output signal-to-noise characteristics, important for mitigating shot noise and contamination effects.

Figures 2-5a and 2-5b show experimental SE energy spectra obtained from the wide-angle toroidal energy analyzer for a test gold sample, with and without specimen/inner cap electrode biasing. A FEI Quanta 3D Focused Ion Beam (FIB)-SEM system was used, with a primary beam voltage of 2 kV and beam current of 40 pA. A deflection plate voltage range of 0 to 15 V was scanned in 0.1 V steps with a dwell time of approximately 0.2 s per point. Five separate waveforms were collected, each from a different area on the specimen in order to reduce contamination effects. Figures 2-5a and 2-5b indicate that the peak signal level rises by around an order of magnitude for the biased case, compared to the unbiased case, and it exhibits considerably less shot noise.

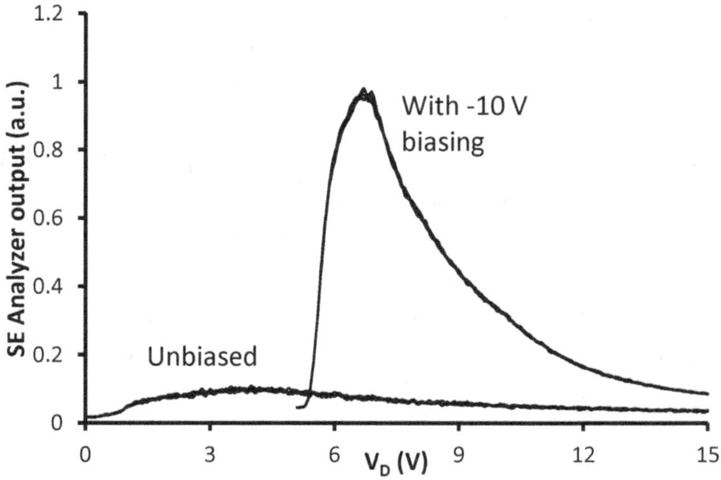

(a)

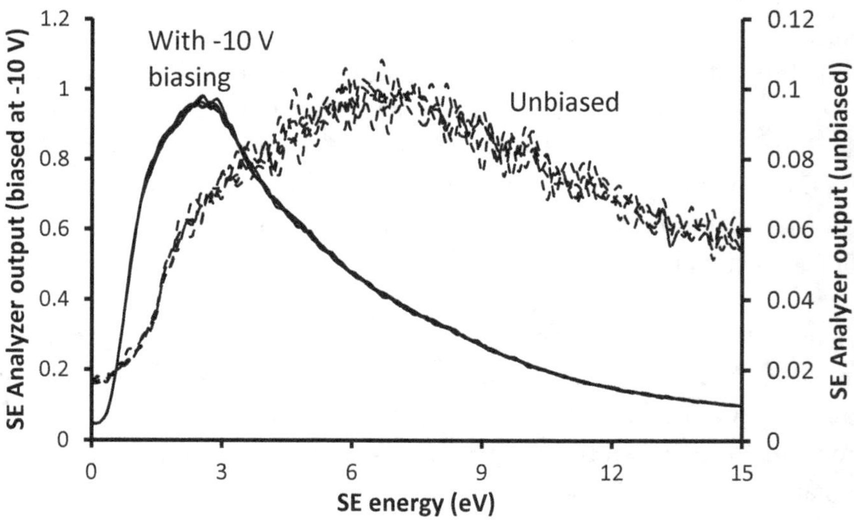

(b)

Fig. 2-5. Five experimental SE energy spectra taken for a bulk test gold sample. Primary beam voltage of 2 kV and current of 40 pA, 150 points with a dwell time of 0.2 s per point. (a) Plot as a function of the analyzer deflector voltage (V_D). (b) Plot as a function of SE energy (eV).

The SE spectral signal for biased specimen/inner cap electrodes shown in Fig. 2-5b has a much more symmetric, sharper and higher peak than the one for the unbiased case, which is obviously much better suited for tracking small changes in the spectral signal shape. It can be used only when relative changes in the SE energy spectral signal are required, which is the case for most applications. On the other hand, it has the disadvantage of no longer directly representing the emitted SE energy distribution, which means that its experimental SE energy spectral data cannot be directly compared to the spectra predicted by theoretical emission models or Monte Carlo specimen-beam interaction simulations.

Table 2-1 presents shot noise data extracted from the spectral signals shown in Fig. 2-5. These results provide a rough guide to the kind of precision that can be obtained for tracking different features of the output SE energy spectral signal, such as its peak signal height, average energy (most probable energy), μ, and standard deviation, σ. The noise characteristics for a more restricted range, a smaller region around the SE spectral peak are also given (from −1.67 to 1.67 eV). Which parameter needs to be tracked from the output analyzer spectral signal, peak signal height, μ or σ, depends on the application.

Table 2-1 Shot noise characteristics of SE energy spectral parameters taken from a test gold sample.

	Full Range			
	Average value	%Noise		
Signal height (unbiased)	0.0971 (a. u.)	4.48	Around the peak (−1.67 to +1.67 eV)	
Signal height (biased)	0.956 (a. u.)	1.01	Average	%Noise
μ (unbiased)	7.83 eV	2.38	6.66 eV	0.157
μ (biased)	5.00 eV	0.312	2.04 eV	0.0775
σ (unbiased)	3.98 eV	0.37	0.623 eV	0.195
σ (biased)	4.39 eV	0.053	0.610 eV	0.141

The highest levels of shot noise (> 1%) occur for the unbiased case when tracking peak signal height and average energy. Specimen/inner conical cap negative biasing brings the shot noise down by a factor of approximately 4.5 for the signal height and 7.6 for the average energy. It is interesting to note that tracking the standard deviation in the unbiased case also gives relatively low shot noise (< 1%). In general, using average energy and standard deviation in the more restricted region around the SE peak has better noise characteristics, where the levels of shot noise all lie below 0.2 %. These results indicate that biasing the specimen/inner cap significantly improves the output spectral signal's noise characteristics. They also suggest that from a noise perspective, changes around the SE signal peak position can be tracked with more precision than using the wider SE energy range.

It is important to note that the experimentally measured shot noise errors for monitoring SE spectral signal height, μ and σ (for the unbiased case), are all significantly larger than the corresponding errors predicted by the analyzer having a finite relative energy resolution of 2.5%, where changes in signal height, μ and σ are all expected to be smaller than 0.024%. Observations of this kind indicate that for the present wide-angle toroidal energy analyzer design, operating under routine experimental conditions (40 pA beam current and a 0.2 s dwell time per spectral point), the precision to which the SE energy distribution can be captured is dominated by shot noise, and not by analyzer energy resolution.

For the case of biasing the specimen/inner cap electrode to −10 V, the percentage shot noise levels limit the precision of tracking changes in the output spectral signal to 1.01 % for the signal height, 0.312% for the average energy μ, and 0.053% for the standard deviation σ, corresponding to the signal-to-noise ratios of 99, 320.5 and 1886.8 respectively. This should be adequate for most applications, however, it should also be noted that the data acquisition time for each parameter is not the same. The signal height, being only one point on the spectrum, has data acquisition times that are 150 times shorter than those required to acquire μ or σ.

Note that there is a background signal in the experimental spectral data shown in Figs. 2-5a and 2-5b, indicated by an initial non-zero current. This signal most likely comes from the generation of tertiary electrons caused by multiple scattering of high energy SEs/BSEs colliding with the inner conical cap electrode and analyzer deflection plates, as illustrated in Fig. 2-6. These multiple collisions occur between the specimen and inner conical cap, generating multiple points of emission from the specimen. An important characteristic of the background signal is that it behaves in a similar way to the main SE spectral signal, in terms of intensity variations caused by changes in primary beam current and the build-up of contamination. It can therefore be used to mitigate errors caused by primary beam current variations and contamination effects in some instances. An example of this will be given in the next chapter.

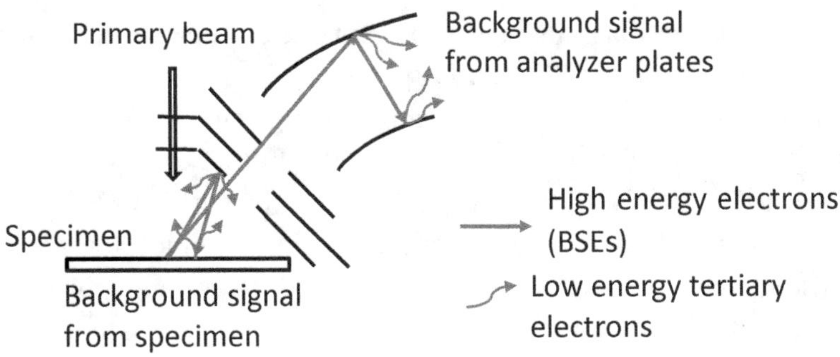

Fig. 2-6. Schematic diagram to illustrate the generation of a background signal caused by multiple high energy electron scattering.

The range of improvement by biasing the specimen/inner cap electrode extends considerably beyond a biasing voltage of −10 V. Figure 2-7 shows experimental results obtained by Hoang *et al.* (2011a) where the SE spectral signal peak height rises by around a factor of 30 for a biasing voltage of −24 V, compared to the initial unbiased signal peak height. This will in turn, lead to an even further reduction of shot noise effects.

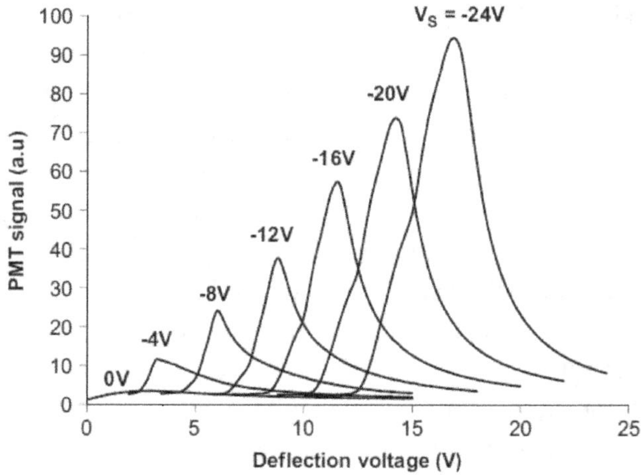

Fig. 2-7. Experimental secondary electron output signals at different specimen/inner cap biasing voltages.

The ray tracing diagrams using Lorentz-2D (Integrated Engineering Software) shown in Fig. 2-8 help explain why biasing of both the sample and inner cap electrode to the negative voltage of −10 V improves the signal-to-noise performance of the analyzer. In this case, the SE energy was set to 2 eV, giving a 2 eV pass energy for the unbiased case, and 12 eV for the biased case. SE trajectories were plot for an energy emission range of 0.3 to 3.3 eV in 0.1 eV steps, and polar angular emission of 45 ±15° in steps of 2°. When the sample is unbiased, the electrons travel in straight trajectories through the conical electrodes (Fig. 2-8a), forming a point of focus at the exit slit (Fig. 2-8c). When the sample is biased, an acceleration field penetrating through the inner conical electrode causes low energy SEs to pre-focus at the analyzer entrance, and increases the angular range entering the spectrometer (see Fig. 2-8b). In the case of 1 eV SEs, the angular range increases to 45 ±15°.

Biasing the sample and inner cap electrode to −10 V leads to a gain of 10 eV in the kinetic energy of SEs as they pass through the analyzer, which in turn leads to a proportional increase in the detected energy bandwidth, resulting in an increase in signal height. However, this also results in electrons defocusing near the exit slit. These effects are demonstrated in Fig. 2-8d: the electrons that pass through the exit slit come from an increased electron energy range of 0.3 to 3.2 eV and an increased polar angular range of 34°–48°. Simulation results predict that the total number of SEs that pass through the exit slit increases by a factor of around 7.69 (at 2 eV) for negative biasing compared to the unbiased case. However, the pre-focusing effect (Fig. 2-8b) caused by the acceleration field for negative biasing also leads to a worse energy resolution, around 5–7% in the 1 to 2 eV SE energy range.

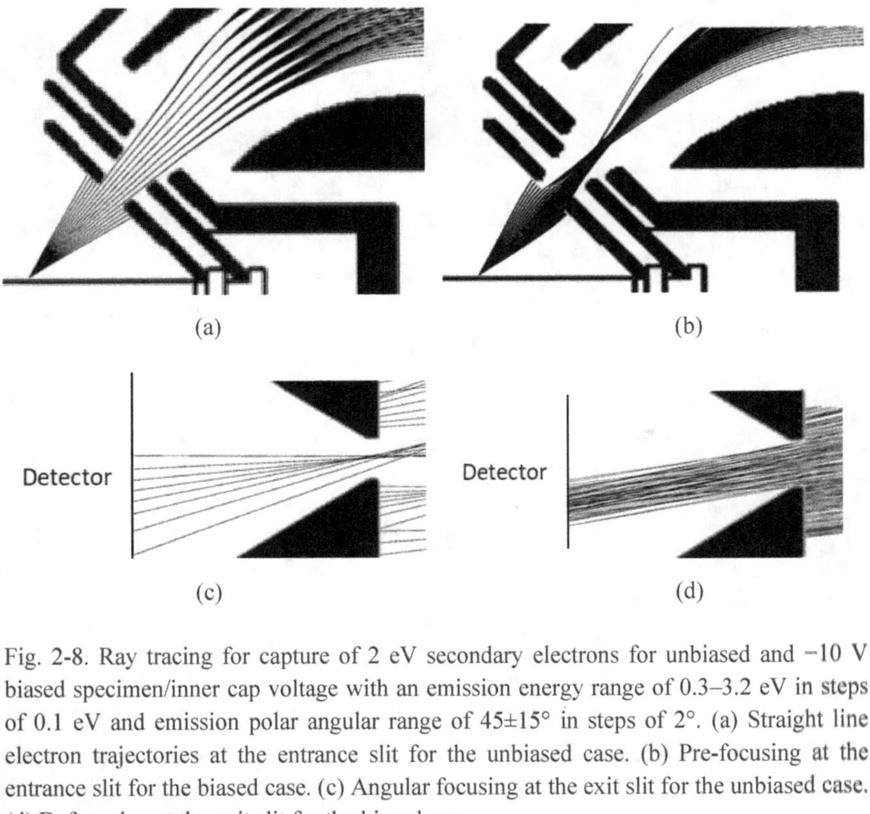

Fig. 2-8. Ray tracing for capture of 2 eV secondary electrons for unbiased and −10 V biased specimen/inner cap voltage with an emission energy range of 0.3–3.2 eV in steps of 0.1 eV and emission polar angular range of 45±15° in steps of 2°. (a) Straight line electron trajectories at the entrance slit for the unbiased case. (b) Pre-focusing at the entrance slit for the biased case. (c) Angular focusing at the exit slit for the unbiased case. (d) Defocusing at the exit slit for the biased case.

Figures 2-9 and 2-10 show some results of ray tracing simulation that were carried out by Hoang *et al.* (2011a) for the wide-angle toroidal energy spectrometer in biased specimen/inner cap conditions. A Chung-Everhart model was used for the emission SE energy distribution. The simulation predicts that the entrance angular spread enlarges considerably for low energy SEs less than 1 eV (Fig. 2-9). This effect, together with the effective increase in detection bandwidth explains why there is more than an order of magnitude increase in SE signal height compared to the unbiased case. Figure 2-10 shows that the simulation predictions of SE energy spectral signals for specimen/inner cap electrode biased voltages of −10 and −14 volts approximately agree with their corresponding experimental SE spectral signal data.

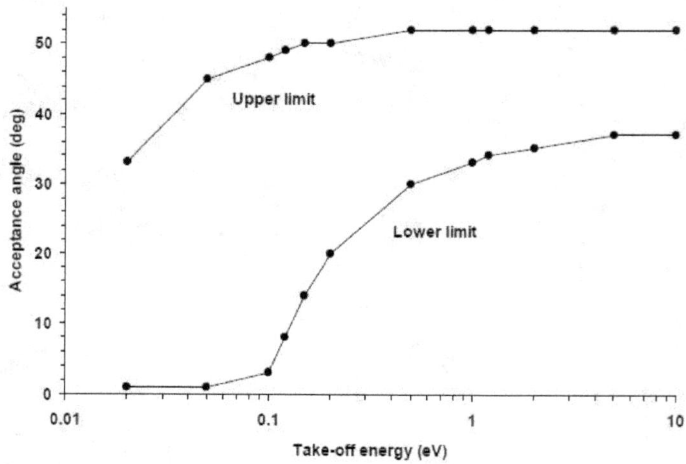

Fig. 2-9. Polar angular collection of different SE energies through the wide-angle toroidal energy analyzer when the specimen/inner cap is biased to −14 V (entrance aperture is set for ± 8° angular spread entry).

Despite there being considerable signal-to-noise advantages to biasing the specimen/inner cap electrode, there are some disadvantages. One problem comes from it having a poorer energy resolution in the low SE energy range. Consider for instance, 1 eV SEs, they pass through the analyzer with an energy of 11 eV (bias of −10 V), which leads to the detected energy width increasing by a factor of 11. This, together with the loss of energy resolution caused by the pre-focusing effect (Fig. 2-8b), leads to an energy resolution of around 0.55 eV (5% of

11 eV), compared to 0.025 eV (2.5% of 1 eV) for the unbiased case. Depending on the kind of contrast mechanism being qunatified, this may lead to a loss of sensitivty in the low energy SE range, in the rising edge part of the output spectral signal. Another problem comes from a larger background signal effect, caused by low energy tertiary electrons being pulled into the analyzer by the acceleration field that exists between the inner cap and outer cap electrodes, adding to the detected current in the low SE energy range.

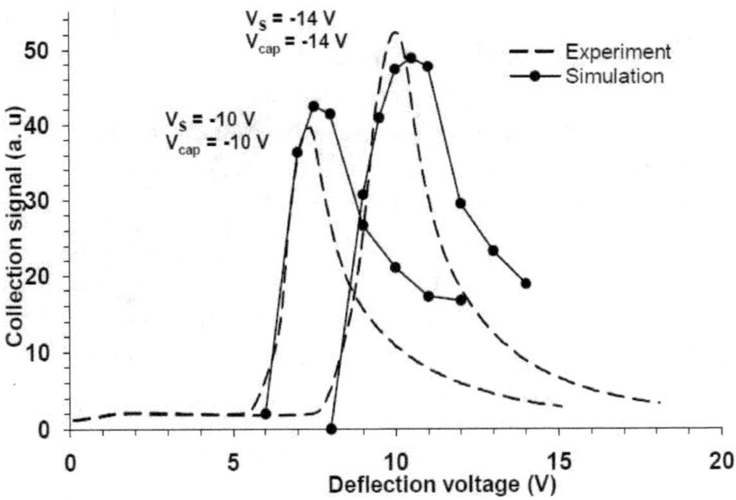

Fig. 2-10. Comparison of experimental and simulated SE output signals for the specimen/inner cap voltages of −10 and −14 volts.

One useful way to characterize the limitations imposed by shot noise is to link the output measured experimental signal-to-noise ratio of a particular parameter, such as peak signal height, μ or σ, to the number of primary beam electrons striking the specimen during spectral signal capture. Let us for instance, first consider measuring the spectral peak height which occurs at the SE energy of E_P, $S(E_P)$. In order to measure this parameter, the pass energy is fixed and it is not necessary to generate the SE energy spectrum. If the primary beam current is I_P, and the total time required to capture $S(E_P)$ is t_{tot} (which includes multiple measurements for signal averaging purposes), then the output signal-to-noise S/N can be related to the primary beam current and total acquisition time by a spectrometer constant k_{sp}

$$\frac{S}{N} = k_{sp}\sqrt{\frac{I_P t_{tot}}{e}} \tag{2.3}$$

where e is the electronic charge of an electron. The above expression assumes that the shot noise follows normal distribution statistics, valid for a large number of detected electrons (say over 100). The expression inside the square root operator represents the total number of primary beam electrons incident on the specimen during the total acquisition time. The spectrometer constant, k_{sp}, can be calculated once the output S/N is measured for a given set of operating conditions. For instance, applying this formula to the data shown in Table 2-1 for a gold specimen, the unbiased SE peak height was measured with a percentage noise of 4.48%, which corresponds to an output signal-to-noise of 22.3 (100/4.48). The primary beam current of 40 pA with a dwell time of 0.2 s averaged over 5 samples gives a total number of 245.65 million electrons involved in the measurement. The spectrometer constant is then calculated to be $k_{sp} = 1.411 \times 10^{-3}$.

The spectrometer constant k_{sp} for tracking the SE peak signal height is useful for many reasons. It serves as a guide on what operating conditions to use in order to achieve a desired output signal-to-noise value. Typically, for a given primary beam current value, the dwell time on the specimen and/or the number of samples used for averaging will be varied. It is also useful for comparing the overall efficiency between different spectrometer-detection systems. In this case, the spectrometer azimuthal capture angle is 100° and the entrance polar angular spread is ±8°, which provides a certain transport efficiency of SEs emitted from the specimen to the detector. If the spectrometer design were changed to capture an angular azimuthal angle spread of 200°, a factor of √2 improvement in the output signal-to-noise ratio can be expected. In general, energy analyzer designs differ significantly in their respective transport efficiencies. Another important parameter involved in the measured spectrometer constant value is the quantum efficiency of the detector, which also varies significantly between different energy spectrometer-detection systems. Therefore, by providing a measured spectrometer constant value for a given set of operating conditions, it is possible to compare the overall signal-to-noise characteristics of different energy spectrometer-detection systems with one another. The higher the spectrometer constant, the better the energy spectrometer-detection system from a shot noise point of view.

It is important to note that the spectrometer constant depends on the kind of parameter being measured. If for instance, the measured parameter is the output spectral signal standard deviation, σ, then from Table 2-1, the output signal-to-noise for an unbiased specimen is 270 (100/0.37), and since the total data acquisition time is 150 s (5 spectra each captured over 30 s). The spectrometer constant for a 40 pA primary beam in this case is 1.396×10^{-3}, slightly lower than the one for tracking the spectral signal peak height. The spectrometer constant depends on the shape of the output spectral signal and SE yield, and will be different for another sample type (such as silicon). If the parameter tracked is the average energy μ, the equivalent output signal-to-noise for the unbiased gold specimen is 42 (100/2.38), which gives a significantly lower spectrometer constant value of 0.217×10^{-3}. This can be improved to 3.293×10^{-3} if a smaller interval around the spectral peak is taken (that has an output signal-to-noise ratio of 100/0.157). The value of the spectrometer constant provides a guide on which parameter to use from a shot noise perspective.

It is also possible to apply the spectrometer constant concept to characterize the signal-to-noise characteristics of the spectral image acquisition mode. If for instance, the total number of pixels in the image is N_{pixel}, then the spectral constant in equation 2.3 will be reduced by a factor of $\sqrt{N_{pixel}}$, necessitating a greater amount of averaging, in order to maintain the output signal-to-noise per pixel to a manageable level. For some specimens, this extra amount of averaging might be available in the image itself, if that is not the case, the extra amount of averaging will need to be acquired by capturing more images.

2.3 Comparison of experimental SE energy spectra to theoretical analytical energy distributions

When comparing theoretical signals to experimental signals, two factors need to be taken as free parameters in order to optimize the fit, namely, SE spectral peak height and energy shift. The SE peak height from experiment depends on many different parameters, such as primary beam current, SE yield, detector response, head amplifier gain, and contamination levels induced by the primary beam, so it is therefore best left as a free parameter when comparing it to an analytical model. It will be represented here in terms of a signal height scaling parameter, S_0. There are also contact differences between the sample and its holder, work function changes caused by beam induced contamination, and the presence of a background signal generated by high energy multiple scattering. All these effects make it difficult to determine the same zero energy starting point for each spectrum and combine to create random small energy shifts. For this reason, energy shift will be treated here as a free offset parameter, E_{off}.

Taking into account the freedom to scale in signal height and shifts in the SE energy position, the theoretical signal takes the following form:

$$S_T(E_P) = S_0 \cdot T(E_P) \cdot N\left(E_P - E_{off}\right) \quad (2.4)$$

The best fit to experimental signals is obtained by optimizing S_0 and E_{off} through minimising the root-mean-square deviation (RMSD) between the theoretical signal $S_T(E)$ and the experimental signal $S_E(E)$:

$$RMSD = \sqrt{\frac{\sum_i^n (S_E(E_i) - S_T(E_i))^2}{n}} \quad (2.5)$$

where E_i represents an output energy channel. The optimization algorithm used here, is the generalized reduced gradient (nonlinear) method present in Microsoft Excel Solver (Lasdon *et al.*, 1978). The signal accuracy was then quantified by the % RMS deviation (% RMSD) relative to the average signal value $\overline{S_E}$:

$$\%RMSD = \frac{RMSD}{\overline{S_E}} \times 100\% \qquad (2.6)$$

The SEM used for all experiments reported here, unless otherwise stated, is the FEI Quanta 3D Focused Ion Beam (FIB)-SEM system, typically operating with 1 to 5 kV primary beam voltages, 40 to 400 pA beam currents, in a SEM specimen chamber vacuum level of approximately 2 to 3 × 10^{-6} Torr. Eight elemental metals (Al, Ti, Cr, Fe, Cu, Ag, Pt and Au) were chosen for comparing experimental SE energy spectral results to their theoretical counterparts (Chung and Everhart, 1974) for a wide range of atomic numbers *Z*, from 13 (Al) to 79 (Au). The initial work function values used in the fit were taken from Linn and Joy (2005), as reported from the 66th edition of the CRC Handbook of Chemistry and Physics, but their values were also allowed to vary in order to find the best fit between theory and experimental data.

The first set of metal samples were in the form of cylindrical pellets, 3 mm in diameter and 3 mm in height. As the work function is highly sensitive to surface conditions, all pellets were cleaned in ultrasound with acetone and isopropanol. The pellets were then inserted into a custom made sample holder, shown schematically in Fig. 2-11. Each metal pellet was paired with a Ti pellet, where the Ti SE energy spectral signal served as a reference, in order to eliminate experimental variations caused by changing the specimen. The theoretical fit to the output spectral signal was made while maintaining the background signal. This was done in order to avoid spectral shift errors created by background signal subtraction, coming from the difficulty of not being able to locate a consistent zero energy point for each spectrum. The spectral results are shown in Fig. 2-12 for a primary beam voltage of 2 kV. A difference signal $g(E) = S_E(E) - S_T(E)$ is also plot on each graph. In general, there is a good fit between the experimental spectral signals and the Chung-Everhart analytical distribution.

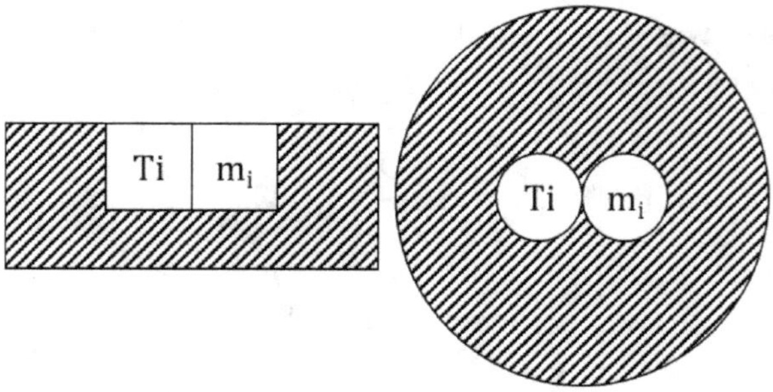

Fig. 2-11. Layout of metal pellets in a custom made sample holder. Each metal m_i is paired with a Ti pellet for spectral signal referencing purposes.

Table 2-2 gives the % RMSD difference for each element with an initial work function value, and an adjusted work function value in order to minimize the % RMSD. There is a definite improvement in the agreement if the work function is allowed to vary. Since the Chung-Everhart distribution is only an overall analytical approximation, with in-built assumptions about the way SEs are generated and escape from the sample, an exact match to experimental SE energy spectral data is not expected. The purpose here is to quantify how closely they do match. Beam induced contamination is another contributing factor for the difference, indicated by the better fit when the work function is a free parameter. There will also be differences caused by the presence of the background signal, a finite spectral sampling interval, and the finite energy resolution of the energy analyzer.

Overall, there is a high level of agreement between the experimental SE energy spectra obtained by the wide-angle toroidal energy spectrometer attachment in the SEM and a Chung-Everhart distribution fit for the 8 bulk metal pellet samples. An average difference of around 4.26% RMSD is obtained, and for most the metal pellet samples, the %RSMD value lies below 5%. These results display a similar level of agreement to the experimental results reported for dedicated UHV chamber electron energy spectral instruments, see for instance the comparison made by Cazaux (2011) for an aluminum sample shown in Fig. 1-3b.

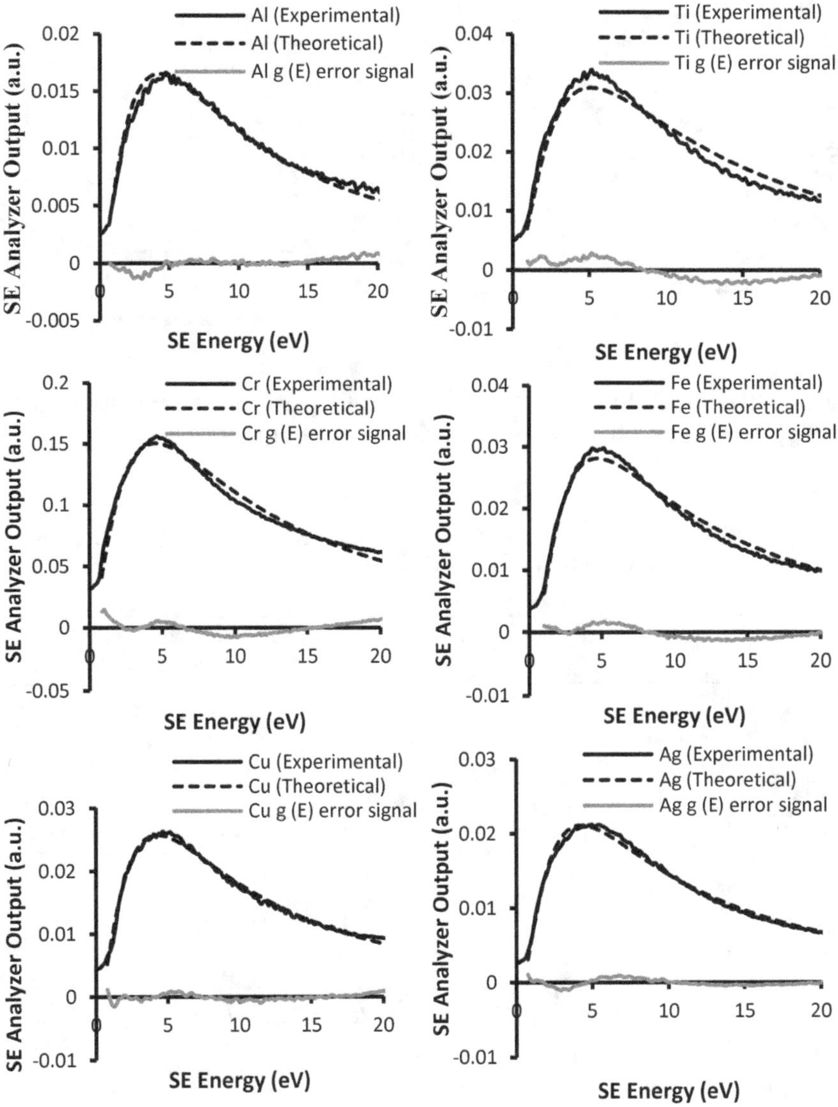
Al (Experimental)
Al (Theoretical)
Al g (E) error signal
Ti (Experimental)
Ti (Theoretical)
Ti g (E) error signal
Cr (Experimental)
Cr (Theoretical)
Cr g (E) error signal
Fe (Experimental)
Fe (Theoretical)
Fe g (E) error signal
Cu (Experimental)
Cu (Theoretical)
Cu g (E) error signal
Ag (Experimental)
Ag (Theoretical)
Ag g (E) error signal
SE Analyzer Output (a.u.)
SE Energy (eV)

(Fig. 2-12)

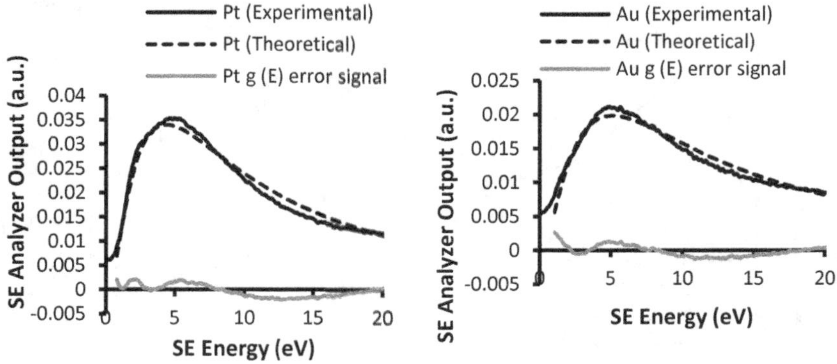

Fig. 2-12. Experimental SE energy spectra fitted with the Chung-Everhart distribution for 8 different bulk metal samples for a primary beam voltage of 2 kV.

Table 2-2. % RMSD difference between experimental spectra and the Chung-Everhart SE energy distribution for 8 different bulk metal samples.

	Theoretical		Fitted with adjusted Work Function	
Element	Work Function (eV)	Fixed Work Function fit %RMSD	Adjusted Work Function (eV)	%RMSD
Al	4.28	5.60%	4.58	3.67%
Ti	4.33	6.56%	4.602	3.80%
Cr	4.5	5.59%	4.48	5.53%
Fe	4.5	5.15%	4.45	3.87%
Cu	4.65	5.95%	4.33	3.74%
Ag	4.26	3.46%	4.28	3.44%
Pt	5.65	6.10%	5.13	4.94%
Au	5.1	6.33%	4.85	5.12%

Since the experimental SE energy spectral signals shown in Fig. 2-12 were obtained with two metal pellets being in the holder at any one time, one of which was always Ti, it is important to examine the SE energy spectral variations for the common Ti pellet in each of the sample pair. This gives information about the kind of variations in the energy spectrum caused by changing the specimen. There will be fluctuations in the specimen chamber vacuum level, primary beam current, and a variation caused by there being a different metal paired with the Ti pellet during each measurement. This common Ti pellet experimental SE energy spectral data is shown in Fig. 2-13, indicating that there are variations in position, height and shape of the output spectral signal. The relatively large variations in these common Ti energy spectral signals suggest that there may be an interference effect coming from the presence of neighboring elements on the specimen. They suggest that additional scattering of tertiary electrons is somehow being created by these neighboring areas.

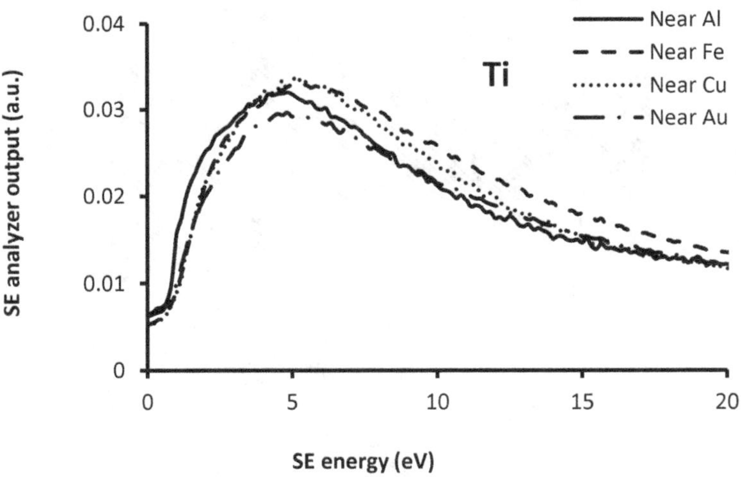

Fig. 2-13. Experimental spectra of Ti pellets when paired with different metals.

Another set of experiments was performed, where metal foil samples were used instead of pellets, and 4 different triangle metal elements were fit into the central area of the sample holder. The 4 elements used in these experiments were Au, Pt, Ag and W. The layout of these foil samples is illustrated in Fig. 2-14a. In order to test out a strategy for reducing the interference signal generated by the presence of neighboring metals, this time the 4 metal elements were partially covered by a thin foil cover made from Al, as illustrated in Fig. 2-14b. The purpose of the Al cover is to reduce the amount of scattering inside the sample holder, since Al has a lower a backscattered/secondary electron yield compared to heavy metals.

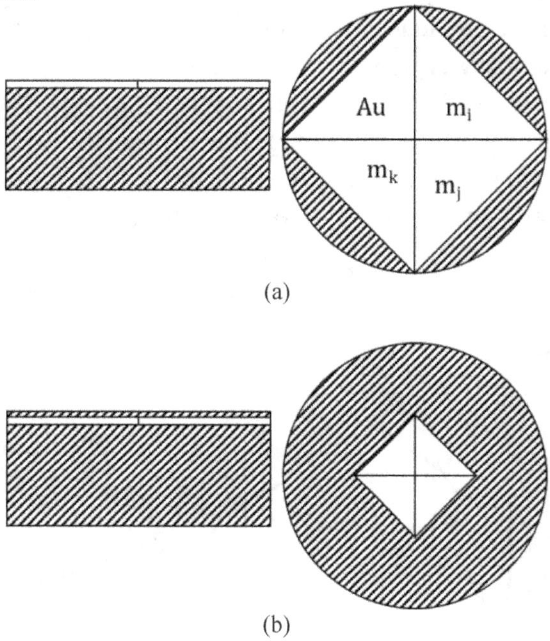

Fig. 2-14. Layout of metal foils in sample holder. (a) Without aluminum cover. (b) With aluminum cover.

Figure 2-15 shows experimental SE energy spectral signals from Au and Pt with and without the presence of the Al cover for a primary beam voltage of 1 kV. The % RMSD of Au and Pt decreased from 6.7% and 12.1% to 4.0% and 2.66% respectively with the presence of the Al cover.

It should be noted that these results are not directly comparable to the experimental results obtained from the pellet samples since they were obtained at a primary beam voltage of 1 kV, while the spectra from the pellet specimens were recorded with a primary beam voltage of 2 kV. In general, better agreement with the Chung-Everhart distribution was obtained for higher primary beam voltages. The match between the 1 kV primary beam voltage spectral signals was consistently found to be worse than the match obtained from the 2 kV spectral signals.

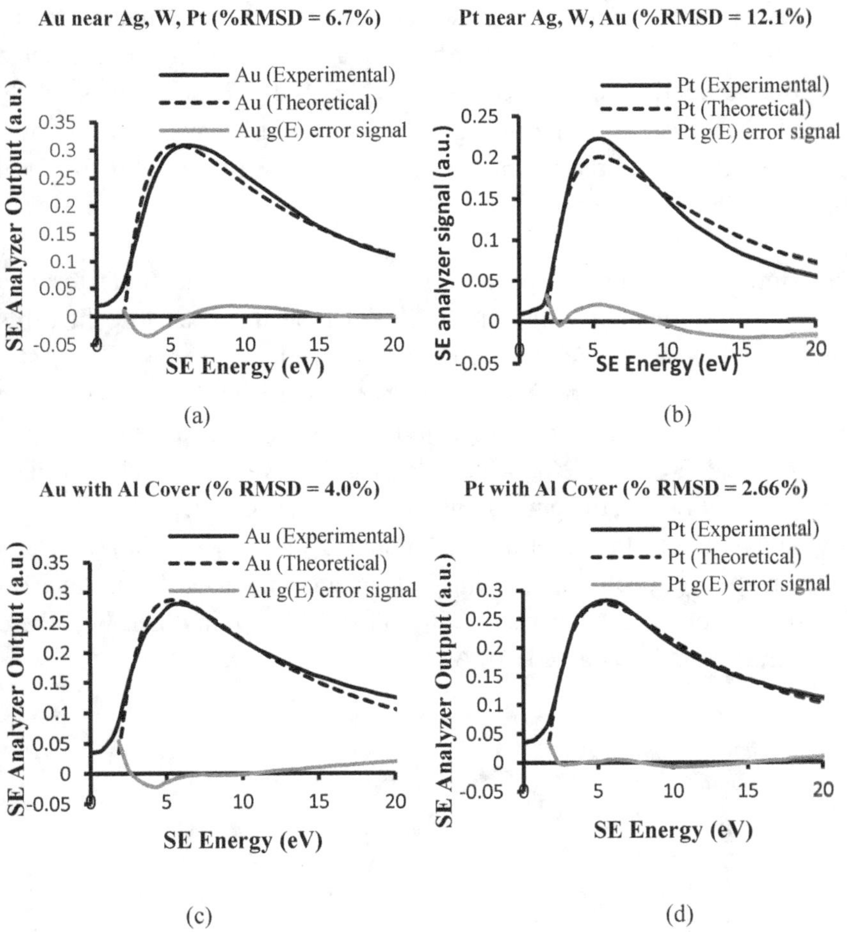

Fig. 2-15. Experimental SE energy spectral signals obtained from foils at 1 kV. Best fit with the Chung-Everhart distribution with the background signal retained. (a) Au foil, with neighboring metals Ag, W and Pt. (b) Pt foil, with neighboring metals Ag, W and Au. (c) Au foil in the same configuration as (a), with Al cover. (d) Pt foil in the same configuration as (b), with Al cover.

The spectral signals shown in Fig. 2-15 display a consistently poor match between theory and experiment for the first points on the rising edge part of the signal. The deviation between theory and experiment is greatest for these first few points. This deviation is probably not related to the finite analyzer relative energy resolution, since those errors are predicted to be in the fractions of a percent range (Fig. 2-4b). They are most likely related to the nature of the background signal, which will have its own characteristic energy distribution, and the finite SE energy step size used (0.167 eV in this case).

Experimental SE energy spectral signals from the wide-angle toroidal analyzer attachment for intrinsic Si also demonstrated a close fit to its corresponding theoretical signal, as shown in Fig. 2-16, taken here at a primary beam voltage of 2 kV. The theoretical signal is based on the Henke *et al.* (1979) model (equation 1.2) using the parameters, χ = 4.05 eV and E_G = 1.1 eV, which are standard values for silicon. The %RMSD of the signal from the theoretical signal is 6.0%. These results are comparable to the level of agreement obtained by specialized UHV electron energy spectral measuring instruments, indicated for instance in the comparison made by Cazaux (2011) for Si shown in Fig. 1-3c, where the overall % RSMD value is 10%.

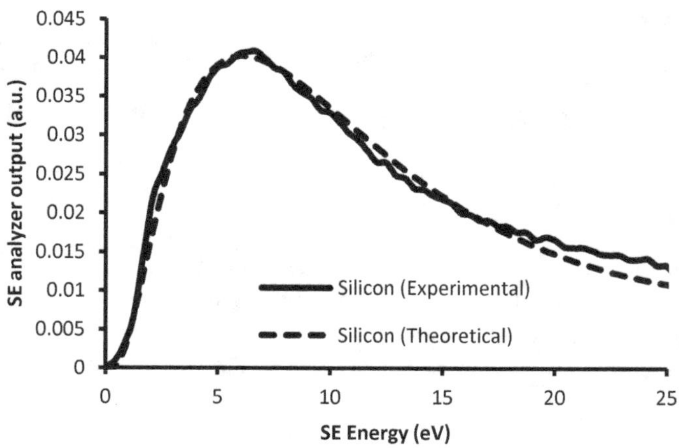

Fig. 2-16. The experimental SE energy distribution taken from a silicon bulk sample compared to its theoretical counterpart at a primary beam voltage of 2 kV.

2.4 Surface field effects and voltage contrast

A common problem for several applications of SE energy spectroscopy is the presence of electric fields. They may exist between the specimen and its surroundings, and locally around the point being probed by the primary beam. This typically happens if the voltage on the entire specimen changes (with respect to its surrounding electrodes), and/or if there is a voltage distribution on the specimen surface. The main difference between these two field distributions is that the former, is a global fringe field, while the latter, is a local surface field effect. The trajectory paths of all secondary electrons, irrespective of where they leave the specimen will be influenced by global fringe fields, while the effects of local surface fields will vary from point to point on the specimen surface.

Local surface field effects have already been mentioned in the context of quantitative voltage contrast as an unresolved problem for Electron Beam Testers (section 1.2, Figs. 1-13 and 1-14). They cause the output S-curve signal of retarding field spectrometers to attenuate and distort. Although their effects can be partially mitigated by applying strong vertical extraction fields (up to 5 kV/mm), the surface field effect grows stronger as feature sizes on the specimen shrink to nanoscale dimensions, requiring unmanageably high compensating extraction field strengths. It is therefore important from the outset to understand the kind of accuracy challenges surface electric fields pose for the wide-angle band-pass toroidal energy analyzer attachment. Background understanding of how specimen global fringe/local surface fields act on the output spectral signals of the wide-angle toroidal energy analyzer is also required when applying it to quantify dopant contrast, or when using it to monitor specimen charge build-up, subjects which that are discussed in the later chapters of this book.

In principle, band-pass SE energy spectrometers should have more opportunity for understanding and correcting the surface field effect than voltage contrast retarding field energy analyzers. This is because the output S-curve signal of retarding field analyzers not only averages out

SE emission energy information, but inherently mixes it together with SE angular emission information. Band-pass energy analyzers on the other hand, provide the opportunity of characterizing surface field effects in terms of how they change the SE energy spectrum.

How both global specimen fringe fields and local surface fields compromise the ability of the wide-angle toroidal energy analyzer attachment to carry out quantitative voltage contrast was investigated by Srinivasan and Khursheed (2015b). They were able to show that although specimen fringe/surface fields significantly change the form of the wide-angle toroidal analyzer output SE energy spectral signal, reliable quantitative voltage contrast information can still be obtained. Their work therefore solves a longstanding problem in the subject of voltage contrast and demonstrates how a voltage contrast measurement system based upon acquiring the SE energy spectrum not only functions successfully in the presence of specimen fringe/surface fields, but can do so without the need to apply any vertical extraction field. Some of the results obtained by Srinivasan and Khursheed (2015b) will be summarized here.

Srinivasan (2014) investigated how the wide-angle toroidal energy analyzer output spectral signals change when the specimen is biased to voltages that are different to the one applied to the inner cap, which creates global fringe electric field distributions between the specimen and inner cap electrode, like the simulated ones depicted in Figs. 2-17a and 2-17b. Figure 2-17a shows simulated fields for the original Hoang *et al.* (2011a) hemispherical cap electrode geometry, while Fig. 2-17b shows it for the conical cap electrode geometry used by Srinivasan and Khursheed (2015b). For the hemispherical cap electrode geometry, a specimen voltage of −10 V and inner conical cap electrode voltage of −11 V are used, and the trajectory path of a 0.5 eV electron (polar launch angle of 45°) is compared to its straight line trajectory path for the field-free case. In the case of the conical cap electrode geometry, direct ray tracing of 0.5 eV energy electrons are plot with an emission polar angular spread of ± 3° for the specimen voltage changing from −10 V to −10.5 V.

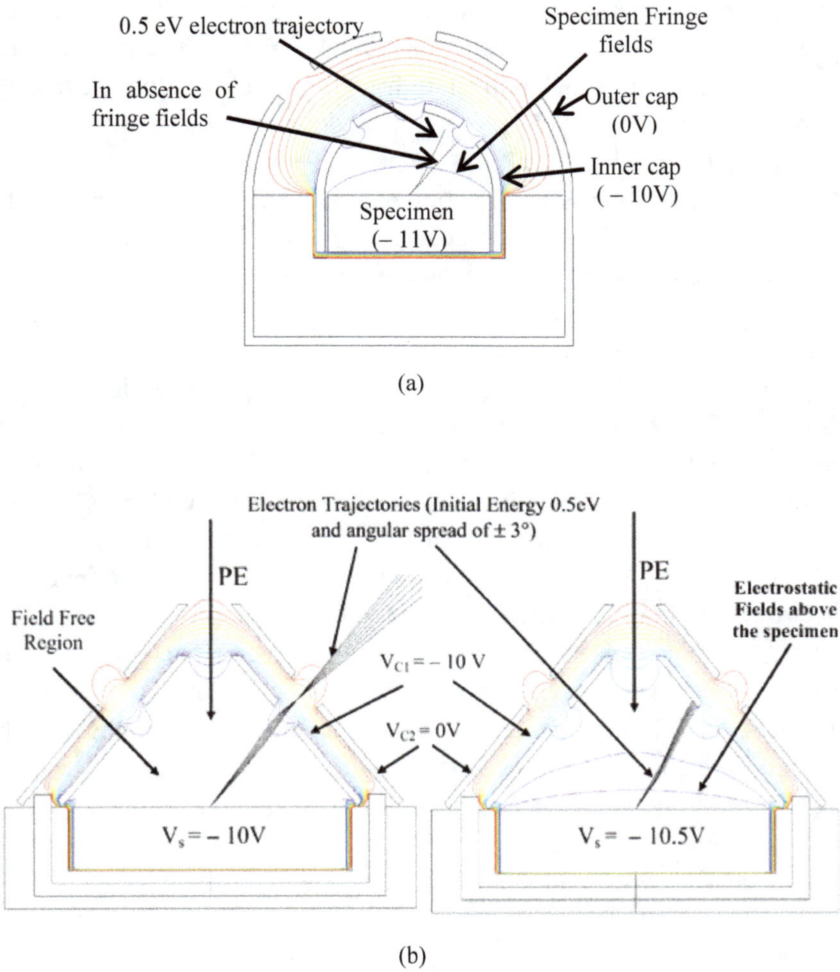

Fig. 2-17. Ray tracing simulation and field distribution of global fringe fields for specimen/inner cap electrode biasing. (a) Original hemispherical cap electrode design by Hoang *et al.* (2011a). Central ray at 0.5 eV ray. (b) Srinivasan and Khursheed (2015b) conical cap electrode design (2015). 0.5 eV emission rays with a ± 3°spread for the specimen voltage changing from −10 to −10.5 V.

The ray tracing simulation shown in Fig. 2-17a predicts that the 0.5 eV electron experiences a vertical accelerating force as it approaches the inner cap electrode, which pulls it up from the path it would have taken if there were no electric field present. This deflection field will cause the electron to enter the analyzer at a steeper angle, most likely resulting in it colliding with the analyzer deflection plates and not reaching the detector at the exit plane. Lower electron energies will naturally be more affected by this deflection field than the higher energy ones.

A similar upward deflection force is created in the case of the specimen changing from −10 V to −10.5 when the inner cap voltage is −10 V, as shown in Fig. 2-17b. For the field-free condition where the specimen voltage is −10 V, 0.5 eV SEs are transmitted into the analyzer, but form a point of focus between the inner and outer comical cap electrodes. This pre-analyzer focusing effect is caused by the acceleration field present between the inner and outer conical caps when the specimen/inner cap electrodes are biased negatively. As already noted (section 2.2), this will degrade the analyzer energy resolution and produce a background signal which will reduce its sensitivity in the rising edge part of the output spectral signal (low energy part). These simulation results predict that the upward deflection force caused by global deflection fields will cause a significant reduction in the output signal height when the specimen voltage is more negative than the inner cap electrode voltage, and that this attenuation will be a function of the fringe field strength.

For the situation where the specimen voltage is positive with respect to the inner cap voltage, not shown in Fig. 2-17, a potential barrier for emitted SEs will be created, and the output signal strength will be expected to drop in the lower energy part of the SE energy spectral range. Overall, these considerations suggest that any departure from field-free conditions between the specimen and inner cap electrode will cause a drop in the output SE energy spectral signal height.

Srinivasan and Khursheed (2015b) went on to obtain experimental SE energy spectral signals from the wide-angle toroidal energy analyzer corresponding to the voltage layout depicted in Fig. 2-17b. They used a *n*-type doped silicon wafer (N_D~10^{18} cm^{-3}) where the specimen voltage (V_S) was varied from –10 V to –13 V, with an inner cap voltage (V_{C1}) set to –10 V, and a grounded outer conical cap voltage (V_{C2} = 0). Their experimental results for the global fringe field free case (specimen and inner cap voltage being the same) are shown in Fig. 2-18a, and are in good agreement with the experimental results obtained previously on a bulk metal sample with hemispherical cap electrodes (Fig. 2-7) by Hoang *et al.* (2011a). The signal height rises steadily and the peak shape becomes sharper and more symmetric as the specimen/inner cap voltage becomes more negative.

The experimental spectral results with the presence of global specimen fringe fields are shown in Fig. 2-18b, and they are quite different to those obtained in field-free conditions. The signal height drops significantly when the specimen voltage is biased to −11, −12 and −13 V, while keeping the inner cap voltage at −10 V. These experimental spectral signals confirm the implications of the earlier simulation predictions (shown in Fig. 2-17), which suggest that there will be a loss in output signal strength due to the presence of an upward deflection fringe field. The signal height drops progressively in proportion to how negative the specimen voltage becomes. On the other hand, as the specimen takes on more negative values, there is also the expected shift to the right, coming from an overall increase in kinetic energy of electrons that pass through the analyzer.

Figure 2-19 shows a plot of the output spectral signal mean, μ (most probable energy), as a function of the specimen voltage difference with respect to the inner cap voltage, ΔV_S, as extracted from the experimental spectral signals shown in Fig. 2-18b. It shows that despite there being a significant drop in the height of the output signals, the spectral mean has a steady monotonic variation with respect to ΔV_S, which is close to being a linear relationship. This means that in principle, quantitative voltage contrast measurements can be made in the presence of global electric fringe fields between the specimen and its surroundings.

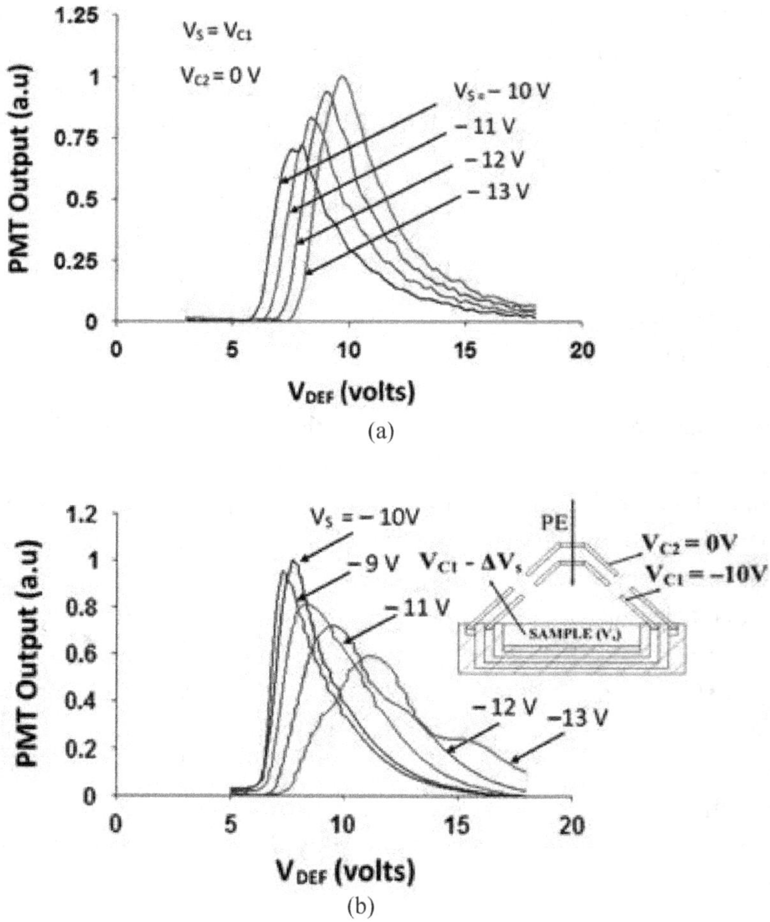

Fig. 2-18. Experimental SE energy spectral signals with and without a global fringe field between the specimen and inner cap electrode. (a) Field free region where both V_S and V_{C1} vary together between −10 V to −13 V. (b) Global fringe field created by V_{C1} = −10 V and V_S varying between −9 to −13 V.

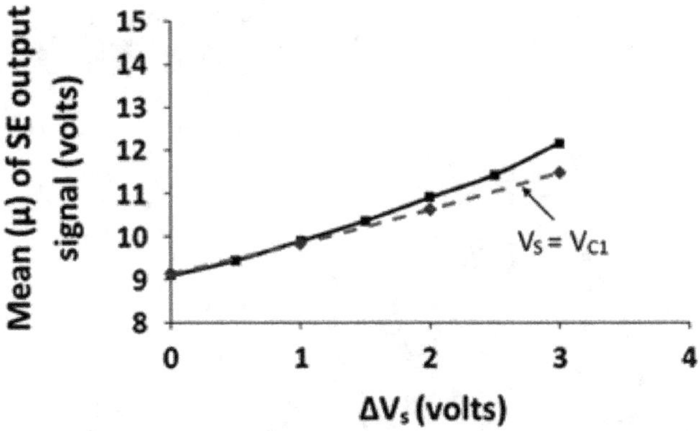

Fig. 2-19. Spectral signal mean in the presence of global fringe fields as a function of specimen voltage difference from the inner cap voltage, $\Delta V_S = V_{C1} - V_S$.

The fact that the spectral signal height is a maximum in Fig. 2-18b when the specimen and inner cap electrode have the same voltage, provides an additional way of estimating the specimen voltage when it is not known, beyond monitoring a change in the output spectral signal mean. In that case, the inner cap voltage can be systematically varied while monitoring the output spectral signal height, and the inner cap voltage that gives the maximum spectral signal peak height will provide an estimate of the specimen voltage. Global fringe field effects may occur for a specimen having a conductive top layer on a non-conductive substrate when irradiated by a high energy primary beam, causing the entire specimen surface to change by the same potential.

Srinivasan and Khursheed (2015b) also demonstrated that the wide-angle toroidal energy analyzer can in principle make quantitative voltage contrast results in the presence of local surface fields on the specimen. Figure 2-20 shows a schematic diagram of a 200 μm diameter enameled copper winding wire mounted on top of a metal base. The outer insulating layer surrounding the wire makes it possible to bias the wire and the metal base to a different voltage, creating local surface fields around the copper wire. The convex curvature of the wire will naturally tend to provide greater angular dispersion for the SEs that leave its surface, an opposite effect to the collimating action of the global fringe field shown in Fig. 2-17.

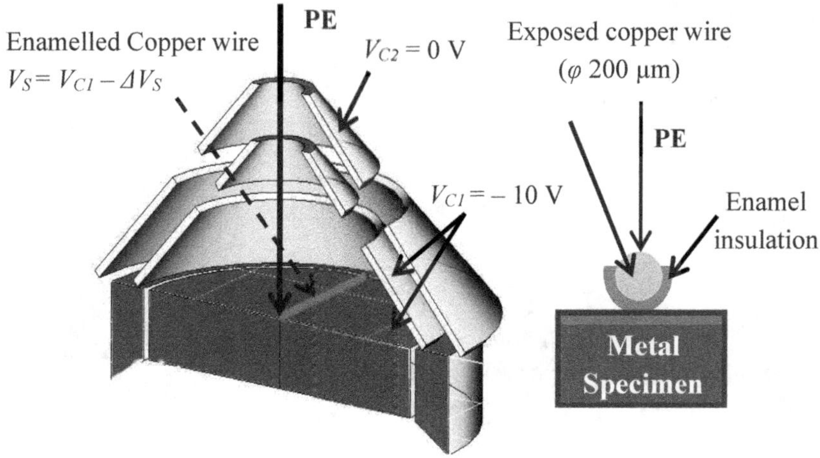

Fig. 2-20. Schematic diagram of the experimental setup to probe surface fields around a copper enamel wire specimen.

The experimental spectral results for the voltage of the copper enameled wire changing are shown in Fig. 2-21. They indicate that as the wire becomes more negative with respect to the metal base/inner cap voltage, varying from −10 to −14 V (while keeping V_{C1} and the metal base biased to −10 V), the SE signal shifts to the right due to an increase in the kinetic energy of the scattered electrons that pass through the analyzer, and that the amplitude of the SE signal goes through a maximum value; it first increases, after which (< −12 V), it decreases. The situation seems to be a combination of two separate field effects: the SE spectral signal height rising as the specimen voltage becomes more negative is most likely caused by the diverging action of local surface electric fields directly above the wire, while the subsequent fall in SE peak height is most likely caused by the influence of global fringe field effects. These results suggest for $V_S > -12$ V, surface micro-fields dominate, while for $V_S < -12$ V, the shape of the output spectral signal is mainly determined by global fringe fields.

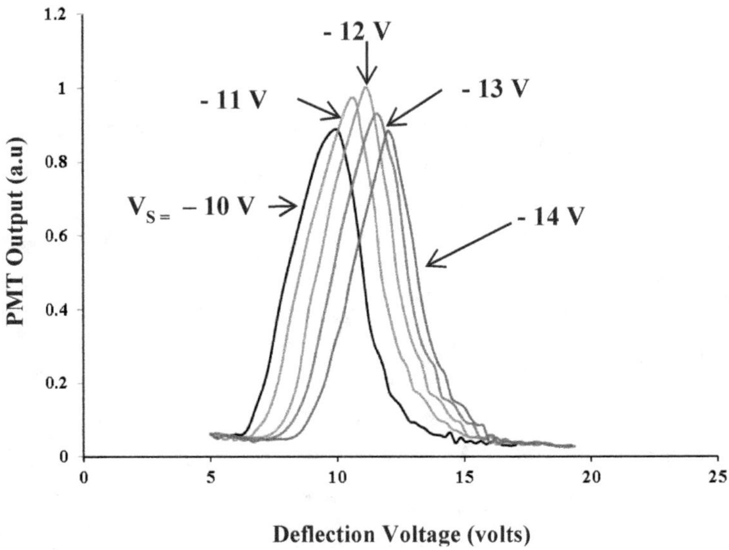

Fig. 2-21. Experimental SE energy spectral results obtained in the presence of local surface fields around a copper enamel wire specimen.

Despite the variations in analyzer spectral signal amplitude, there is still a consistent shift in the peak position as the specimen voltage is biased more negatively. This is demonstrated by Fig. 2-22, which plots the analyzer spectral signal mean value with change in specimen potential, indicating that it increases monotonically. These experimental results therefore show that the wide-angle toroidal energy analyzer can be used to quantify specimen voltage changes even in the presence of local surface fields. The non-linear dependence of signal mean value with specimen voltage will of course, need to be taken into account.

The SE energy spectral results shown in Fig. 2-21, taken together with the signal mean graph depicted in Fig. 2-22, provide information both about the kind of surface fields present at the specimen surface, and a way of tracking surface voltage changes.

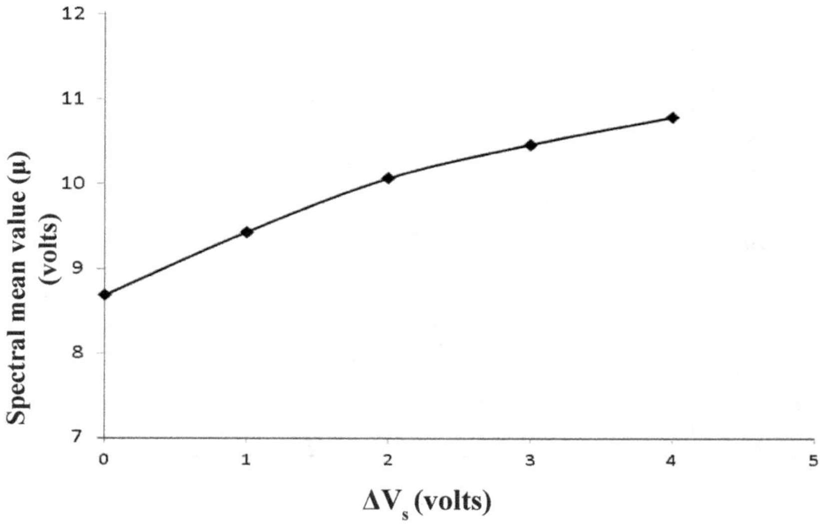

Fig. 2-22. Plot of the spectral mean signal value μ for local surface fields around a copper enamel wire sample as a function of the wire voltage difference with respect to its surroundings (ΔVs) from 0 to 4 V.

Srinivasan and Khursheed (2015b) also investigated surface field effects around doped semiconductor specimens. In this case, they setup a linear potential gradient on the surface of a moderately doped silicon sample ($N_D \sim 10^{18}$ cm^{-3}) by a button cell, as shown in Fig. 2-23. The two electrodes of the button cell are connected to the two far ends of the silicon wafer using metal wires. The distance between the two ends of the specimen is 6 mm, and the applied potential across it is 1.5 V. The entire setup is biased to –10 V, giving a voltage of –8.5 and –10 V at the ends of the specimen.

The case of a potential gradient being present on the surface of a doped semiconductor specimen is quite different to the other global fringe/surface fields examined so far. In this case, there is a local transverse electric field across the specimen surface, which means that SEs will be influenced by this electric field even before they leave the specimen surface.

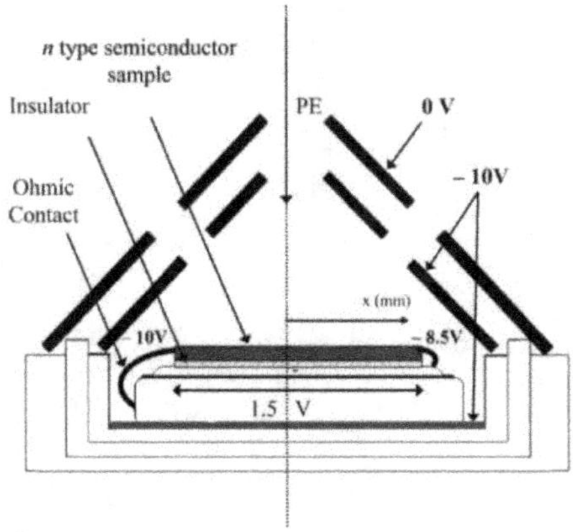

Fig. 2-23. Experimental layout of a potential gradient across the surface of a moderately doped *n*-type semiconductor sample.

Experimental SE spectral results were obtained from the specimen at consecutive points on a straight line along the x direction starting from the center of the specimen, and are shown in Fig. 2-24. The distance between each point of measurement on the specimen was approximately 120 microns, and assuming that the specimen voltage varies linearly across its surface, there is a voltage change of approximately 30 mV between each measurement point. Fig. 2-24 shows that as the potential along the surface of the sample becomes more negative, there is a distinct shift in the SE analyzer signal: the amplitude increases and shifts to the right. It is interesting to note that in this case, there is very little shift in the rising edge part of the spectral signals, most of the contrast seems to come from the trialing edge part of the spectral signal. This would seem to suggest that there is some additional filtering effect of the low energy electrons occurring at the specimen surface.

As before, the specimen voltage change ΔV_S can be tracked by calculating and plotting the output signal mean value μ, and is shown in Fig. 2-25. The plot of μ against the change in potential ΔV_S (proportional to the distance moved along the x direction) is, almost linear, indicating that in principle, small local changes of voltage on the surface of a doped semiconductor specimen can be measured by SE energy spectroscopy.

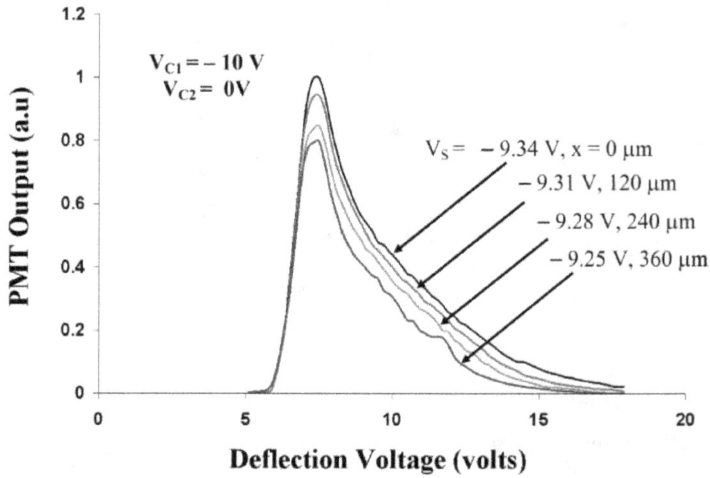

Fig. 2-24. SE experimental spectral results for different points on the surface of a moderately doped silicon *n*-type specimen having a potential gradient on its surface.

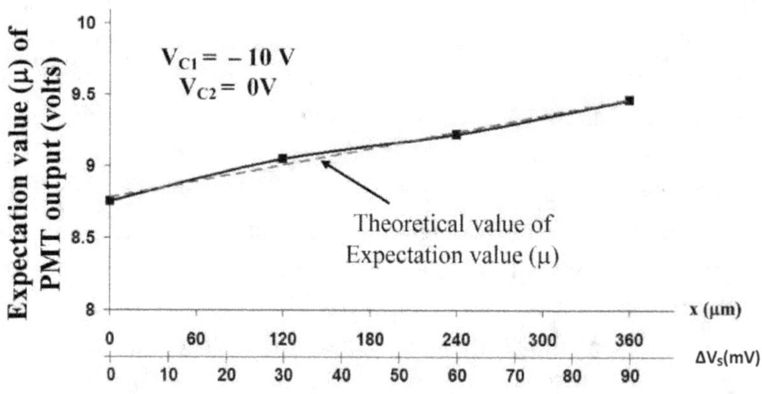

Fig. 2-25. Plot of spectral expectation value (mean µ) as a function of position on the specimen on the surface of a moderately doped silicon *n*-type specimen having a potential gradient on its surface.

The close fit to the theoretical straight line in Fig. 2-25 confirms that the method of monitoring the expectation value (mean μ) of the SE analyzer signal in order to measure small voltage changes on the semiconductor specimen seems to be a good one. An estimate of the voltage measurement accuracy can be obtained by calculating the standard deviation between the dotted and the solid line, translating it into an uncertainty in voltage measurement by using the $\mu/\Delta V_S$ gradient. For the experimental conditions and results shown in Fig. 2-25, the voltage accuracy is calculated to be 5.1 mV, corresponding to an average signal-to-noise ratio of 7.3. These experimental results therefore demonstrate that voltage variations on the surface of a semiconductor specimen can be measured to millivolt accuracy by the wide-angle toroidal energy analyzer attachment.

In this section, SE energy spectra obtained from the wide-angle toroidal energy spectrometer have been described for the purposes of tracking specimen surface voltage changes in the presence of global fringe/local surface electric fields. The results seem to suggest that the energy analyzer attachment can both monitor specimen voltage changes, and provide useful information about the kind of electric fields that are present. Although the presence of global fringe/local surface electric fields cause SEs to deviate from their usual trajectory paths, producing complex changes in the output spectral signal height and shape, its most probable energy (average energy) continues to have a simple monotonic relationship to specimen potential, making it still possible to reliably track voltage changes that occur on the specimen surface. This is a significant advance over the performance of previous retarding field voltage contrast systems, in which the measurement errors caused by local surface field effects was an unsolved long standing problem.

Chapter 3

Quantitative Material Contrast

Energy-Dispersive X-ray Spectroscopy (EDS) is an elemental chemical microanalysis technique routinely performed in SEMs. It functions by detecting and analyzing emitted X-rays from the sample during bombardment by the SEM primary electron beam. The frequency/yield of each characteristic X-ray peak provides quantitative information about elemental composition in the sample. This makes EDS a powerful attachment for carrying out quantitative material analysis in the SEM. However, it also has significant limitations, particularly with respect to operating at low primary beam voltages (LVSEM) and acquiring high spatial/depth resolution information (in the nano-meter range).

Standard EDS is usually carried out at relatively high primary beam voltages (> 6 kV), necessary to produce X-rays generated from inner shell atomic orbitals (K, L, and M lines) of the material being probed. This in turn, generates a large interaction volume, limiting the spatial and depth resolution to be in the micron range (O'Connor *et al.*, 2003). This effectively makes EDS a bulk measurement technique. Even for modern low-kV EDS techniques, the accuracy of material identification is reduced, due to there being fewer emitted characteristic X-ray peaks at low energies (Burgess *et al.*, 2013; Burgess *et al.*, 2017). These peaks tend to overlap with one another, and overlap with the carbon contamination Kα line. Surface contamination also reduces the height of its spectral peaks compared to the background signal (Li-Lung, 2005). The low-kV EDS attachment also needs to be placed very close to the sample, which obstructs other analytical tools in the SEM. These kinds of limitations, present in most EDS attachments for the SEM, mean that there is a possible complementary role that can be played by SE energy spectroscopy, which can operate at low primary beam voltages (down to say 0.5 kV) with nano-meter spatial/depth resolution.

This chapter demonstrates that quantitative material analysis can be carried out in a SEM by using the energy spectrum of its scattered electrons in a variety of different ways. One highlight is the discovery that bulk density of states information can be extracted from the combination of SE energy spectral data captured at two different primary beam low voltages (0.5 and 1 kV). Another significant finding is the existence of a linear relationship between atomic number and SE energy spectral peak height for elemental metals. The chapter demonstrates how material contrast information can be enhanced by tracking changes in the shape of the SE energy spectral signal, both in its rising edge, and in its falling edge, and uses energy spectral techniques for understanding and mitigating carbon contamination. It also demonstrates how experimental SE energy spectra generated by a Ga^+ ion primary beam can be captured in a dual beam FIB-SEM instrument, and how they differ from the SE energy spectra produced by a primary electron beam.

Most of the experimental results presented in this chapter have not been reported before, and it is important to emphasize here, the central role played by the wide-angle toroidal energy analyzer attachment, particularly its high signal-to-noise characteristics as described in the preceding chapter. It is the use of a SE energy spectrometer specifically designed for the SEM that has made quantitative material analysis possible, demonstrating that SE energy spectroscopy has the potential to become a useful technique for LVSEM.

3.1 Quantitative material contrast using SE energy spectral peak shape for elemental metal samples

A quantitative material contrast technique is introduced in this section, based upon tracking variations in SE energy spectral peak shape for different elemental metals. Experimental SE energy spectral signals from metal foil samples were obtained with a 2 kV and 40 pA primary beam, and six of them are shown in Fig. 3-1a. The metal foils used here and their layout in the sample holder are similar to those discussed in the preceding chapter. Metals spanning a wide range of atomic numbers were selected, and two parameters were extracted from each SE energy spectral signal: the peak height, and the Full-Width-Half-Maximum (FWHM).

The SE spectral peak heights were found to rise linearly with increasing atomic number, illustrated by the blue data in Fig. 3-1b. The R-squared value of the experimental SE peak height linear fit is 0.9172, indicating that there is a strong linear correlation between SE energy spectral peak height and sample atomic number. This is the first time a simple quantitative relationship between the SE spectral peak height and atomic number has been found.

Normalizing the SE energy spectral signal by taking the ratio of the signal to the background spurious signal, i.e. the signal output $S_E(0\ \text{eV})$ measured at $E_P = 0$ eV, was found to provide greater linearity. The normalized experimental signal to the background $S_{BN}(E_p)$ was calculated by

$$S_{BN}(E_P) = \frac{S_E(E_P)}{S_E(0\ \text{eV})} \tag{3.1}$$

The SE peak height normalized by the background is shown in Fig. 3-1b, indicated by the red data.

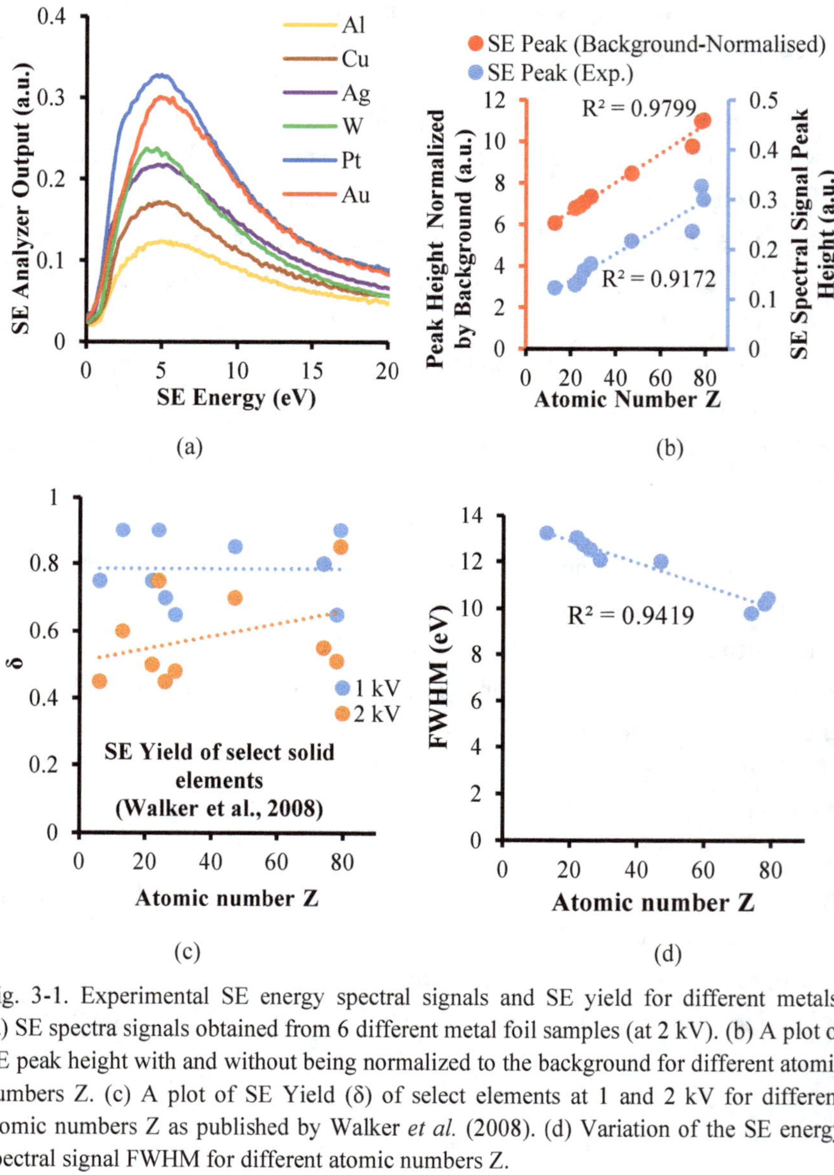

Fig. 3-1. Experimental SE energy spectral signals and SE yield for different metals. (a) SE spectra signals obtained from 6 different metal foil samples (at 2 kV). (b) A plot of SE peak height with and without being normalized to the background for different atomic numbers Z. (c) A plot of SE Yield (δ) of select elements at 1 and 2 kV for different atomic numbers Z as published by Walker *et al.* (2008). (d) Variation of the SE energy spectral signal FWHM for different atomic numbers Z.

Figure 3-1b shows that the SE peak height normalized by the background signal has an improved R-squared fit value of 0.9799. The better linearity of the technique most likely comes from the fact that the magnitude of the background signal is affected by some of the same kinds of errors that affect the SE peak height, caused by primary beam current variations and the formation of a carbon contamination layer on the specimen surface. Since the background signal comes mainly from tertiary electron generated by multiple scattering of high energy SEs and BSEs on the analyzer electrodes, the success of the background normalization techniques suggests that there is a contamination layer also on the analyzer electrodes which builds up at approximately the same rate as the primary beam induced contamination layer on the specimen surface.

It is important to note that the linear dependence between atomic number and SE energy spectral height as reported here, is very different to the kind of results obtained previously by SE emission studies carried out in the SEM, which were based upon recording variations in SE yield (Insepov *et al.*, 2011; Lye and Dekker, 1957; Pawley, 1984; Walker *et al.*, 2008). Figure 3-1c shows SE yield data of this kind for the primary beam voltages of 1 kV and 2 kV that were obtained from Walker *et al.* (2008). Unlike the present SE energy spectral peak height data, the SE total yield data does not show any clear quantitative relationship with atomic number. The main reason for this is that the bulk signals from previous SE yield data collected SEs emitted over a wide range of angles and energies, whereas each point from the SE energy spectrum is detected over a much smaller energy range (2.5% of the SE energy) and polar angular entrance spread (16°). The SE yield effectively represents an integrated SE energy signal, making it less sensitive to material contrast variations. More about the material contrast mechanisms behind the SE energy peak height variation with atomic number is given in section 3.3, where the SE energy spectrum is captured together with the BSE energy spectrum.

Figure 3-1d shows that there is also a linear relationship between the Full-Width-Half-Maximum (FWHM) of the spectral signals shown in Fig. 3-1a and atomic number. In this case, the FWHM of the spectral signal decreases with increasing atomic number. The R-squared value is 0.9419, not as linear as the background normalized SE peak height dependence, but better than the SE peak height signal without normalization. The general change of SE energy spectral signal shape with increasing atomic number therefore seems to be one of a signal that steadily increases in height and becomes narrower. The fact that both these parameters vary linearly with increasing atomic number means that SE energy spectroscopy provides a relatively straight forward way of estimating the atomic number of elemental metal regions on a specimen's surface.

For the purposes of identifying different metal regions on the sample, tracking the SE energy spectral peak height is naturally much faster than monitoring changes in the FWHM, since it only involves a single point on SE energy spectrum. An additional reference metal, perhaps built into the specimen holder, is obviously required. In this way, the linear dependence shown in Fig. 3-1b can be used to provide a fast and simple way to identify an elemental region's atomic number on the sample. It may therefore, in some situations, provide complementary information to EDS, particularly for low primary beam voltages (0.5 to 2 kV). However, it can clearly only provide an approximate value, and it will not be viable for distinguishing between metal regions that have close atomic numbers. In both Figs. 3-1b and 3-1d, the data for Pt (Z = 78) and Au (Z = 79) lie very close together, and are difficult to distinguish. In fact, from Fig. 3-1a, it would appear that the peak height for Pt, which includes the background, is a little higher than the one for Au. The linear trend with atomic number is therefore useful only as a first approximate material identification strategy. Fortunately, there is more material contrast information contained within other parts of the SE energy spectral signal which can be used to enhance material identification further, as will be demonstrated in the next section.

3.2 Enhanced material contrast through tracking gradient variations in the spectral signal rising edge

In this section, enhanced material contrast information required to distinguish close atomic number elements (such as Pt and Au) is obtained by capturing gradient variations in the SE energy spectral signal rising edge.

SE energy spectral signals from 8 metal pellets (described in chapter 2) are normalized by height and shown in Fig. 3-2a. In this way, variations in height are eliminated and only small changes in shape are analysed. Changes in the gradient of the normalized SE energy spectral signals were found to be most sensitive to material contrast over the rising edge part of the spectral signal, typically in the 1 to 2 eV energy range. This is illustrated in Fig. 3-2b, where the spectral signals of Pt and Au, are clearly separated, unlike the situation for SE peak height or FWHM material contrast where they are diificult to distinguish (closeness of the last two points in Figs. 3.1b and 3.1d). The spectral experimental results seem to indicate that different materials give rise to a different rising edge signal shape. To enhance this contrast further, the SE energy spectral signal is differented and its gradient is plot as a function of energy over the 1 to 2 eV range, the result is shown in Fig. 3-2c. The differentiated SE energy spectral signals from all 8 elemental metals are clearly distinguaishable. This is especially true in the case of Pt and Au, where their differented SE spectra are significantly separated and not overlapping. Unlike the situation for tracking SE energy spectral signal height or FWHM, the gradient signals for Pt and Au are more separated with respect to the signals from other metals, indicating that the information contained in the shape of the rising edge part of the spectral signal is of a very different kind, complementary to the information that can be extracted by detecting changes in SE energy spectral height or FWHM. This method therefore represents a way of futher enhancing material contrast from the SE energy spectral signal.

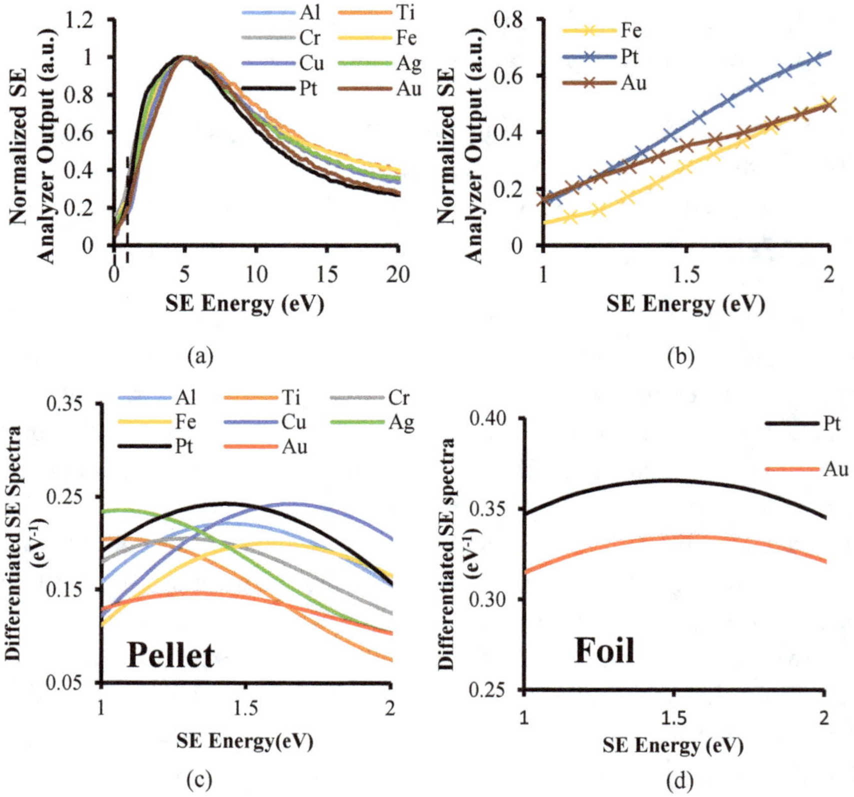

Fig. 3-2. Rising edge SE energy spectra obtained from metal samples. (a) SE energy spectral signals normalized by SE peak height obtained from metal pellets. (b) Normalized SE energy spectral signals for Fe, Pt and Au pellets between 1 and 2 eV. (c) Differentiated SE spectra between 1 and 2 eV from metal pellets. (d) Differentiated SE spectra between 1 and 2 eV of the Pt and Au foil samples.

The reliability of this differentiated rising edge approach is confirmed by obtaining similar experimental SE energy spectral signal gradient variations from the Pt and Au metal foil samples, shown in Fig. 3-2d. Like the metal pellet samples, the experimental spectral signals from Au and Pt metal foil samples when differentiated in the 1 to 2 eV energy range are clearly separable. The %RMSD between the Pt and Au curves from the metal pellet and metal foil samples were found to be 62.5% and

11.7 % respectively. This consistency demonstrates that it is possible to use gradient information extracted from the SE spectral signal rising edge in order to distnguish between elements having close atomic numbers.

3.3 Material contrast through full-range energy spectra

In the previous two sections, it was shown that quantitative material contrast could be obtained from SE energy spectral signals in different ways. This section shows how useful additional material contrast information can be obtained by comparing both SE and BSE portions of the full-range electron energy spectral signal, where the detected energy of scattered electrons goes up to the primary beam energy. This was achieved through increasing the analyzer deflection voltage range by using a low-ripple high-voltage power supply (TREK 677B) that is able to ramp voltages from 0 to 2000 V. The deflection voltage interval, was set to gradually increase with the deflection voltage value, i.e. $\Delta V_{Di} = k\, V_{Di}$. This was done for several reasons. Firstly, since the SE peak energy is one to two orders of magnitude narrower than the BSE peak, the energy interval between sampled points on the full-range spectrum needs to progressively increase in order to keep data acquisition times and beam induced contamination to a manageable level. Secondly, since the energy width of detected electrons increases with electron pass energy (as discussed in the preceding chapter), a pass energy interval increasing by the same proportion is better suited at capturing the broader nature of the BSE peak shape. This strategy makes it possible to collect the full-range electron energy spectrum including both low energy SE and high-energy BSE spectra up to the primary beam energy with sufficient sampling points in a reasonable acquisition time. The number of sampling points used in the full-range electron energy spectra was 130 for a primary voltage of 1 kV, and 155 points for the 2 kV spectra, with acquisition times of 35 and 40 s respectively. The primary beam current was kept to around 40 pA. Metal foil samples were used, where Au was used as a reference foil in each sample holder, and the results are shown in Fig. 3-3. In addition to the results from metal foils, the full range energy spectrum of highly oriented pyrolytic graphite (HOPG) is also included. The HOPG is a grade 2-type sample obtained from SPI Supplies with grain sizes of ~1 mm.

A wide variety of useful material information can be extracted from the full-range energy data shown in Fig. 3-3 by comparing the shapes and heights of the SE and BSE energy spectral peaks. There are some obvious similarities and differences. Both SE and BSE peak heights rise with increasing atomic number, but BSE peak widths become progressively narrower, while SE peak widths do not appear to change significantly (on this full-range scale). Au and Pt are distinguishable in both SE and BSE peaks at 1 kV (Fig. 3-3a), but distinguishable only in the SE peak at 2 kV (Fig. 3-3b). More quantitative information can be obtained by plotting peak height or width as a function of atomic number. To start with, SE and BSE peak height variation for both primary beam voltages of 1 and 2 kV are plot, and shown in Figs. 3-4a and 3-4b. The SE peak heights vary linearly with atomic number at both 1 and 2 kV, consistent with the experimental data shown in Fig. 3-1b. The BSE peak height, on the other hand, increases exponentially with atomic number. This is the first time any quantitative dependence of BSE peak height with atomic number has been reported. Previous research had mainly been limited to recording how the total backscattered electron yield varied with atomic number (Antolak and Williamson, 1985; Archard, 1961; Arnal *et al.*, 1969; Reimer and Tollkamp, 1980; Sternglass, 1954), with a general consensus that the yield initially increases, and is then followed by a region having either a plateau, or values of the yield decreasing. This is for instance, illustrated in Fig. 3-4c which depicts experimental results reported by Sternglass (1954).

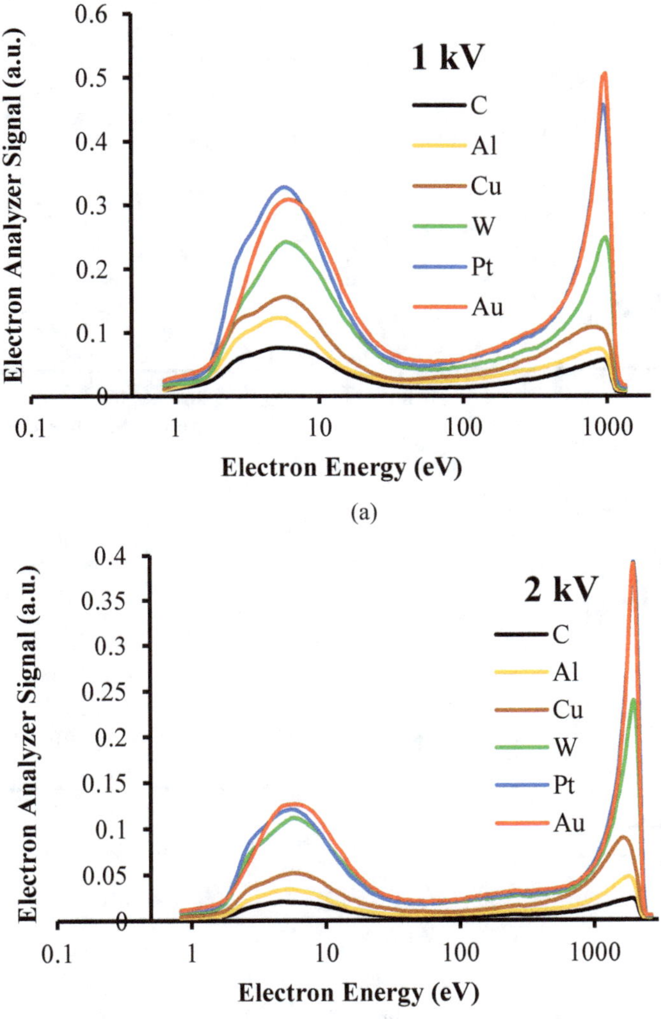

(a)

(b)

Fig. 3-3. Full range electron energy spectra obtained from metal foils and highly oriented pyrolytic graphite (HOPG) at the primary beam voltages of (a) 1 kV (b) 2 kV.

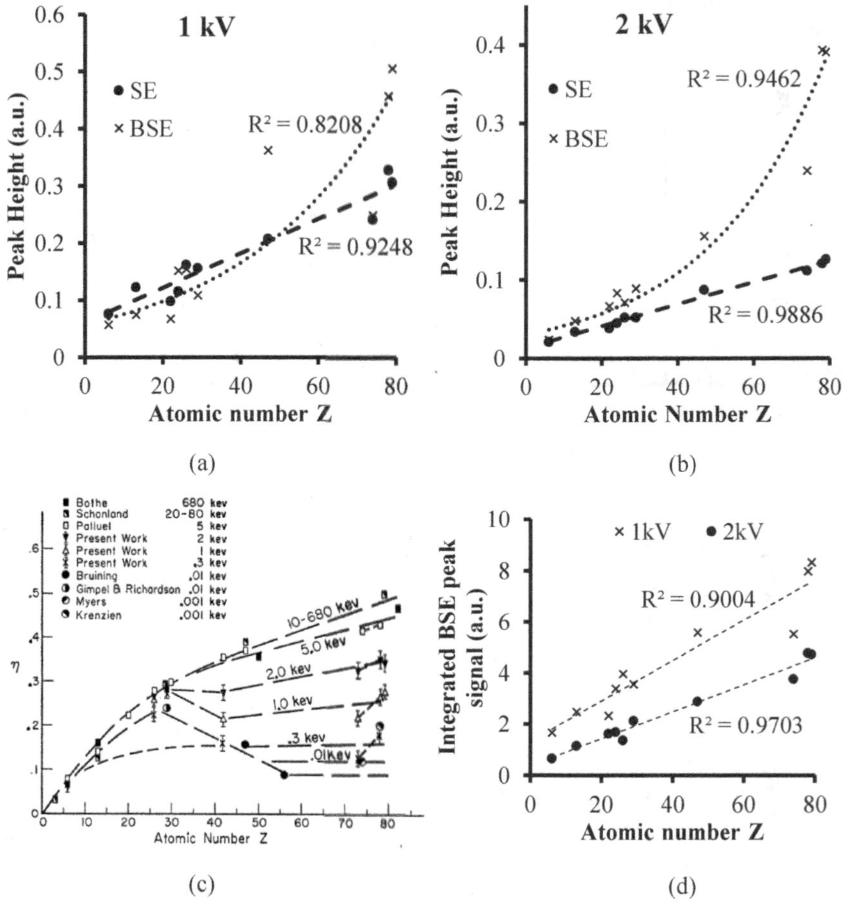

Fig. 3-4. SE and BSE peak height, and BSE yield as functions of atomic number and different primary beam voltages. (a) SE and BSE peak height obtained with a 1 kV primary beam. (b) SE and BSE peak height obtained with a 2 kV primary beam. (c) BSE yield η obtained by Sternglass *et al.* (1954) at different primary beam energies. (d) Integrated BSE peak signal (0.4 to 1 keV and 1.4 to 2 keV). Copyright 1954 by American Phyiscal Society. Reprinted with permission.

In order to make a more meaningful comparison with the previous published BSE yield data, the area under the BSE energy spectral peak was integrated. This parameter will be referred to here as an "Integrated BSE peak signal". Although this is a bulk BSE signal, it is still significantly different from the previous BSE yield signals. It is extracted from the experimental spectra data shown in Fig. 3-3, which was obtained by the wide-angle toroidal energy analyzer attachment capturing electrons emitted in the 45 ± 8° polar angular range. Previous BSE yield results were typically obtained from hemispherical collectors (0 to 90° polar angles (Assa'd and Gomati, 1998; Farhang *et al.*, 1993; Reimer and Tollkamp, 1980). Another important difference is that the present Integrated BSE peak signal was integrated over a selected energy range, from 0.4 to 1 keV for the 1 kV primary beam data, and from 1.4 to 2 keV for the 2 kV primary beam data, suitably adjusted so that it represents the area under each BSE spectral peak. Previous BSE yield signals were obtained by excluding SEs from the output signal, typically filtering out electrons below 50 eV (consistent with the usual SE cut-off definition), and were therefore formed from a much wider range of electron energies.

Figure 3-4d shows that the Integrated BSE peak signal plot as a function of atomic number is linear for both the primary beam voltages of 1 and 2 kV, with a high degree of linearity in the 2 kV case (R-squared value of 0.9703). This gives better agreement with the previously reported BSE yield data shown in Fig. 3-4c, providing a closer fit for the low atomic number range, but still differing significantly for the higher atomic number range. This discrepancy may be due to the very different electron emission energy and angular collection ranges inherent in the previous BSE detector systems that tracked yield as a function of atomic number, compared to the comparatively small angular/energy emission range captured by the present band-pass toroidal energy analyzer attachment.

As with the SE energy spectral peak height data, the discovery of a simple quantitative relationship between BSE energy spectral peak height data (exponential) and Integrated BSE peak signal (linear) with atomic number demonstrates the advantages of using a band-pass energy spectrometer attachment in the SEM, instead of relying on SE or BSE yield data.

The full range electron spectral signals shown in Fig. 3-4 also provides data that can better help understand how SE and BSE energy spectral signals might be related to one another. Since both the Integrated BSE and SE peak height data rise up linearly together with atomic number, the results suggest that changes in SE peak height are primarily caused by secondary electrons that are generated by backscattered electrons, known as SE2s (Drescher *et al.*, 1970; Joy, 1987; Peters, 1982). This is a significant result, and is further strengthened by the fact that the degree of linear correlation between the SE peak height and the Integrated BSE peak data is approximately the same for both the primary beam voltages of 1 and 2 kV (the R-squared value being between 0.96 to 97 in both cases), as shown in Fig. 3-5a. This is strong evidence for the linear dependence of SE peak height with atomic number coming directly from SE2s, the secondary electrons generated from backscattered electrons. The same amount of loss in linearity observed in the SE peak height variation for the 1 kV beam with atomic number (compared to the 2 kV beam), also occurs for the Integrated BSE spectral signal variation with atomic number. Monitoring variations in the SE energy spectral height for different metals is therefore a way of tracking overall changes occurring in BSE scattering (in the polar angular 45° ±8° range).

It is interesting to note that the FWHM of the BSE energy spectral peak also gives an approximately linear dependence with atomic number, as shown by Fig. 3-5b. Although decreasing with atomic number and less linear than the SE peak height or Integrated BSE peak signal, it provides another useful cross-check for extracting quantitative material contrast information from the full-range energy spectrum. For LVSEM applications, both the SE energy spectrum and BSE energy spectrum as a way of obtaining an atomic number estimate of an elemental region on the specimen can be used. For applications where the primary beam voltage is higher than 3 kV, generation of the BSE energy spectrum will be impractical, due to electrical spark-over limitations inside the present wide-angle toroidal energy analyzer attachment.

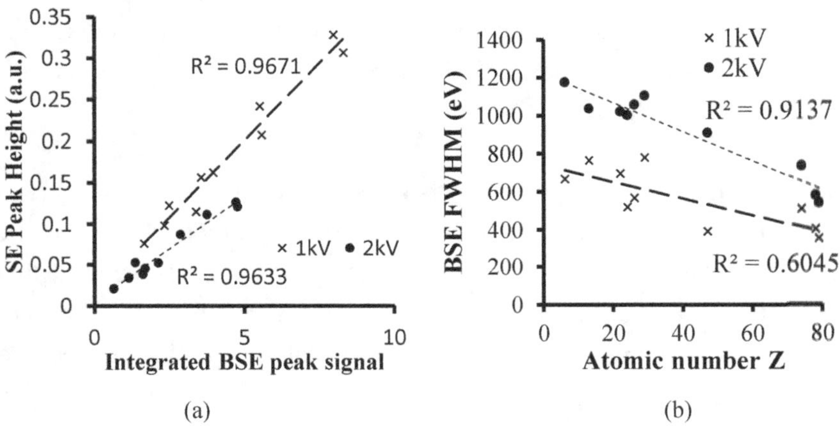

Fig. 3-5. SE-BSE energy spectral peak material contrast characteristics. (a) SE peak height correlation to the integrated BSE total peak signal for different materials. (b) BSE energy spectral peak FWHM variation with atomic number.

There is more information that can be extracted from the experimental results shown in Figs. 3-4a and 3-4b that has not been mentioned so far. For both the SE and BSE spectral signals, the scattering of the results about their respective fitted trends is smaller at the primary beam voltage of 2 kV than at 1 kV: at 1 kV the R-squared values for SE and BSE peak heights are 0.9248 and 0.8208 respectively, and they improve to 0.9886 and 0.9462 for 2 kV. The reduced scattering about the fitted trend with increased beam energy most likely comes from at least two reasons. Firstly, at a lower beam energy, the backscattering of electrons is no longer solely dominated by nuclear coulomb field scattering, which increases with atomic number, instead, inelastic collisions with outer-shell electrons also contribute, resulting in the backscattered distribution being less dependent on atomic number (Sternglass, 1954). This translates into a greater deviation from the fitted trends for both SEs and

BSEs at lower primary beam voltages (1 kV in this case). Secondly, there is an increased interaction with whatever lies at the sample surface (a contamination layer/native oxide) at lower beam energies, which becomes more apparent at 1 kV. Since these surface effects are irregular and independent of atomic number, they show up as further deviations from the fitted linear trend for SE peak heights, and an exponential one for BSE peak heights.

The full-range energy spectral results tracking the comparative change in SE and BSE peak heights, taken together with the gradient changes in the SE energy spectral signal rising edge (shown in Fig. 3-2), suggest that different parts of the SE energy spectral signal correlate to the different ways SEs are generated. SE energy spectral peak height changes seem to be indirectly related to BSE scattering, while variations in the rising edge SE energy spectral signal shape most likely comes from inelastic collisions with outer shell electrons, surface states, and surface ad-layers.

3.4 Quantitative material contrast mapping for close atomic numbers

An alternative to enhancing material contrast information through monitoring the SE spectral signal rising edge shape, is to bias the specimen/inner cap electrode negatively. In this case, extra sensitivity to material contrast seems to appear in the spectral peak height and shape in the falling edge part of the spectral signal. This approach is both faster and better adapted for material contrast mapping. Biasing the specimen/inner cap electrode negatively, has the advantages of providing better output signal-to-noise characteristics, and producing sharper and more symmetric SE energy spectral signals, as already noted in the preceding chapter.

A simple test sample consisting of platinum and gold foils on an aluminium base plate was assembled in order to test material contrast mapping for close atomic number metals, as shown in Fig. 3-6. The foils were pushed towards one another until the gap size between them was in the tens of microns range.

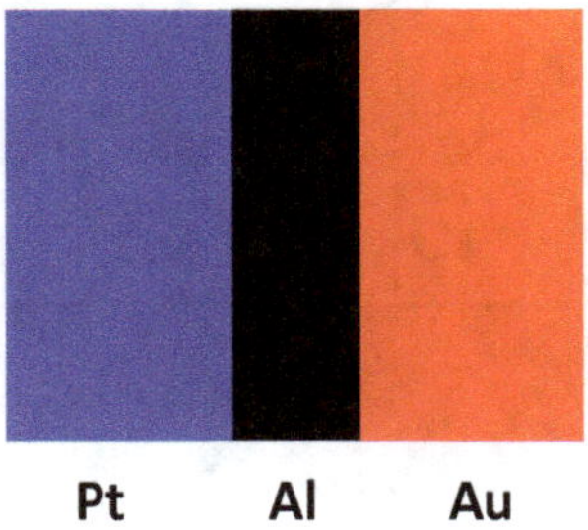

Fig. 3-6. A Pt/Au foil sample on an Al baseplate to test quantitative material contrast mapping for close atomic number metals.

Figure 3-7 shows experimental SE energy spectral signals taken from points located on the Pt and Au sides of the test sample for which the inner cap electrode and sample were biased to −10 V for a primary beam voltage of 5 kV. These experiments were carried out on the FEI NOVA NanoSEM™ 230, with a primary beam current of 130 pA and a 15 s frame scan time. Material contrast information appears both in terms of a peak height difference (Fig. 3-7a) and overall shape change (Fig. 3-7b): the Au spectral signal has a higher peak signal height than the Pt signal, while the Pt spectral signal has a broader shape. Most of this material contrast shape change information appears in the falling edge part of the signal. Figure 3-8 shows experimental SE energy spectra to illustrate how both the height and shape material contrast between Au and Pt also exist for a primary beam voltage of 1 kV. Although the precise nature of the contrast differs from the 5 kV spectra, the Au spectral signal still has the higher peak value, while the Pt spectral signal still has the broader shape.

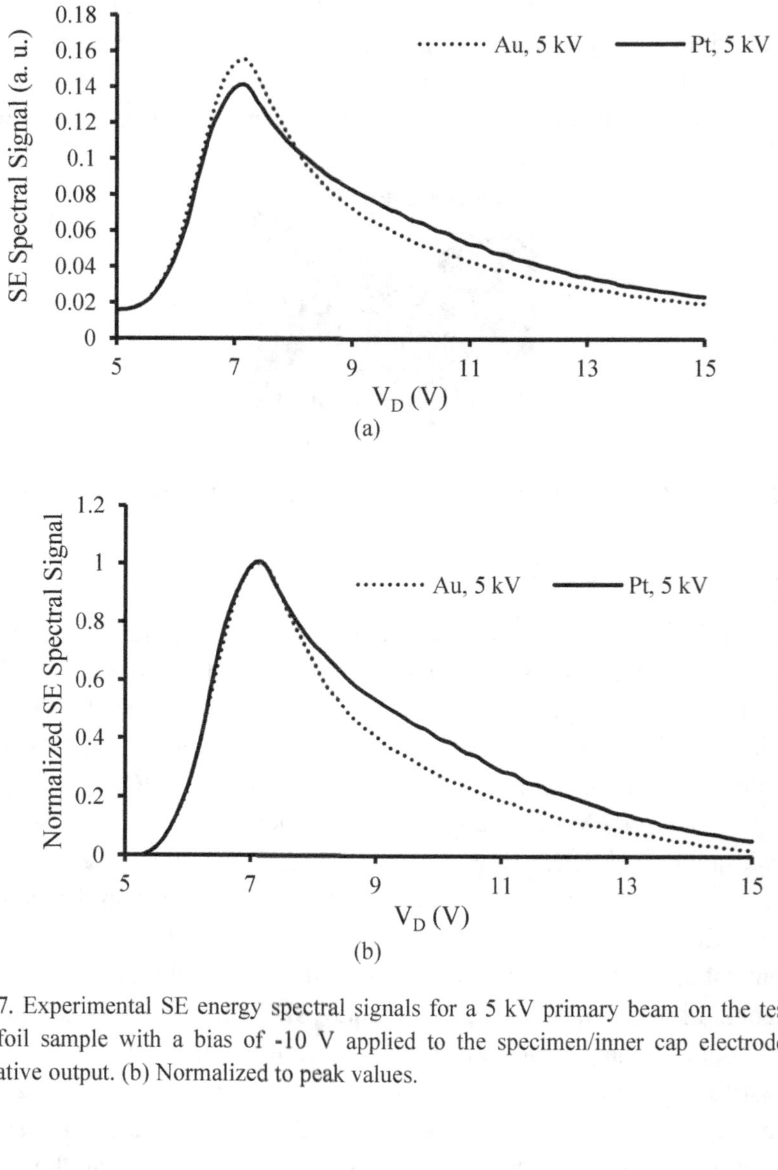

Fig. 3-7. Experimental SE energy spectral signals for a 5 kV primary beam on the test Pt/Au foil sample with a bias of -10 V applied to the specimen/inner cap electrode. (a) Relative output. (b) Normalized to peak values.

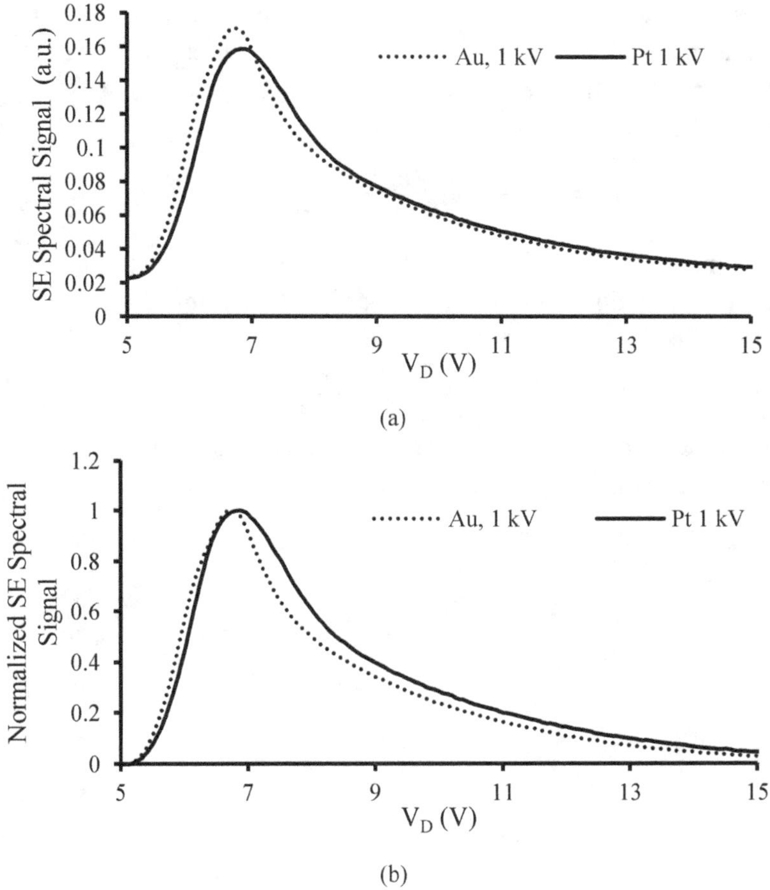

Fig. 3-8. Experimental SE energy spectral signals for a 1 kV primary beam on the test Pt/Au foil sample with a bias of -10 V applied to the specimen/inner cap electrode. (a) Relative output. (b) Normalized to peak values.

In order to quantify the spectral signal change of shape for every pixel in the image, four images were acquired at different analyzer deflection voltages, as shown in Fig. 3-9. The image I_{0ij}, where the subscripts *i* and *j* refer to image pixel indices, is acquired at $V_D = 7$ V, I_{1ij} at $V_D = 8$ V, I_{2ij} at $V_D = 9$ V, and I_{3ij} at $V_D = 10$ V, corresponding to the SE energies of $E_0 = 2.55$ eV, $E_1 = 4.22$ eV, $E_2 = 5.9$ eV, and $E_3 = 7.57$ eV respectively. Material contrast information is first extracted from the intensity of the band-pass image close to the SE spectral peak value, I_{0ij} image, and additional material contrast information is then obtained by calculating the standard deviation over the other three band-pass images σ_{ij} with respect to the first image, by,

$$\sigma_{ij}^2 = \frac{I_{1ij}(E_1-E_0)^2+I_{2ij}(E_2-E_0)^2+I_{3ij}(E_3-E_0)^2}{I_{0ij}+I_{1ij}+I_{2ij}+I_{3ij}} \tag{3.2}$$

A combination spectral image, P_{ij}, uses the extra material contrast information contained within the standard deviation image by translating it into an effective height change, given by,

$$P_{ij} = I_{0ij} - k\sigma_{ij} \tag{3.3}$$

where the parameter *k* in this expression is a constant that maximizes the material contrast information present in the final image.

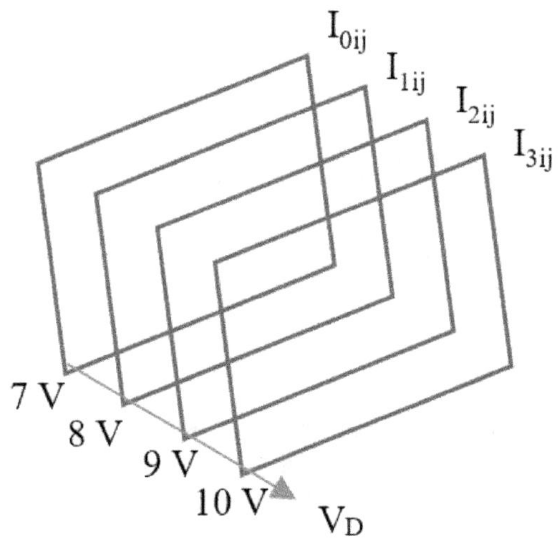

Fig. 3-9. Spectroscopic image acquisition of four images along the falling edge part of the output spectral signal.

Figure 3-10 acquires experimental material contrast spectroscopic imaging data for a primary beam voltage of 5 kV on the Pt/Au foil test specimen, and compares it to its corresponding SE detector image and EDS data obtained by the windowless Oxford Instruments Ultim®

Extreme system, an EDS system that has significantly better sensitivity than conventional EDS systems at low primary beam voltages. The total acquisition time for both EDS and spectroscopic imaging was 16 minutes. A histogram of pixel intensity is also shown alongside the SE detector and combination spectroscopic image. The SE detector image shown in Fig. 3-10a has very little visible variation of pixel intensity between Pt and Au regions within the sample, which is also reflected by the Au and Pt peaks in its histogram directly overlapping with one another. On the other hand, the combination spectroscopic image P_{ij} shown in Fig. 3-10b displays good material contrast separation, the blue color contains 97.9% of all Pt pixels, and the red color contains 99% of all Au pixels for $k = 0.8$. This high level of material contrast mapping comes from the fact that the Pt and Au histograms of the spectroscopic image are well separated and distinct. The color coding intensity threshold value was placed in the small overlapping region between the Pt and Au histograms. For these 5 kV primary beam voltage conditions, the corresponding windowless EDS system is also able to give a distinctive color coded elemental map of the specimen, as shown in Fig. 3-10c.

The component band-pass image I_{0ij} and standard deviation image σ_{ij} used to produce the spectroscopic combination image P_{ij} shown in Fig. 3-10b, are given in Fig. 3-11. They clearly demonstrate the advantage of using the combination imaging strategy to form P_{ij}. Before the combination prcocedure, only 74.2% of all Pt pixels are coded blue while only 59.7% of all Au pixels are coded red in the band-pass image I_{0ij}. In the σ_{ij} image, the material contrast separation is 71.3% for Pt and 70.4% for Au. After forming the combination image, the material contrast separation goes up to over 97% for both Pt and Au. The small region of overlap is most likely caused by shot noise and topographical effects.

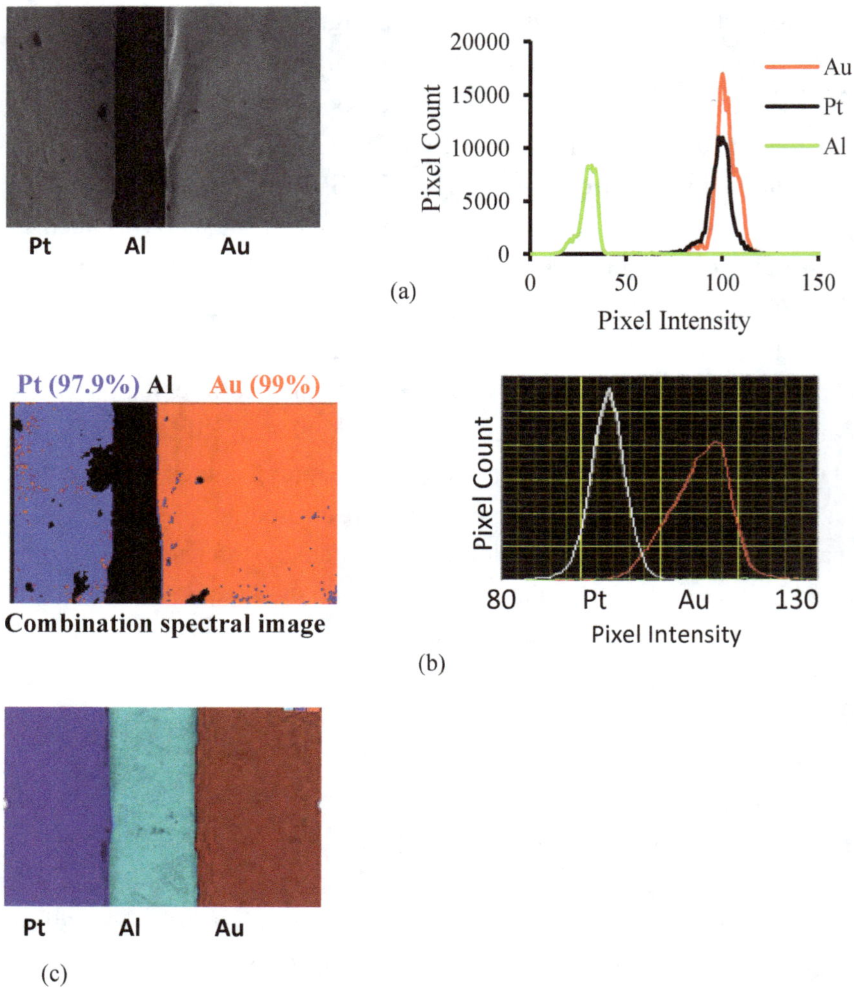

Fig. 3-10. Quantitative material contrast mapping of Pt/Au foils on Al at a primary beam voltage of 5 kV. (a) Conventional SEM topographic image with pixel intensity histogram. (b) Combination spectral image with pixel intensity histogram ($k = 0.8$). (c) EDS windowless low energy mapping system image.

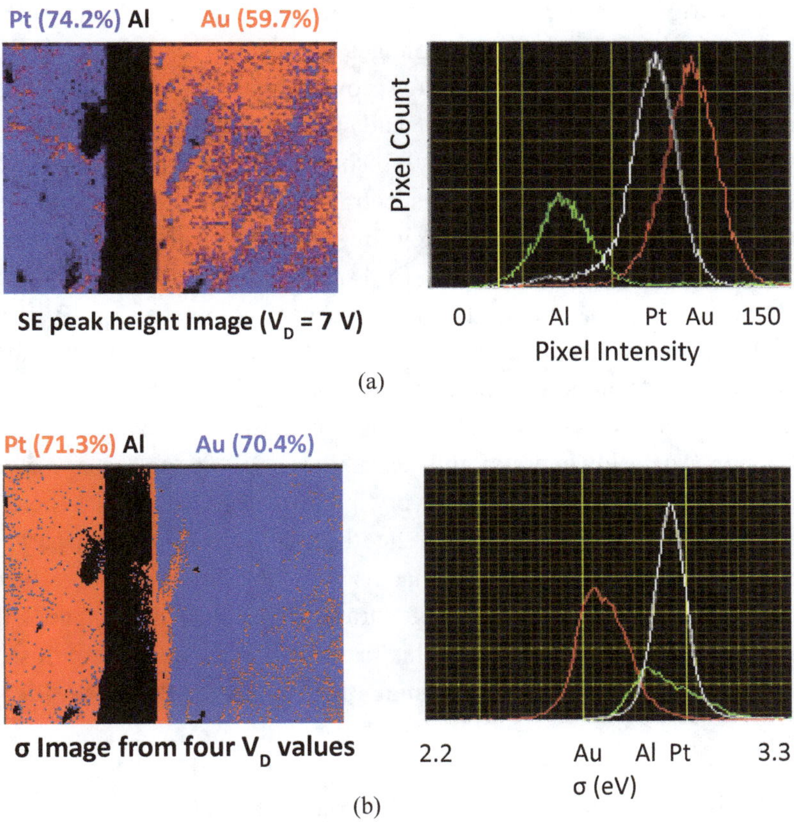

Fig. 3-11. Component images to the secondary electron spectroscopic image at a primary beam voltage of 5 kV. (a) Band-pass contrast image at a SE energy of 2.55 eV ($V_D = 7$ V) with pixel intensity histogram. (b) Mapping σ (std) extracted from images captured at deflection voltages of −7, −8, −9 and −10 V with pixel histogram. Note that standard deviation of pixels in the Al region have been deliberately assigned low values in the color coded image so the Al region appears black, to keep it consistent with the SE peak height image.

The SE energy spectroscopic elemental mapping technique is particularly important for low primary beam voltages (< 5 kV). This is because EDS systems are generally not able to provide material contrast at low primary beam voltages (Wuhrer and Moran, 2016). This point is illustrated by the experimental results shown in Fig. 3-12, where a SE detector image and a SE energy spectral combination image obtained with a primary beam voltage of 1 kV are compared to the corresponding EDS map obtained by the Oxford Instruments Ultim® Extreme windowless system acquired at a primary beam voltage of 2.5 kV. Like the 5 kV primary beam voltage results, the SE detector image shows almost no visible material contrast, while the SE spectral combination image has near 100% material contrast separation (99.6% for Pt and 98.1% for Au). However, in this case, the EDS image (at a 2.5 kV primary beam voltage), displays very poor material contrast, there is almost no discernible visible difference between the Pt and Au parts of the sample. Figure 3-13 shows that as with the 5 kV primary beam voltage case, the combination image strategy significantly enhances the final material contrast, only in this case, better material contrast is initially obtained in the band-pass image I_{0ij} (91.9% for Pt and 83.4% for Au).

In forming the final spectral combination image at a primary beam voltage of 1 kV, an additional step of using the average standard deviation values in the Pt and Au regions was found to be effective in reducing the influence of topographic noise. After the combination image step, the standard deviation of each pixel was tested to see if it was closer to either the average Pt value or the average Au value: if the pixel had been assigned a blue color (identified as Pt) but had a standard deviation value closer to the average Au value, its color was then turned to red. Likewise, if a pixel had been assigned a red color in the combination image (identified as Au), but its standard deviation value was closer to the average Pt value, then its color was changed to blue. This second step improved the metal identification accuracy from 93% to 99.6% for Pt, and from 88% to 98.1% for Au.

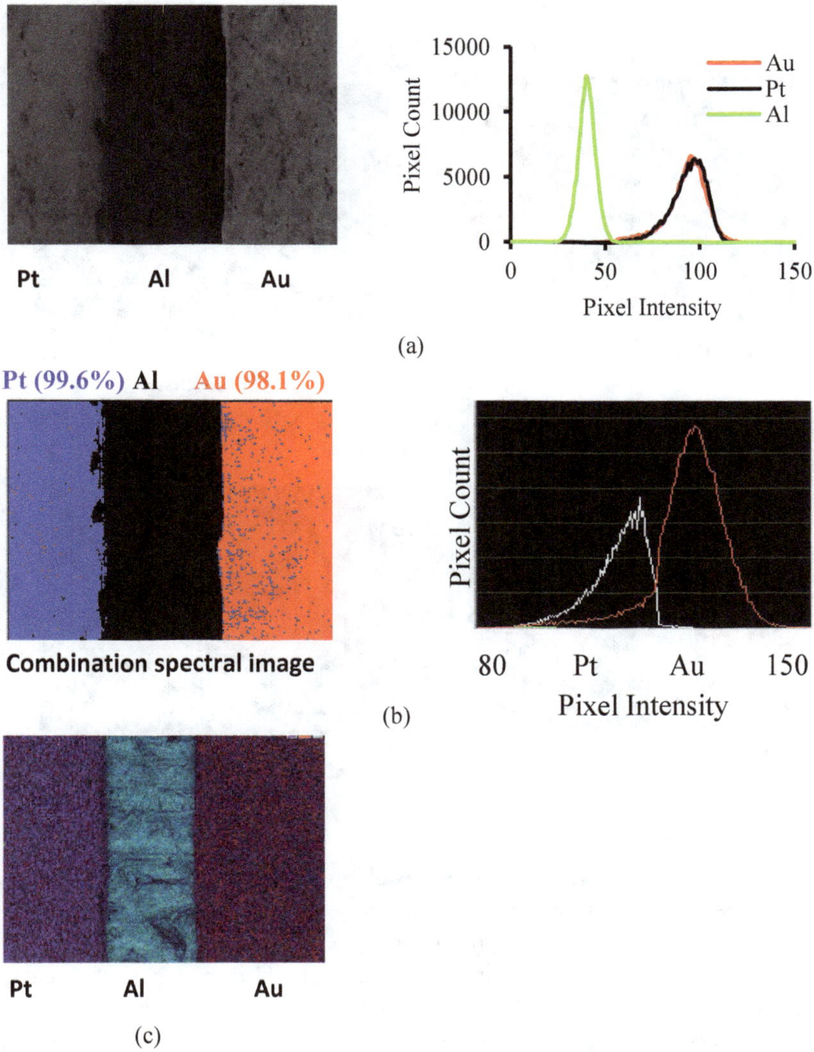

Fig. 3-12. Quantitative material contrast mapping of Pt/Au foils on Al at a primary beam voltage of 1 kV. (a) Conventional SE detector image with pixel intensity histogram. (b) Combination spectral image with pixel intensity histogram k = 0.5. (c) EDS windowless low energy mapping system image at a 2.5 kV primary beam voltage.

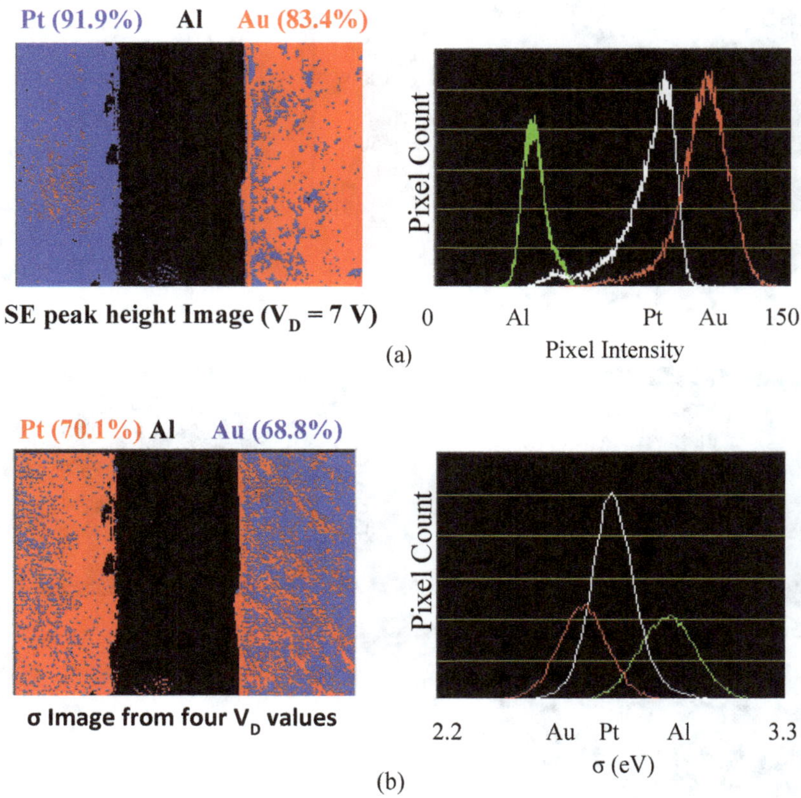

Fig. 3-13. Component images to the secondary electron spectroscopic image at a primary beam voltage of 1 kV. (a) Band-pass contrast image with pixel intensity histogram at a SE energy of 2.55 eV (V_D = 7 V) (b) Mapping σ extracted from images captured at deflection voltages of −7, −8, −9 and −10 V with pixel σ histogram. Note that standard deviation of pixels in the Al region have been deliberately assigned low values in the color coded image so the Al region appears black, to keep it consistent with the SE peak height image.

Overall, the experimental results obtained at a primary beam voltage of 1 kV indicate that near 100% material identification between platinum and gold can be made with SE energy spectroscopy by biasing the specimen/inner cap electrode negatively (−10 V). Under the same conditions, almost no material contrast is obtained by a windowless low energy EDS system or by the conventional SE detector. These results illustrate how SE energy spectroscopy might become a useful companion technique to EDS for LVSEM. Obviously more metals with close atomic numbers need to be tested, such as nickel and copper, to confirm that the same approach is still effective for lower atomic number metals.

Extending SE energy spectroscopy to even lower primary beam voltages still seems to provide a significant amount of material contrast, despite surface/topographical effects playing a greater role. This is illustrated in Figs. 3-14 and 3-15, where the results for a 0.5 kV primary beam voltage are shown. There is clearly more mixing of surface effects/topography in the SE spectral height image shown in Fig. 3-15a, particularly on the Au side of the specimen, and much greater overlap in the standard deviation Pt and Au histograms, as shown in Fig. 3-15b. In this case, the standard deviation image cannot be used to enhance material contrast in the SE spectral signal height image. However, there is both material and surface effects/topography information contained within the SE spectral signal height image, which has a 96.2% identification of the Pt region, and 79.7% identification of Au. It is interesting to note that there are more areas of intermixing within the Au region than on the Pt side of the specimen, suggesting that more variations through surface effects/topography occur for Au than Pt at a primary beam voltage of 0.5 kV.

A similar analysis was tried with the specimen/inner cap electrode unbiased, but extracting material contrast information in that case was not possible since the analyzer spectroscopic images were found to be too noisy. Beyond the signal-to-noise advantage, the biased case produces a different SE energy spectral signal shape, one where the SE peak energy position is located at a much lower energy (around 2 eV compared to 5 eV for the unbiased case), which seems to enhance its material contrast sensitivity for close atomic number metal samples.

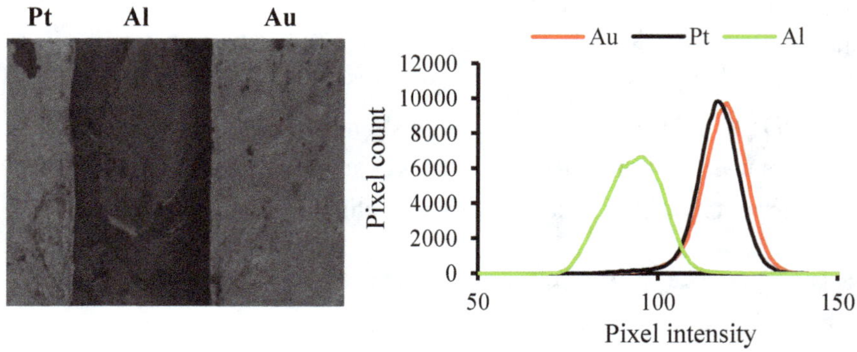

(a)

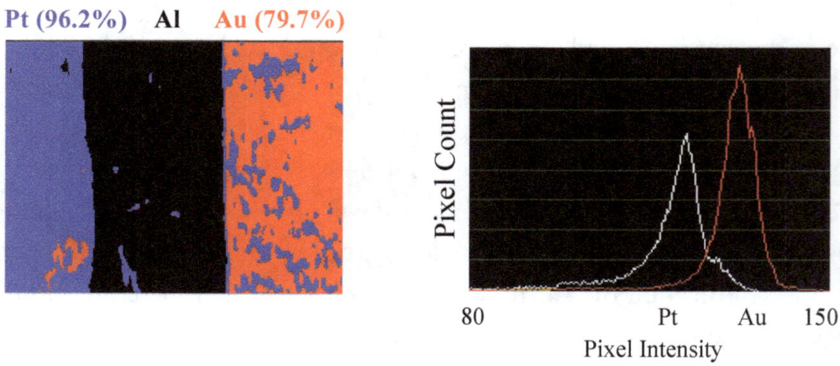

(b)

Fig. 3-14. Quantitative material contrast mapping of Pt/Au foils on Al at a primary beam voltage of 0.5 kV. (a) Conventional SEM topographic image/pixel intensity histogram. (b) Combination spectral image/pixel intensity histogram (k = −0.1). Pt in the histogram is denoted by white, and Au is represented by red.

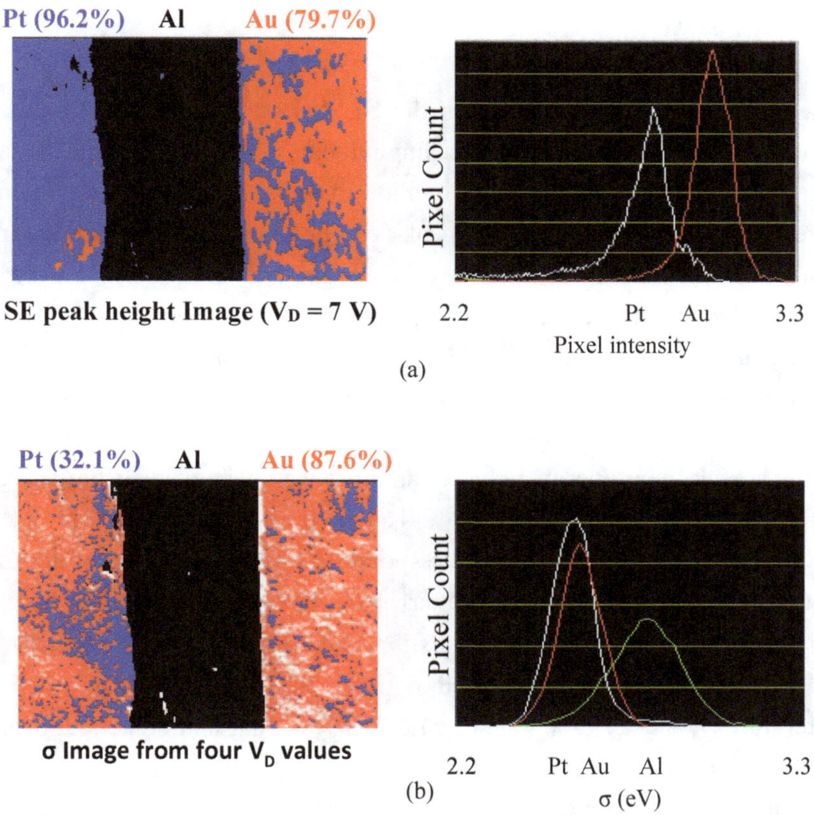

Fig. 3-15. Component images to the secondary electron spectroscopic image at a primary beam voltage of 0.5 kV. (a) Band-pass contrast image with pixel intensity histogram at a SE energy of 2.55 eV (V_D = 7 V). Pt in the histogram is denoted by white and Au is represented by red. (b) Mapping σ (std) extracted from images captured at deflection voltages of −7, −8, −9 and −10 V with pixel σ histogram, Pt in the histogram is denoted by white and Au is represented by red. Note that standard deviation of pixels in the Al region have been deliberately assigned low values in the color coded image so the Al region appears black, to keep it consistent with the SE peak height image.

3.5 Detection of the Auger hydrocarbon contamination peak in the SEM

Reliable material information in the SEM depends upon being able to monitor the level of primary beam induced contamination during any measurement. In this section, a method to track carbon contamination is presented using Auger electron peak detection. The experimental setup is identical to that used for the full-range electron energy spectrum acquisition described earlier. In Fig. 3-16a, the experimental full range electron energy spectra of Cu, HOPG and Au are shown. An Auger peak is observed to consistently appear around 270 eV on each of these backscattered electron spectra, and can be identified as the carbon KVV Auger peak. The energy range of interest was narrowed to be between 200 to 400 eV, as shown in Fig. 3-16b. The pass energy interval is kept to around 4 eV (close to the corresponding energy detection width ΔE), small enough to resolve the Auger peak. The presence of this peak on metals as well as HOPG indicates that the contamination present in the SEM chamber is carbon based, similar to the findings in previous literature (Postek, 1996; Vladar *et al.*, 2001; Vladar and Postek, 2005).

Further analysis investigating the nature of the contamination was performed by obtaining the differentiated electron energy spectra over the same energy range of 200 to 400 eV. Taking the derivative of the Auger peak removes the backscattered electron background, and enhances peak position and peak width information. The differentiated spectra, displayed in Fig. 3-16c, show significant differences between the metal sample Auger peaks and that of HOPG. The differentiated Auger peaks from Cu and Au are very similar, with local maxima at 252 eV and minima at 295 eV, as opposed to the Auger peak from HOPG, where the local maximum and minimum are located at 247 eV and 312 eV respectively.

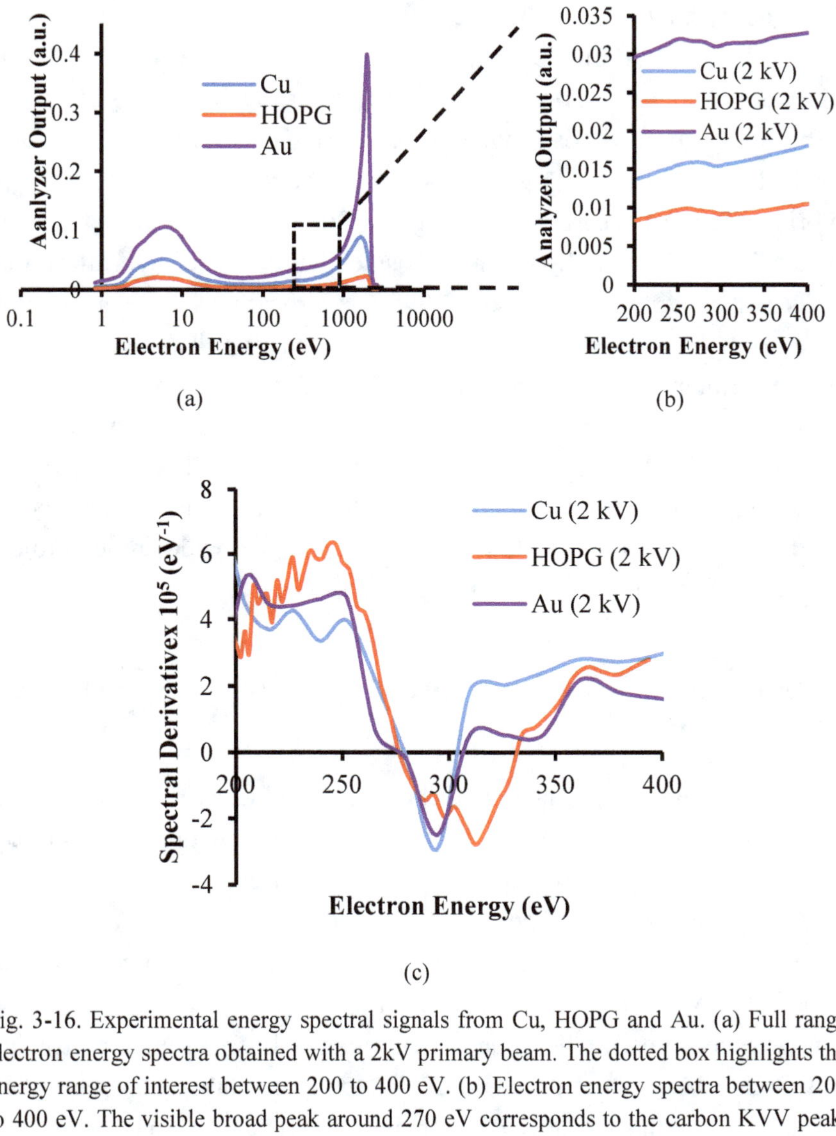

Fig. 3-16. Experimental energy spectral signals from Cu, HOPG and Au. (a) Full range electron energy spectra obtained with a 2kV primary beam. The dotted box highlights the energy range of interest between 200 to 400 eV. (b) Electron energy spectra between 200 to 400 eV. The visible broad peak around 270 eV corresponds to the carbon KVV peak. (c) Differentiated electron energy spectra between 200 to 400 eV.

The difference between the local minimum and the local maximum energies, known as the D-value, is used to determine the Auger line width, a parameter known to be an indicator of the chemical state of the carbon being probed (Lascovich *et al.*, 1991; Lascovich and Scaglione, 1994). In particular, the D-value is known to be proportional to the % sp^2 hybridization of the carbon (Lesiak *et al.*, 2018), which is expected to be higher for HOPG than hydrocarbons, since HOPG is sp^2 hybridized due to its planar structure (Jackson, 1995). Hydrocarbons have an overall lower % hybridization depending on hydrogenation, The agreement of the experimental spectra shown in Fig. 3-16 to these theoretical considerations about the nature of the hydrocarbon contamination Auger peak in relation to the HOPG Auger peak, suggests that useful Auger peak information can be obtained for the purposes of tracking beam induced contamination in the SEM by using the wide-angle toroidal electron energy analyzer attachment.

3.6 Mitigating contamination effects by spectral background normalization

The background normalization technique introduced in section 3.1 to reduce errors when plotting SE spectral peak height with atomic number is employed here to mitigate contamination effects. Experimental SE energy spectral signals from Au foils with different scan areas were collected, as shown in Fig. 3-17a, where the 1×Area unit refers to a scan area of 10 by 6 μm. The overall signal height decreases steadily as the scan area becomes smaller, due to beam induced contamination growing with higher beam current densities. The experimental signals were then normalized by the background signal level $S_E(E_p = 0)$, and plotted in Fig. 3-17b. These normalized signals, unlike their non-normalized counterparts, showed minimal difference with respect to each other, the average difference between the 4 normalized signals is 1.2%. This indicates that the background normalization technique provides a way of mitigating contamination effects on experimental SE energy spectral signals for the Au sample.

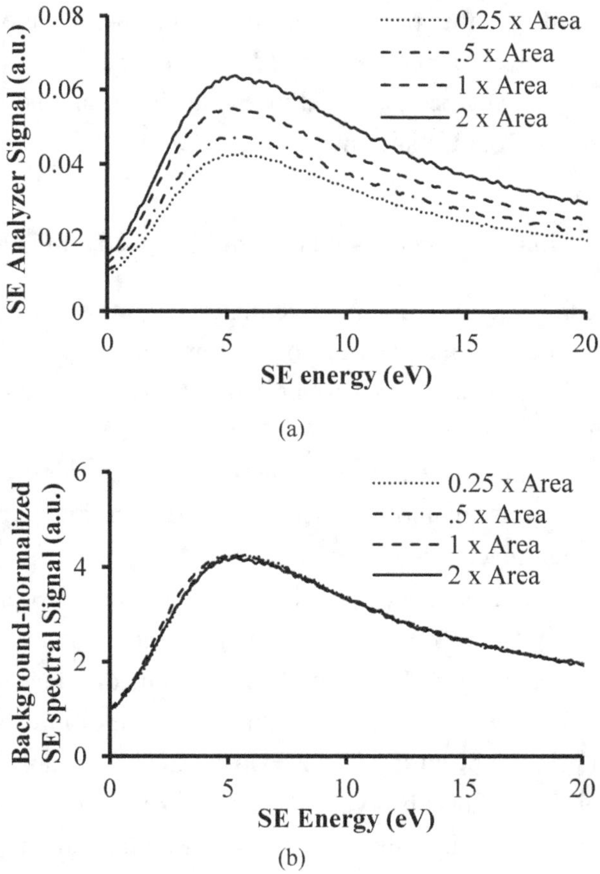

Fig. 3-17. Experimental SE energy spectral signals obtained from Au foils, with identical scan times but different scan areas. (a) Experimental signal as collected. (b) Experimental signals normalized by background $\boldsymbol{S_E(E_P = 0)}$.

Note that the background signal falls by approximately the same proportion as the SE spectral signal height in Fig. 3-17a; this is why normalizing it to the background signal is an effective way of mitigating contamination effects. This result suggests that the build-up of contamination on the inside surfaces of the inner conical cap and analyzer plates caused by high energy multiple electron beam scattering occurs in a similar way to the contamination build-up on the specimen surface caused by the primary beam. More investigation of these contamination effects is required in order to understand how much of it comes from scattering occurring within the analyzer, and how much of it comes indirectly from the specimen.

3.7 Gallium ion beam generated SE energy spectra for metals

In the preceding sections, quantitative material contrast using SE energy spectroscopy required the use of a focused primary electron beam inside a SEM. In this section, SE energy spectroscopy is carried out by using a focused primary gallium ion beam (Ga^{+}) inside a Focused Ion Beam (FIB) instrument. Secondary electron imaging by a gallium ion beam inside a FIB instrument is commonly used for specimen imaging/navigation during the micromachining of nanostructures, and is an integral part of how a FIB instrument functions. There is therefore, the opportunity of also carrying out SE energy spectroscopy in FIB instruments. However, a FIB SE detector image differs significantly from the one generated by the primary electron beam of a SEM. One important difference is that the gallium ion primary beam inside a FIB instrument inevitably sputters the specimen surface while observing it, and unlike the case with electron beam irradiation, the specimen surface is constantly changing. Although parameters such as the gallium ion beam current are deliberately reduced in order to minimize this effect during SE image acquisition, it is an additional feature of FIB SE energy spectroscopy, and it is therefore inherently more imprecise than SEM SE energy spectroscopy.

Another important difference between FIB SE energy spectroscopy and its SEM counterpart lies in its respective interaction volume inside the specimen. The interaction volume of a primary ion beam is predicted to be much smaller than the one created by a primary electron beam for the same energy. This is because a primary ion beam has a much smaller backscattered yield than a primary electron beam, making FIB SE images more surface sensitive, similar to low voltage SEM SE images. Monte Carlo simulations highlighting this difference have shown there to be a similarity in interaction volume size for a 30 kV gallium ion beam and a 1.5 kV electron beam (Ishitani and Tsuboi, 2006). Since ion beams have a small backscattering yield, their SE energy spectra are predicted to be much narrower. Joy and Griffin (2011) have speculated that separation of different components to the SE energy spectrum inside a SEM might be used to improve its image resolution/signal to noise characteristics, if some form of energy filtering can be performed to separate SE1s from SE2s. They suggest capturing SEs over a 5-6 eV energy window around the SE peak position. They presented ion-induced SE energy spectra of Ag, Cr and InP, suggesting that better material contrast would be obtained by "SE1 spectroscopy", because it is unique to the element or compound being probed, and it would therefore be well suited for microanalysis applications, an alternative to EDS. This prediction can be experimentally tested out here, by acquiring SE energy spectra inside a FIB instrument.

Figure 3-18 illustrates the experimental set-up used for the collection of experimental SE energy spectral signals using the Ga^+ ion beam inside a FEI Quanta 200 3D dual beam focused FIB/SEM instrument. The wide-angle toroidal energy analyzer attachment was installed on the specimen stage, which was tilted to 52° so that the analyzer central axis was aligned to that of the ion-beam column, a routine procedure required for the Ga^+ ion mode of operation inside the FIB-SEM. Axis alignment was done by using low magnification SE detector imaging. Experimental spectral signals were collected using a primary Ga^+ beam voltage of 5 kV with a 40 pA ion beam current on metal pellet samples. This relatively low beam voltage/current was used in order to minimize sputtering effects on the metal surface during spectral signal data acquisition.

Spectral signals typically took 30 s to acquire for each spectrum, and consisted of 150 points taken in even steps of 0.17 eV.

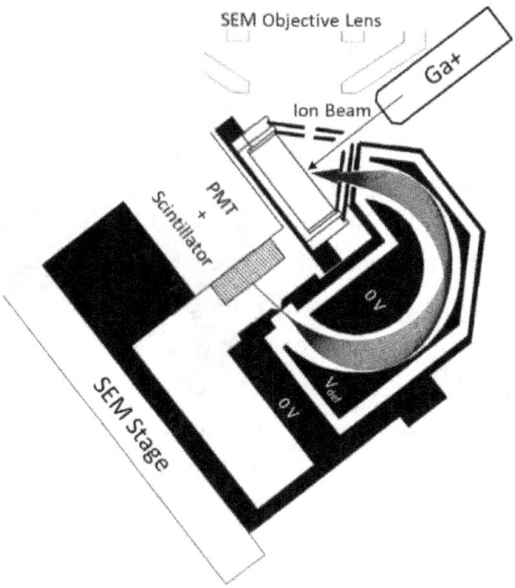

Fig. 3-18. Experimental set-up for Ga+ ion-beam-induced SE spectral signal acquisition on the specimen stage of a FEI Quanta 200 3D dual beam focused ion beam/SEM instrument.

Experimental ion-induced SE (iSE) spectral signals were collected and are shown in Fig. 3-19, together with their electron beam-induced SE (eSE) counterparts for 6 different metal pellet samples. The average energy/FWHM of the iSE and eSE energy spectral signals were extracted and are presented in Table 3-1. The iSE energy spectral signals possess a much narrower SE peak width than their eSE counterparts, and have smaller average energy values. This is in general agreement with SE energy spectra obtained by primary ion beams in other ion beam instruments (Joy and Griffin, 2011; Suzuki *et al.*, 2008). They are also in accord with Monte Carlo simulations that predict the backscattered ion yield to be much lower than the one for primary electron beams, producing a narrower output spectral signal (Ishitani and Tsuboi, 2006). The present experimental spectral results provide more quantitative information about this for the 6 different metal specimens investigated. The average energy of the gallium ion beam SE energy spectral signals is typically a factor of two lower than their electron beam generated counterparts, while the FWHM is a factor of around 3 times smaller.

While the signal shapes differ significantly, the relative peak heights of both iSE and eSE spectra follow a similar trend. This suggests that the ion generated SE spectra also contain material contrast information related to atomic number, like the electron beam generated ones. This material contrast is more evident among the heavier metal samples (W, Pt, and Au), compared to the lighter metal samples (Al, Cu and Ag). This may be due to the heavier elements being less affected by the primary ion beam's sputtering effects. The FWHM of the iSE energy spectral signals were obtained and plot as a function of atomic number in Fig. 3-20. Compared to similar results obtained with the electron beam in Fig. 3-1d, the relation to atomic number is much weaker and the scattering about the fitted line is much larger (R-squared value of 0.536). Therefore, the iSE generated SE spectral signals are not as suitable for quantitative material identification as the electron beam induced ones.

The weaker linear relationship obtained for the ion-induced SE energy spectral height/FWHM with atomic number, is similar to the results obtained earlier in this chapter for the lower primary beam voltage of 1 kV, compared to 2 kV in the electron beam induced case, indicating greater sensitivity to specimen surface effects. However, the fact that there is still an approximate linear dependence would suggest that the distinction between SE1s and SE2s is not as clear cut as previously thought, and that nuclear coulomb field scattering events are also important for SE1 emission. Imprecision caused by specimen surface sputtering may also be another contributing factor causing deviation from a linear relationship with atomic number.

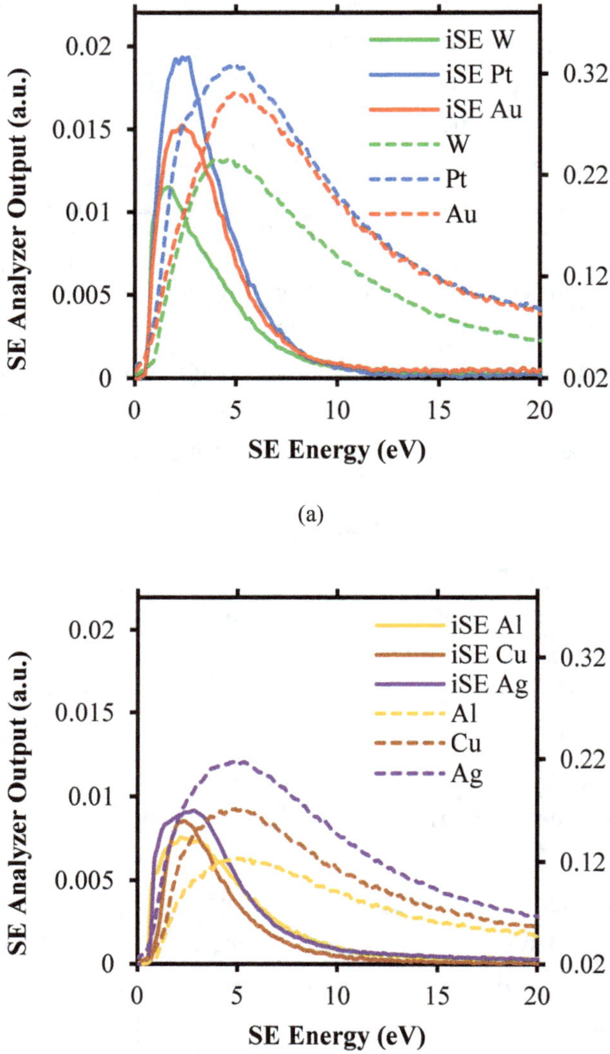

(a)

(b)

Fig. 3-19. Ion-induced SE (iSE) signals with their electron-induced SE (eSE) signal counterparts for different elements. (a) W, Pt and Au. (b) Al, Cu and Ag.

Table 3-1. Average Energy (μ) and Full-Width at Half-Maximum (FWHM) of ion-induced SE (iSE) and electron-induced SE (eSE) spectral signals for various metal samples.

	μ (eV)		FWHM (eV)	
	iSE	eSE	iSE	eSE
Al	5.599	10.578	5.033	13.208
Cu	4.404	10.093	3.336	12.077
Ag	5.314	9.686	4.461	12.004
W	4.719	9.438	3.613	9.780
Pt	4.071	9.593	3.576	10.196
Au	5.019	9.830	3.336	10.427

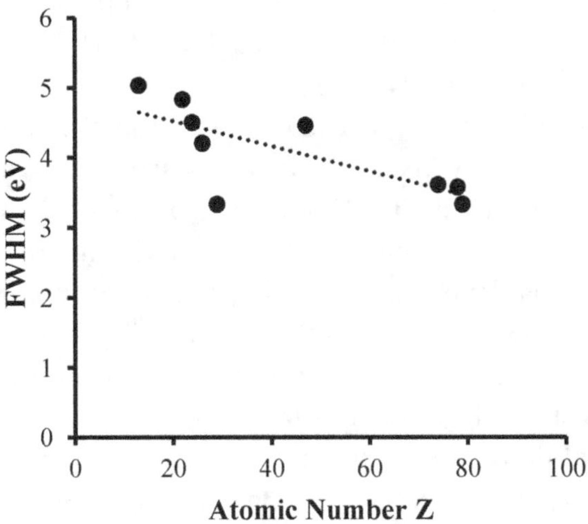

Fig. 3-20. FWHM of iSE energy spectral signals for different metal elements.

3.8 SE spectral fine features and the bulk density of states

A close examination of the full range spectra shown in Fig. 3-3 reveals the presence of fine spectral features in the 0 to 10 eV range, reproduced here in Fig. 3-21. There seems to be a sharp rising edge followed by a lower gradient linear shoulder occurring just before the SE peak for all the metals examined. This effect seems greater for the lower primary beam voltage of 1 kV. Given that previous research work carried out in dedicated UHV SE electron energy spectral systems had correlated SE energy fine structure spectral features of metal specimens to bulk density of states (DOS) information (Willis and Christensen, 1978; and Shafer *et al.*, 1981), a natural starting point would be to subtract out the background cascade and compare the residual fine structure spectral features to peaks in the corresponding theoretical DOS data.

The experimental setup which captured the SE energy spectra shown in Fig. 3-21 was carried out using the wide-angle toroidal energy analyzer attachment inside the HV specimen chamber of a FEI Quanta 200 3D dual beam focused ion beam/SEM instrument, where a primary electron beam of 1 and 2 kV was focused on to a metal sample, normally incident to it. The previous UHV chamber SE energy spectral research work was obtained by an unfocused 100 eV primary energy electron beam that was directed at a W (100) specimen at an angle of 45°. In order to overcome these very different layout and operating conditions, the approach taken here is to normalize the heights of the 1 kV and 2 kV SE energy spectral signals, differentiate them, and then subtract them away from one another, so that residual small changes between the 1 and 2 kV SE energy spectral derivative shapes are obtained. The rationale for doing this is based upon the observation that the fine SE energy spectral features seem to be progressively more pronounced as the primary beam voltage is lowered, suggesting that subtraction of SE spectral signals may be a way of suppressing the background SE cascade influence.

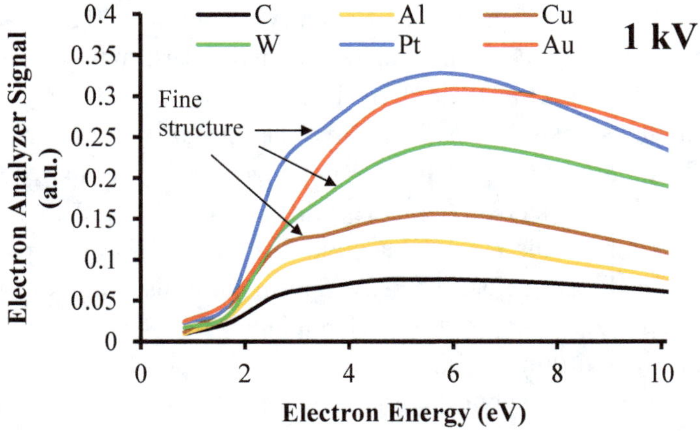

(a)

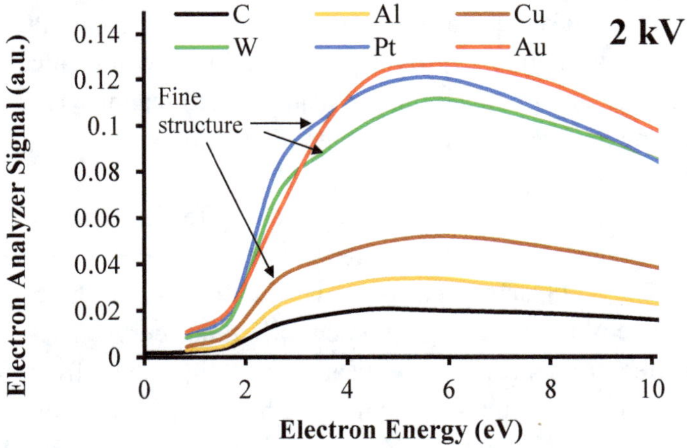

(b)

Fig. 3-21. Experimental SE energy spectra for different metal specimens illustrating the presence of fine structure features. (a) 1 kV primary beam voltage. (b) 2 kV primary beam voltage.

Figure 3-22 shows spectral data after carrying out the differentiation and subtraction procedure for the Cu, Au, W and Pt test metal foil samples, the SE energy spectra, their derivatives, and the residual signals. The differentiated signals in Fig. 3-22 show that the SE spectra for the metal samples examined have a sharper rising edge at 1 kV compared to 2 kV, which shows up as a positive gradient difference in the 0–2 eV range (the only exception to this is the W sample). The linear shoulder part of the signal (immediately before the peak) appears as a discontinuous bump in the trailing edge part of the differentiated signal between 3–5 eV, followed by a turning point in the 4–5 eV range. The first and most obvious observation about the gradient difference curves is that they are very different in shape for each metal. This is particularly useful for distinguishing between materials that have close atomic numbers, such as Au and Pt. The Pt gradient difference signal for instance, has a broader shaped negative peak to the one for Au, and crosses the axis at different energies. This makes it easy to distinguish between the Pt and Au samples, and it therefore provides an important complement to the change in SE energy peak height data, which cannot be used for distinguishing materials having close atomic numbers.

Close examination of the SE energy gradient difference curves shown in Fig. 3-22 reveals that there is a general resemblance in shape to the corresponding DOS data. This is shown in Fig. 3-23, which plots the absolute value of the SE energy gradient difference curve. The DOS data was extracted from previous publications: DOS_1 for W from Rai and Thapa (2014), DOS_2 for Cu from Galashev and Ivanichkina (2019), DOS_3 for Au from Krasavin *et al.* (2018) and DOS_4 for Pt from Krupski *et al.* (2015). The agreement is more in terms of an overall trend, rather than a precise match for the spectral peaks. A precise match cannot be expected from the data shown in Fig. 3-23 since there are not enough SE energy data points, only a total of 7 for Cu and W, 11 for Au, and 9 for Pt.

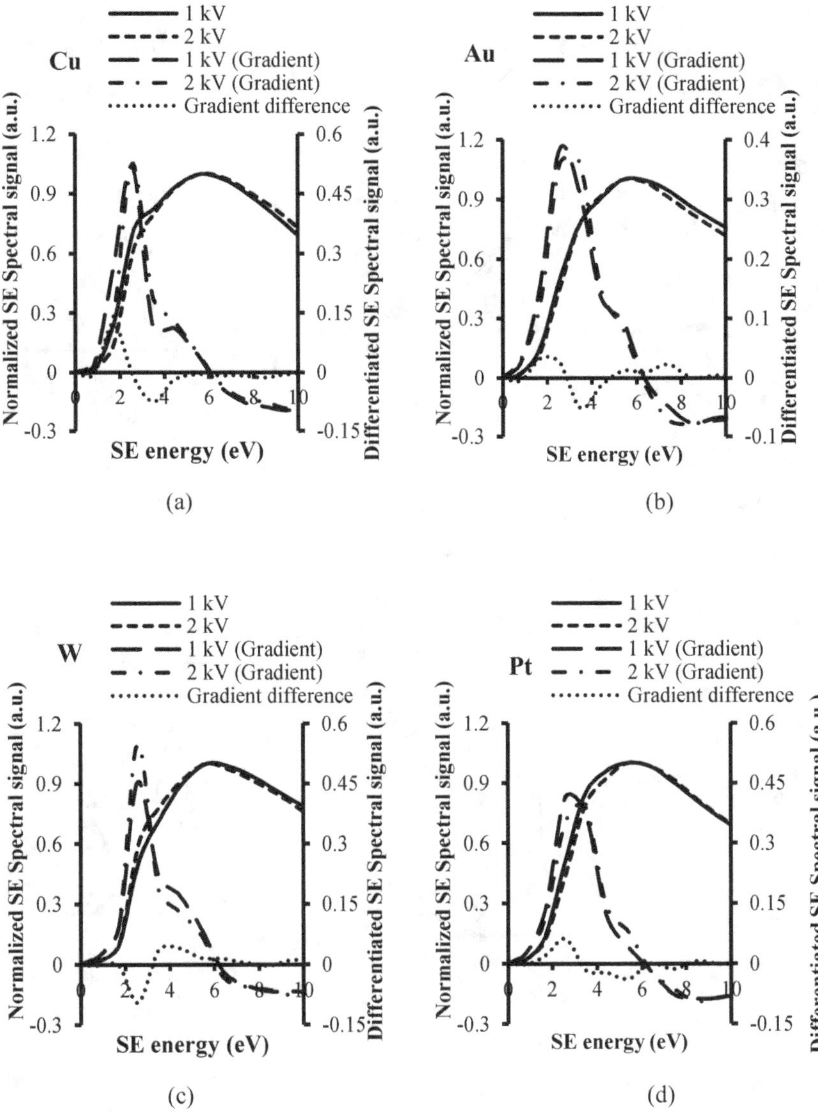

Fig. 3-22. SE energy spectral signals and the differentiated SE energy spectral signals (gradient) obtained at 1 and 2 kV, and the difference in the gradients between 1 kV and 2 kV for metal foils. (a) Cu, (b) Au, (c) W, (d) Pt.

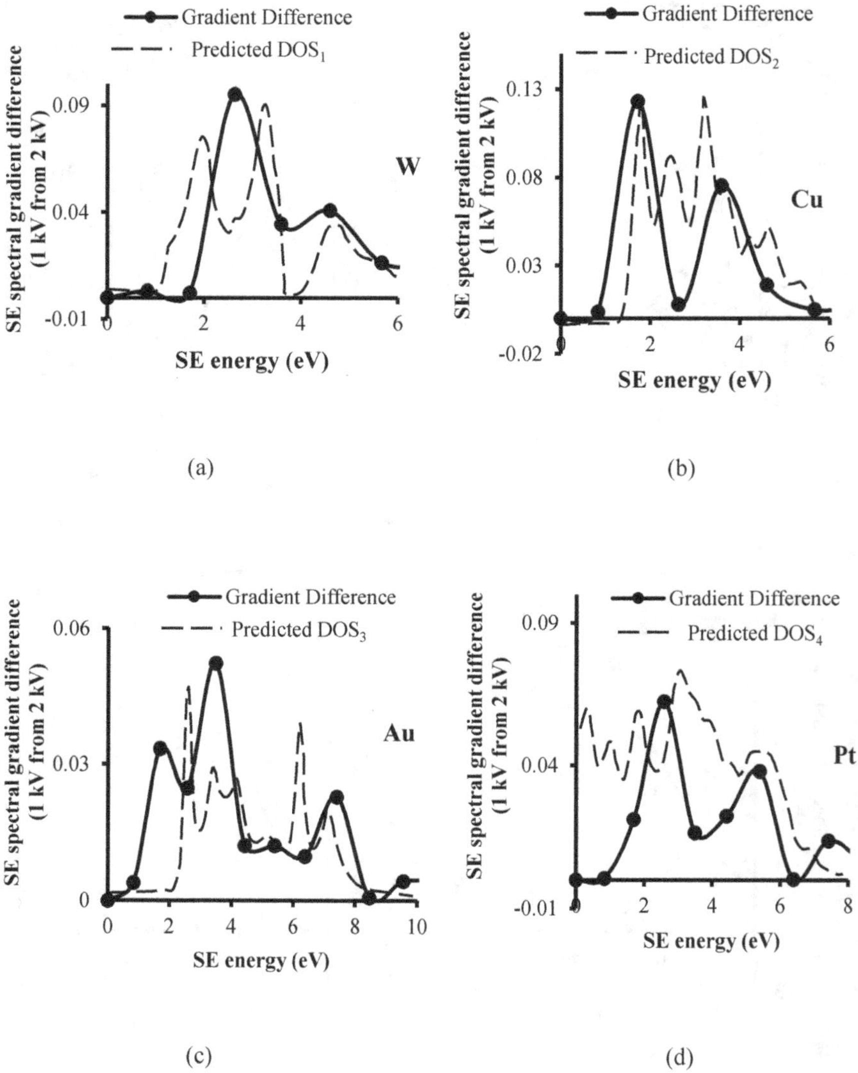

Fig. 3-23. Absolute value of the SE energy spectral gradient difference between 1 and 2 kV signals and their comparison to theoretical bulk density of states (DOS) data. DOS_1 from Rai and Thapa (2014), DOS_2 from Galashev and Ivanichkina (2019), experimental DOS_3 from Krasavin *et al.* (2018), and DOS_4 from Krupski *et al.* (2015). (a) W, (b) Cu, (c) Au, (d) Pt.

Significantly larger spectral fine features were obtained in the SE energy spectrum at the lower primary beam voltage of 0.5 kV, shown alongside the 1 kV SE spectrum in Fig. 3-24a for W. The difference signal between these two spectra is also shown on the same graph. The SE energy step size was decreased to 0.1667 eV, increasing the total number of points in this graph to 36, and the analyzer pass energy was scanned at a rate of 4 pt/s. Five spectral signals were acquired over 5 by 5 μm areas on the specimen surface. Each spectral signal was obtained from a different 5 by 5 μm area in order to minimize beam induced contamination effects. The primary beam current at 1 kV was 44 pA, and it was lowered to 22 pA at 0.5 kV. It is interesting to note that the 0.5 kV spectral signal not only has larger fine structure feature amplitudes, but its overall background gradient significantly differs from the one in the 1 kV spectrum. The spectral derivative of the 0.5 kV signal and a cubic polynomial fit extracted from it are shown in Fig. 3-24b.

Figures 3-25a and 3-25b show the residual signal left after subtracting the background SE cascade contribution from the 0.5 kV spectral gradient signal in two different ways. The first, shown in Fig. 3-25a, subtracts the cubic polynomial fit, this is similar to the method used by previous SE energy spectroscopy work carried out in dedicated UHV energy spectral systems, which used analytical representations of the background SE cascade (Willis and Christensen, 1978; and Shafer *et al.*, 1981). Figure 3-25b shows the result of the present approach, where spectral gradient distributions are subtracted from one another at different primary beam voltages, the 1 kV spectral gradient signal from the 0.5 kV one in this case. The match to the corresponding DOS theoretical data is significantly better for the present method, which seems to capture the correct position and relative amplitudes of all three main peaks in the tungsten DOS data. This experimental result suggests that the multiple scattering effects present in the 1 kV SE energy spectrum, are very similar in character to those present in the 0.5 kV spectral signal.

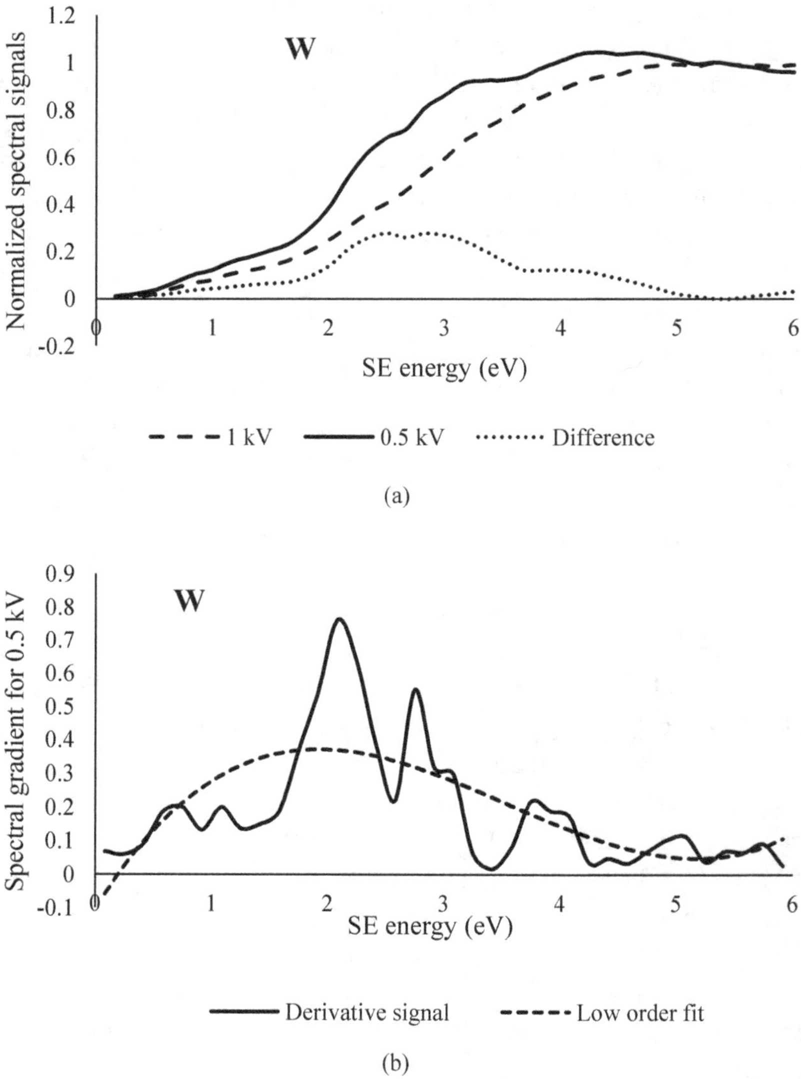

Fig. 3-24. SE spectral fine features for W at the primary beam voltages of 0.5 and 1 kV. (a) Normalized spectra for 1 and 0.5 kV and their difference signal. (b) Spectral derivative for the 0.5 kV signal with a low order cubic polynomial fit.

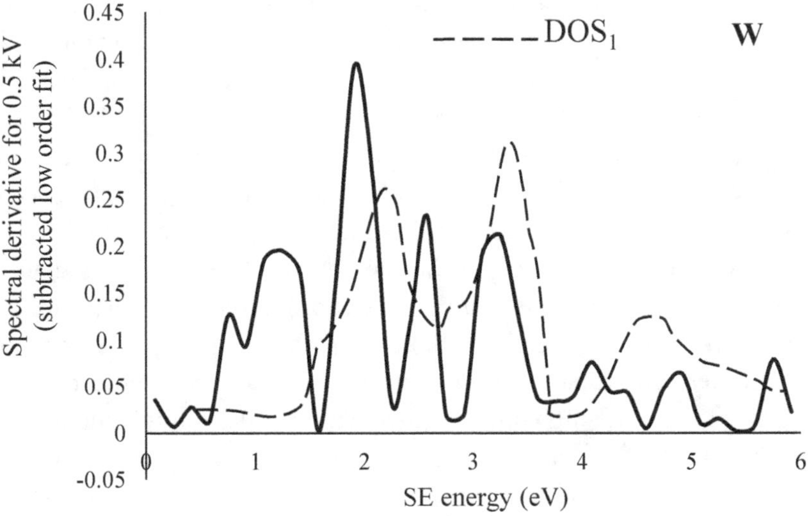

(a)

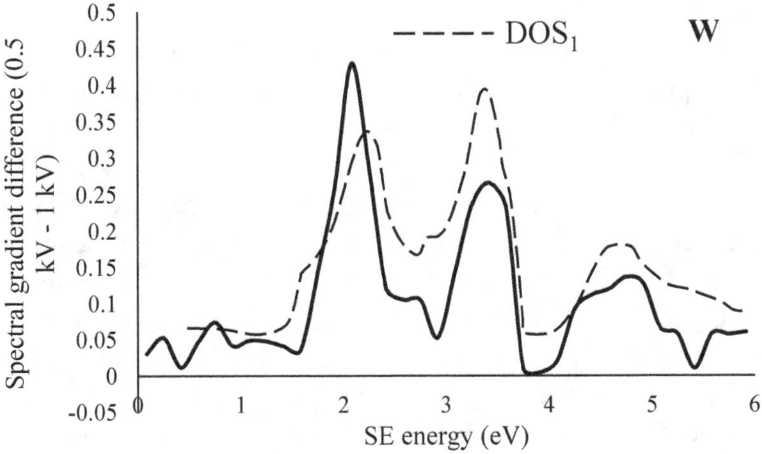

(b)

Fig. 3-25. Subtracting the SE background contribution from the 0.5 kV spectral signal for W. DOS_1 from Rai and Thapa (2014). (a) Subtracting the low order polynomial (cubic) fit. (b) Subtracting the 1 kV spectral signal.

Further investigation showed that in order to improve the match to previous DOS data, a faster pass energy scan rate in combination with more spectral signal averaging is required (while moving to a fresh area on the specimen for each spectrum). Figures 3-26 to 3-29 show experimental SE energy distributions and SE energy gradient difference signals for 12 spectra averaging, while using a pass energy scan speed of 8 pts/s for the W, Cu and Au samples, and a 16 pts/s scan speed for the Pt sample. The better match to the theoretical DOS data is evident by comparing the faster scanned/higher averaged data for the tungsten sample shown in Fig. 3-26b, to the one in shown in Fig. 3-25b, which was recorded with a pass energy scan rate of 4pt/s and averaged over 5 spectral signals. Although not presented here, similar improvements were also obtained for the other metal samples. It should be noted that the match in the case of Au was made to DOS data obtained from experimental XPS results (Krasavin *et al.*, 2018); this is because previously published DFT simulation results seem to give a much poorer match in that case.

The higher pass energy scan rate of 16 pt/s was required in the case of the Pt sample, and will most likely also be required to improve the match further for the Cu sample. The reason for the better match probably comes from the fact that faster spectral data acquisition speeds reduce the influence of beam induced contamination, while greater spectral averaging reduces the effects of shot noise. The reason for the mismatch in the very low energy part of the Pt data (< 0.5 eV) may be due to the fact that the energy window of detection is proportional to the pass energy, requiring even higher levels of signal averaging in that part of the spectrum. It may also be related to the fact that the sample surface needs to undergo some form of pre-cleaning treatment. More accurate DOS theoretical data and a wider range of experiments are required before any conclusions can be made.

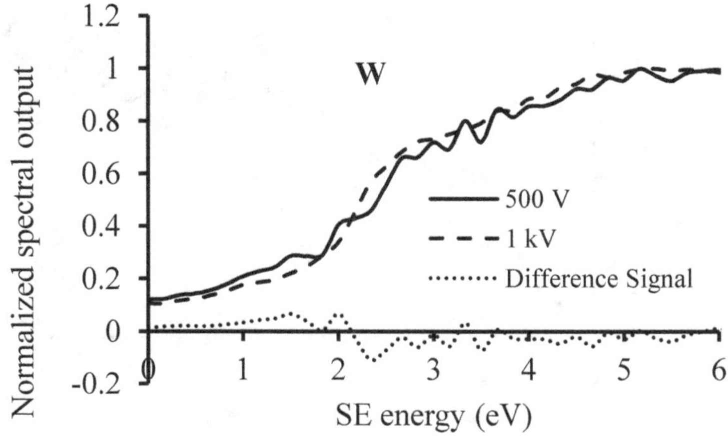

(a)

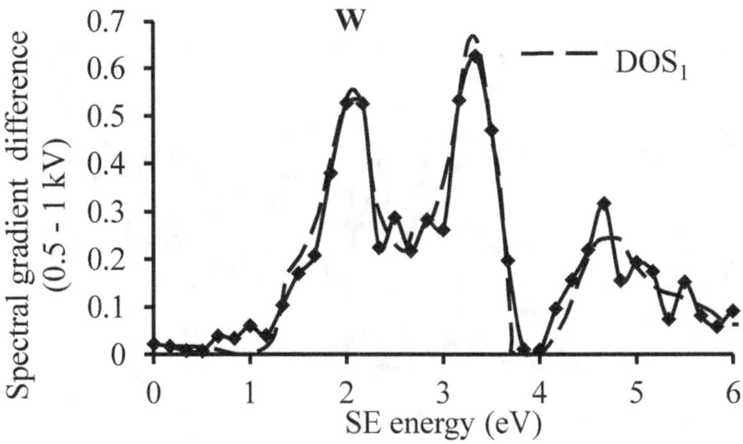

(b)

Fig. 3-26. Extraction of bulk DOS information from SE energy spectral experimental data for a W specimen. The data was averaged over 12 spectra and a pass energy scan speed of 8 pts/s was used. DOS_1 data from Rai and Thapa (2014). (a) Normalized SE energy spectra. (b) Spectral gradient difference signal (0.5 – 1 kV).

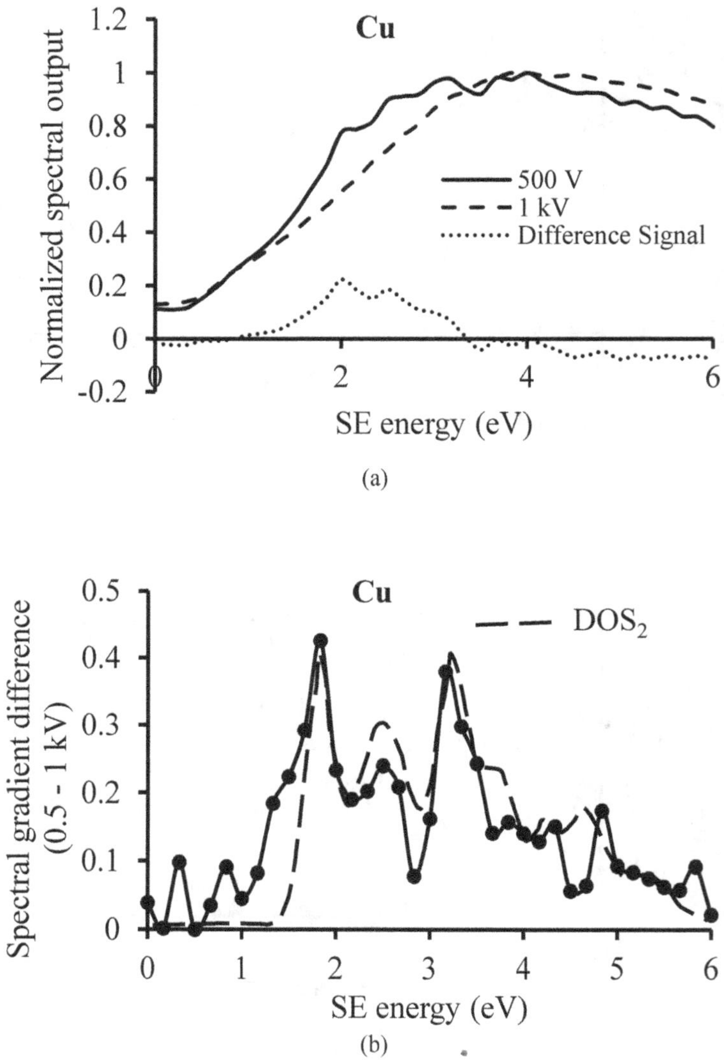

Fig. 3-27. Extraction of bulk DOS information from SE energy spectral experimental data for a Cu specimen. The data was averaged over 12 spectra and a pass energy scan speed of 8 pts/s was used. DOS$_2$ data from Galashev and Ivanichkina (2019). (a) Normalized SE energy spectra. (b) Spectral gradient difference signal (0.5 – 1 kV).

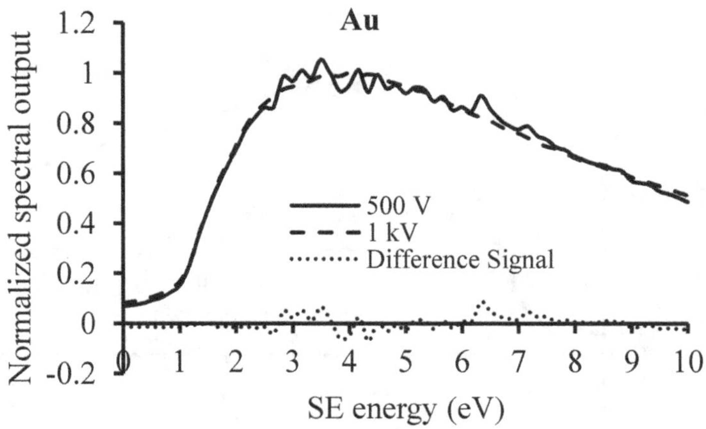

(a)

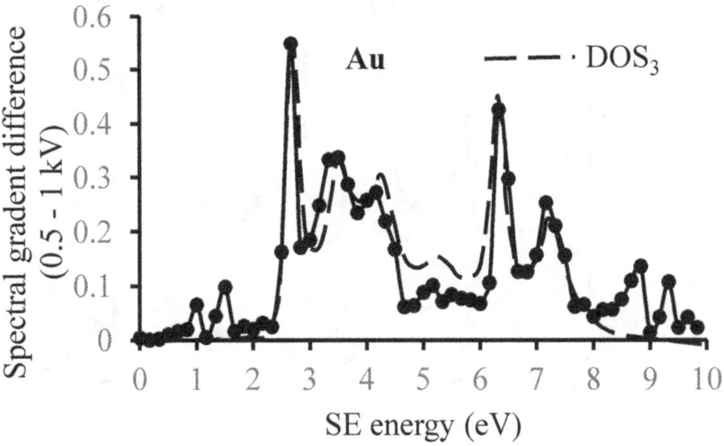

(b)

Fig. 3-28. Extraction of bulk DOS information from SE energy spectral experimental data for an Au specimen. The data was averaged over 12 spectra and a pass energy scan speed of 8 pts/s was used. Experimental DOS_3 data from Krasavin *et al.* (2018). (a) Normalized SE energy spectra. (b) Spectral gradient difference signal (0.5 – 1 kV).

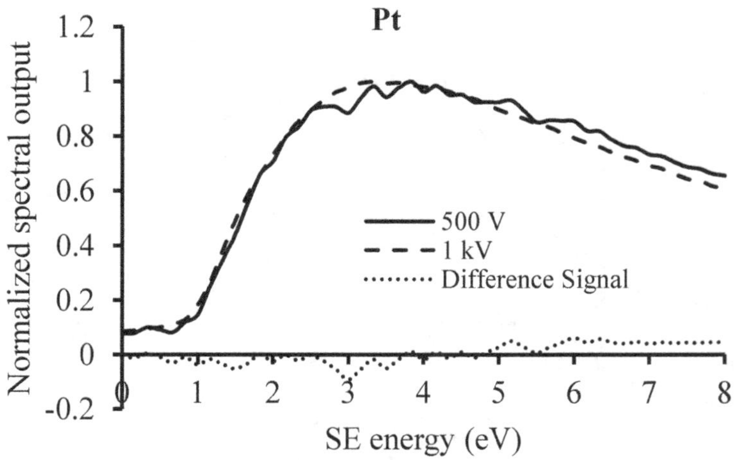

(a)

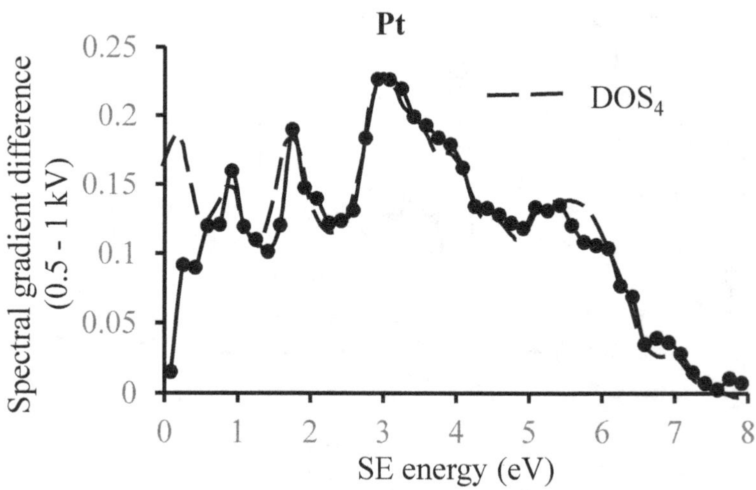

(b)

Fig. 3-29. Extraction of bulk DOS information from SE energy spectral experimental data for a Pt specimen. The data was averaged over 12 spectra and a pass energy scan speed of 16 pts/s was used. DOS_4 data from Krupski *et al.* (2015). (a) Normalized SE energy spectra. (b) Spectral gradient difference signal (0.5 – 1 kV).

The observation that valence band information is embedded into the lower energy part of the SE energy spectral signal sheds some light on the earlier experimental observations reported in sections 3.2 and 3.3. It helps explain why quantitative material contrast can be obtained from the rising edge part of the SE energy spectral signal, in addition to extracting information from tracking SE peak height or FWHM variations. The mechanism responsible for this can now be identified as interactions with valence band electrons. They cause small variations in gradient to appear along the rising edge of the SE energy spectral signal, which were not identified as fine spectral features in the earlier results since they were masked by relatively high levels of contamination. This also helps explain why the deviations of experimental SE energy spectra from the analytical Chung-Everhart distribution seem to grow as the primary beam voltage is lowered (observed in section 2.3).

The preliminary SE energy spectroscopic bulk density of states work reported here naturally raises many questions. What are its accuracy limitations? How do the results change for different area scan sizes on the specimen surface? Does its accuracy improve if two other low primary electron beam voltages are used? Can the method be used as a finger-printing quantitative material identification method for LVSEM? Are there any advantages in biasing the specimen and inner cap electrode negatively? Why are valence band fine spectral features absent from experimental data reported by previous UHV electron energy spectral systems that used the same metal samples (W, Cu, Au and Pt) and same primary beam voltage (0.5 kV) as used here (Ding *et al.*, 2008)? Will the method work for semiconductors such as silicon? Can the method be extended to become an angle-resolved SE energy spectroscopy technique to acquire electronic band-structure information by tilting the specimen through different polar angles while using an array of detectors in the azimuthal direction? How does the density of states SE energy spectroscopy technique compare with angle-integrated photo electron spectroscopy? Future work needs to answer these kinds of questions.

Overall, the preliminary experimental results shown in Figs. 3-26 to 3-29 represent a breakthrough for SE energy spectroscopy inside the SEM. It demonstrates that valid information about the bulk density of states (DOS) of a metal sample can be obtained by inelastic collisions with valence band electrons, which show up as fine spectral features in the SE energy spectrum at low primary beam voltages (< 2 kV). A good match to bulk DOS data can be obtained by subtracting the differentiated SE spectral distributions taken at two different low primary beam voltages from one another (the 1 kV one from the 0.5 kV one). An accurate match to previous theoretical/experimental bulk DOS data for 4 different metal samples has been successively demonstrated under normal SEM operating conditions. The method is inherently spatially resolved, and can in principle combine with the high imaging resolution capability of the SEM. These results, together with the other quantitative material contrast methods presented in this chapter, indicate that SE energy spectroscopy has the potential to become a quantitative material analysis technique for LVSEM.

Chapter 4

Dopant Profiling and Semiconductor Characterization

This chapter demonstrates that SE energy spectroscopy in the SEM is able to provide reliable dopant profiling and characterize semiconductor samples in a variety of different ways. Experimental results are presented to show that SE energy spectroscopy can make *in-situ* native oxide thickness measurements, map surface/space-charge distributions, and carry out local conductivity measurements on bulk silicon samples. Most of these results were recently reported by Han *et al.* (2019), and it is the first time a SEM has been used to extract information of this kind from SE energy spectra.

The need to carry out accurate and reliable dopant profiling is critical to both the Failure Analysis of semiconductor devices as well as the development of new nanostructured materials. Quantitative dopant profiling is an important characterization requirement for devices such as solar cells, and has been consistently featured in International Technology Roadmaps for Semiconductors, where 2D lateral profiling of dopant concentration variations need to be measured with errors typically below 5%. In materials science research, introducing dopants and controlling dopant concentration plays a central role in developing new types of materials such as graphene (Wang *et al.*, 2012).

In the Failure Analysis of semiconductors, the most widely utilized methods for dopant profiling, such as Secondary Ions Mass Spectrometry (SIMS) and the Electrochemical Capacitance-Voltage (ECV) technique, are well known to have limited 2D lateral profiling capability: ECV is inherently a 1-dimensional (1D) depth technique, while 2D-SIMS has only obtained sub 100 nm lateral resolution in recent years (Seebauer and Barlaz, 2016).

For a period of more than two decades, the method of using Scanning Electron Microscopes (SEMs) to make quantitative dopant concentration measurements has been cited to be a promising Integrated Circuit Failure Analysis tool, due to its inherent 2D profiling capability, nanometre spatial resolution, high scan rate, minimal specimen preparation requirement, and wide accessibility. However, it is still not reliable enough to make dopant profiling measurements on real semiconductor devices such as solar cell samples. Quantitative SEM dopant contrast measurements have so far been limited to probing idealized test *pn*-junction specimens, where only step dopant concentrations have been quantified (Perovic, 1995; Kazemian *et al.*, 2006; Kazemian *et al.*, 2007; Volotsenko *et al.*, 2010; and Chee, 2018).

Lack of reliability and accuracy for SEM dopant contrast has been attributed to a variety of different causes, which include the presence of a native oxide layer (Dapor *et al.*, 2009), electron beam induced contamination, and the influence of transverse electric fields above the sample surface, also referred to as "patch fields" (El-Gomati *et al.*, 2005; Elliott *et al.*, 2002; Mika and Frank, 2008; Schönjahn *et al.*, 2002). In addition, dopant contrast mechanisms are not well understood. Some have proposed detecting shifts in secondary electron kinetic energy (Kazemian *et al.*, 2006; Schönjahn *et al.*, 2002), while others have favored using brightness changes (contrast) in the SE detector output signal (Chee, 2016; Elliott *et al.*, 2002; Volotsenko *et al.*, 2010; Xu *et al.*, 2013). There is no agreement about which one is more accurate, how they relate to one another, or what underlying mechanisms in the semiconductor link them together. Both in-built space-charge and surface band-bending are thought to play an important role (Perovic *et al.*, 1995; Schönjahn *et al.*, 2002; Volotsenko *et al.*, 2010), but it has not yet been possible to track them simultaneously and understand how they relate to one another.

Where brightness changes in the SE detector image have been used, puzzling contrast reversal effects have consistently been obtained when parameters such as the image scan speed and primary beam voltage were changed (El-Gomati *et al.*, 2005; Kazemian *et al.*, 2004; Mika and Frank, 2008). Attempts at detecting a shift in SE kinetic energies encountered problems of scattered electrons from grids/electrodes being superimposed on to the final output cumulative SE energy distribution signal, severely distorting it and making it difficult to find a linear region from which energy shifts could be extracted (Kazemian *et al.*, 2006). These difficulties have already been mentioned in chapter 1, in the context of past research work being over-reliant on using in-built SE detector systems.

The main objective of this chapter is to investigate what kind of improvements in SEM semiconductor characterization can be obtained through the use of an electron energy band-pass spectrometer attachment specifically designed for the SEM, one that has enough precision to track small changes in the SE energy spectral signal shape. The work builds upon the earlier high signal-to-noise energy shift results reported from the use of the wide-angle toroidal spectrometer (Hoang *et al.*, 2011a; Srinivasan and Khursheed, 2015b), and demonstrates how it can overcome longstanding dopant profiling difficulties, as well as create new ways of characterizing semiconductor samples inside a SEM.

4.1 Doped stripe test samples

P-doped stripes on *n*-doped substrate and *n*-doped stripes on *p*-doped substrate test semiconductor samples were provided by the ISI research institute, Brno, Czech Republic (Mika *et al.*, 2010). The *p*-doped stripes on *n*-doped substrate sample, as illustrated in Fig. 4-1, comprises of repeating patterns of 4 Boron-doped stripes of concentrations between 10^{16} cm^{-3}–10^{19} cm^{-3} on a lightly doped *n*-type substrate (10^{15} cm^{-3}). The *n*-doped stripes on *p*-doped substrate test sample is similar in geometrical layout to the Phosphorus-doped stripes, defined in lightly doped *p*-type substrate, and also has similar dopant concentrations. The experimental results presented in this chapter were obtained in the FEI XL30 FE-SEM, unless otherwise stated.

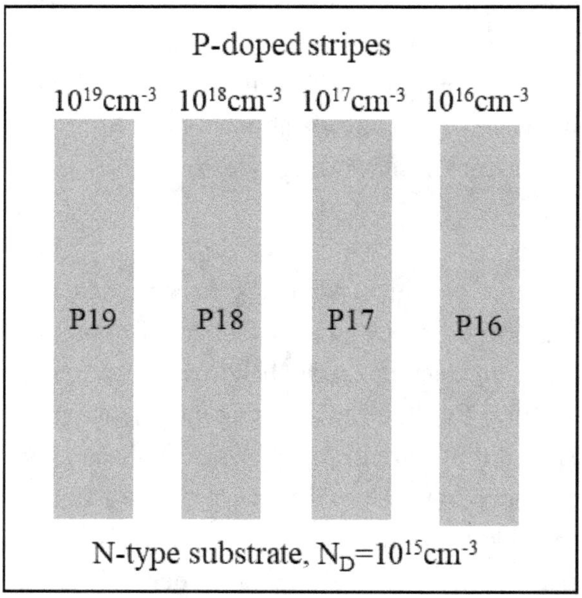

Fig. 4-1. Schematic diagram of a test sample consisting of *p*-doped striped regions in *n*-type substrate. Each stripe is labeled based on its dopant concentration.

The test samples were cleaned with an ultrasonic cleaner in acetone and isopropanol sequentially for 10 minutes each, before removing oxide layers using the Oxford Plasma Pro Cobra 100 Deep RIE. The etched samples were then transported in a vacuum sealed box and inserted into the SEM. The samples with thicker native oxide layers were prepared by immersing the etched sample in fresh ultrapure water for 12 hours, and leaving them in open air atmosphere for different periods of time before inserting them into the SEM. The native oxide layer thickness was then measured using a Semilab SE-2000 Spectroscopic Ellipsometer.

Lateral dopant profiles were obtained in this sample by capturing a series of band-pass energy filtered images, as illustrated in Fig. 4-2. The layout for the sample was orientated such that the dopant distribution along the stripe was constant in one direction. A series of images were taken and stored along this direction, where each frame was moved to a different area on the specimen in order to reduce beam induced contamination. The images were taken for a field of view of 5 μm × 3 μm. Line-scans were averaged within a frame to reduce the effect of shot noise. In this work, one frame was recorded for each of 15 different pass energy values between 0 eV and 17 eV, and 242 lines were averaged within a frame.

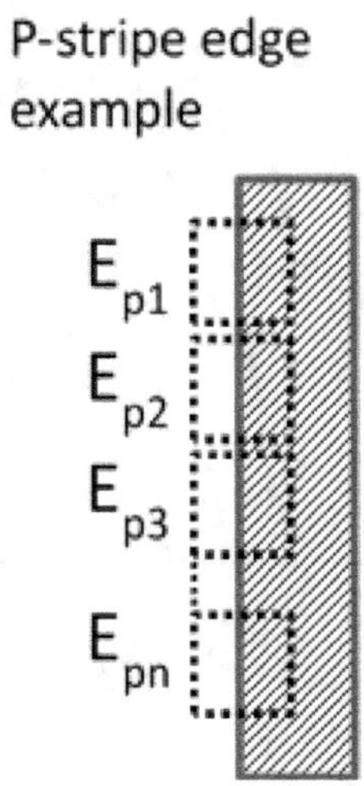

Fig. 4-2. Schematic of image acquisition along the direction of constant dopant distribution, with each image taken at a different pass energy to obtain the SE energy spectral signal for each pixel.

Another important feature of the experimental setup used to capture all SE energy spectral signals is that both the specimen and the inner conical cap electrode above it were biased to −10 V. As explained in section 2.2, this is an effective strategy for improving the signal-to-noise of the output signal. Figure 4-3 shows the output SE energy spectral signals for an intrinsic silicon sample where the specimen and inner cap electrode are first unbiased, and then subsequently biased to −10 V. The signal-to-noise ratio of the spectral signal peak height is 8.5 for the unbiased sample signal and is improved to 42.8 for the biased sample/inner cap electrode spectral signal. For the spectral standard deviation (σ), biasing improves the signal-to-noise ratio from 198 to 1741. For the spectral average energy (most probable energy), the signal-to-noise ratio increases from 140 to 540. Overall, the signal-to-noise ratio of the output SE energy spectral signal's parameters such as peak height, standard deviation and average energy improve by more than a factor of five by using the specimen/inner cap electrode negative biasing strategy.

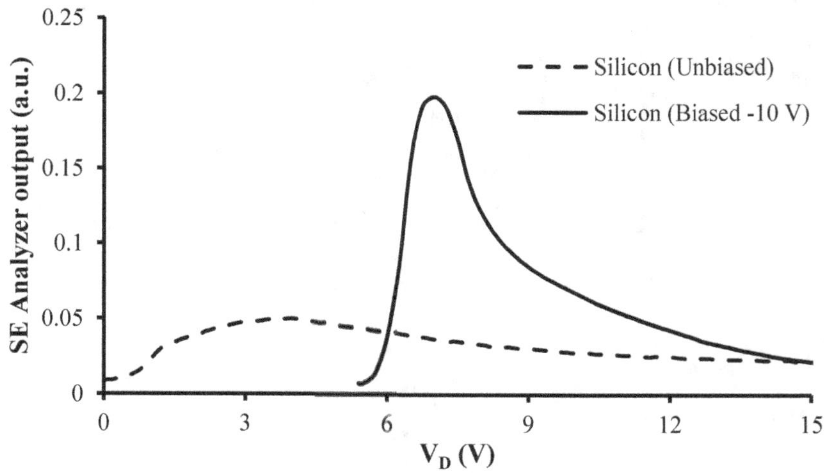

Fig. 4-3. Unbiased and biased (−10 V) specimen/inner cap electrode spectra for an intrinsic silicon sample inside a SEM as a function of deflector plate voltage VD.

4.2 SE energy spectral signal shape dependence on probing position, primary beam voltage, and native oxide thickness

In this section, SE energy spectral signals were investigated for their sensitivity to different probing positions, primary beam voltages and native oxide thicknesses. Figure 4-4 shows SE dopant contrast energy spectral signals taken from the test *p*-doped striped semiconductor sample of varying dopant densities in an *n*-doped substrate. This data was obtained for points on and around a 10^{19} cm^{-3} *p*-stripe region having native oxide thicknesses of 1.75 nm and 3.91 nm with the primary beam voltages of 2 and 5 kV. Figure 4-4a–d show that the point of maximum average SE energy shift is obtained for point C, located close to the *p*-stripe edge, while for points inside the *p*-stripe region, such as point B, there is mainly a change in height. For the thin oxide thickness of 1.75 nm, there is a higher level of dopant contrast (both in terms of height and average shift) in the 2 kV signals compared to the 5 kV signals, but no observable difference in shape, while for the 3.91 nm thick native oxide, the 5 kV signals become much broader and have greater dopant sensitivity than the 2 kV signals.

The results shown in Fig. 4-4 indicate that changes in the shape of the SE energy spectral signal contain a wide variety of useful information about the sample. They show for instance, how both changes in SE peak position height/average energy shift can be simultaneously tracked as a function of probing position on the sample. The fact that there is only a height change for points inside the *p*-stripe, such as point B in Fig. 4-4b and Fig. 4-4d, and that this height change is constant across the strip interior, confirms the validity of the trapped charge model causing band-bending at the Si-SiO_2 interface, an effect that is present across the entire width of the *p*-stripe. On the other hand, the average energy shift is localized to the *pn*-junction region only, suggesting that it is correlated in some way to in-built space-charge regions.

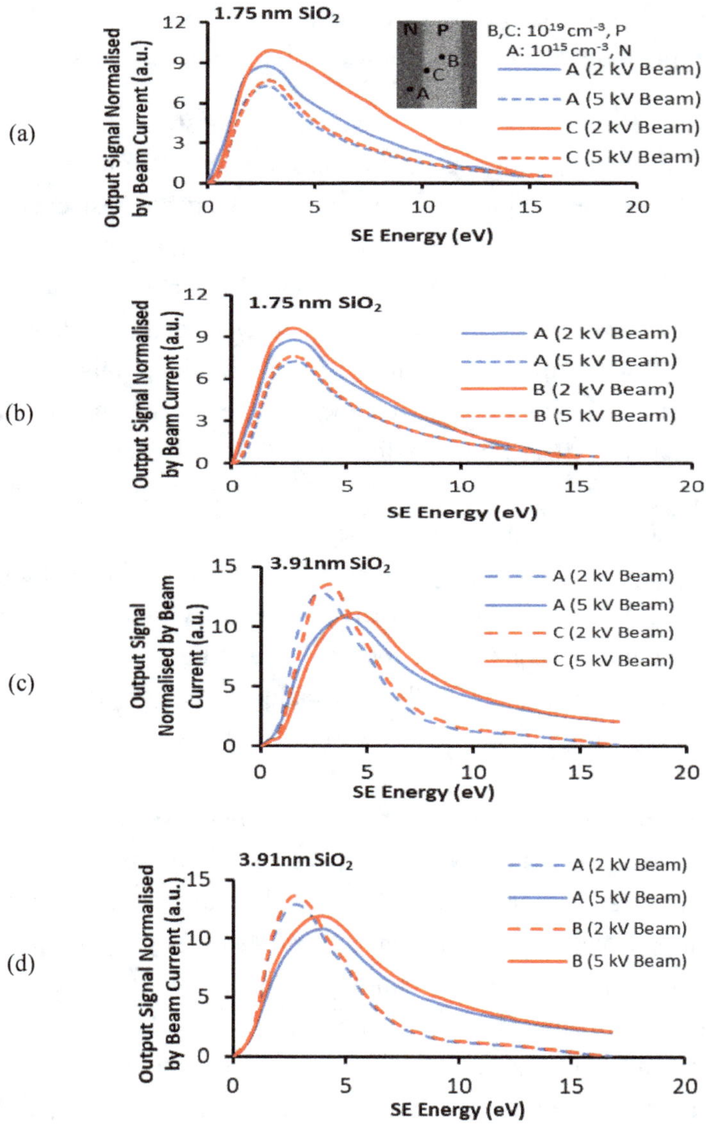

Fig. 4-4. Experimental SE energy spectral signals for different beam voltages, native oxide thicknesses and probing positions on a p-stripe in n-substrate test specimen. (a) From points A and C with 1.75 nm oxide thickness. (b) From points A and B with 1.75 nm oxide thickness. (c) From points A and C with 3.91 nm oxide thickness. (d) From points A and B with 3.91 nm oxide thickness.

This is the first time position-sensitive information for the average energy shift has been obtained, and it is very different to the one predicted by standard one-dimensional *pn*-junction theory which calculates there to be a negative in-built potential across the width of the *p*-stripe region. In the past, a shift in SE energies was related to an in-built *pn*-junction potential, but variations of that potential as a function of position were not reported (Kazemian *et al.*, 2006). It should also be noted that this kind of information is only available here because variations in SE peak height and the overall average energy shift can be tracked separately from overall changes in the signal shape. Most previous dopant contrast measurement systems were not able to do this since they relied on SE detector image brightness changes only, and were therefore not able to distinguish between brightness contrast caused by an SE energy spectral shift or brightness contrast caused by a change in overall SE intensity.

4.3 Dopant contrast measurements from p-stripes of different concentrations

In order to quantify and track changes in the SE energy spectral signal's shape caused by different dopant mechanisms, a Band-Pass Contrast function (C_{BP}) evaluated at the SE energy E_j, $C_{BP}(E_j)$, and an average energy shift $\Delta\mu$ are calculated, given by the following expressions:

$$C_{BP}(E_j) = \frac{\left(I_p(E_j) - I_n(E_j)\right)}{I_n(E_j)} \tag{4.1}$$

and

$$\Delta\mu = \left(\frac{1}{I_{p0}}\right)\sum I_p(E_j)E_j - \left(\frac{1}{I_{n0}}\right)\sum I_n(E_j)E_j \tag{4.2}$$

where I_{p0} is the total current from all energy points on the *p*-type output spectral signal $I_p(E_j)$ and I_{n0} is the total current from all energy points on the *n*-type output spectral signal $I_n(E_j)$. The subscript *j* runs from 1 to *n*, the number of sampled energy points in the output spectral signal.

The band-pass contrast parameter $C_{BP}(E_j)$ is more flexible than the bulk contrast parameter, $C = (I_{p0} - I_{n0})/I_{n0}$ representing brightness changes in the SE detector image used previously, since it can be evaluated for different SE energies. Also, the average energy shift $\Delta\mu$ is more flexible than the measured voltage shifts from linear parts of the cumulative energy spectrum (S-curve) (Kazemian *et al.*, 2006) used earlier, since it takes into account energy shifts across the entire SE spectrum energy range.

The secondary electrons in a SEM that contribute to the SE detector signal come from a wide range of emission energies and angles, limited by an upper cut-off energy E_F. This is true for conventional Everhart-Thornley detectors as well as through-the-lens detector systems. The contrast in SE signal between *p*-type and *n*-type regions on a sample in SE energy terms is therefore a bulk signal contrast effect, as illustrated in Fig. 4-5a. On the other hand, the output SE signal obtained from the wide-angle angle toroidal spectrometer has a much narrower electron energy and angular range. Its detected energy range is illustrated in Fig. 4-5b by a narrow shaded area. The band-pass contrast parameter $C_{BP}(E_j)$ used in this work is therefore much more sensitive to changes in the SE spectrum caused by the presence of *p*-type and *n*-type regions on the sample than the previous bulk contrast parameter C, as schematically illustrated in Figs. 4-5a and 4-5b. In addition, the SE energy E_j at which the contrast parameter $C_{BP}(E_j)$ is evaluated can be changed, which further helps in enhancing the dopant contrast and minimises errors that come from effects such as beam induced contamination.

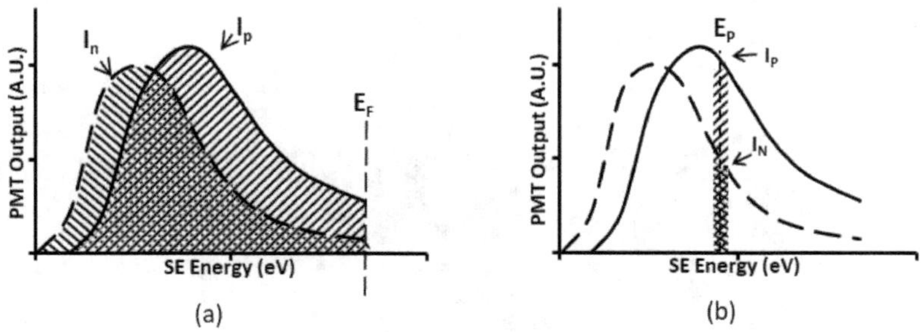

Fig. 4-5. Schematic diagram illustrating the difference between SE detector and band-pass energy dopant contrast: (a) Bulk SE detector signal. (b) Band-pass spectral signal.

The improved sensitivity of band-pass contrast over a conventional SE detector bulk contrast signal can be illustrated experimentally for the *p*-doped stripe test sample, and is shown in Fig. 4-6. A primary beam voltage of 2 kV with a 40 pA beam current was used. The band-pass filtered SE image obtained with the electron energy analyzer set to capture 3.4 eV SEs (close to the SE peak position) shows significantly more contrast between the *p*-doped stripes and the substrate than the corresponding SE detector micrograph. In particular, the two lowest doped concentration stripes P16 and P17 are hardly visible in the SEM micrograph, but can be seen clearly in the band-pass filtered image. A more quantitative comparison was performed by obtaining their respective pixel-averaged horizontal line-scans, shown in Fig. 4-6c. The band-pass contrast of all 4 stripes was found to be higher than the bulk contrast signal obtained from the SE detector. In particular, the band-pass contrast of P17 is 2-3 times the SE detector bulk contrast. These results demonstrate that dopant contrast sensitivity can improve by a factor ranging from 1.5 to 3 times by using SE band-pass spectral signal, compared to the conventional SE detector bulk signal. This is in general agreement with previous results, such as those generated from the narrow angle toroidal energy analyzer attachment (Rau and Tagachenkov, 2013), and the low-pass energy filtered images obtained from the ad-hoc use of a SEM TLD detector (Chee, 2016, 2018).

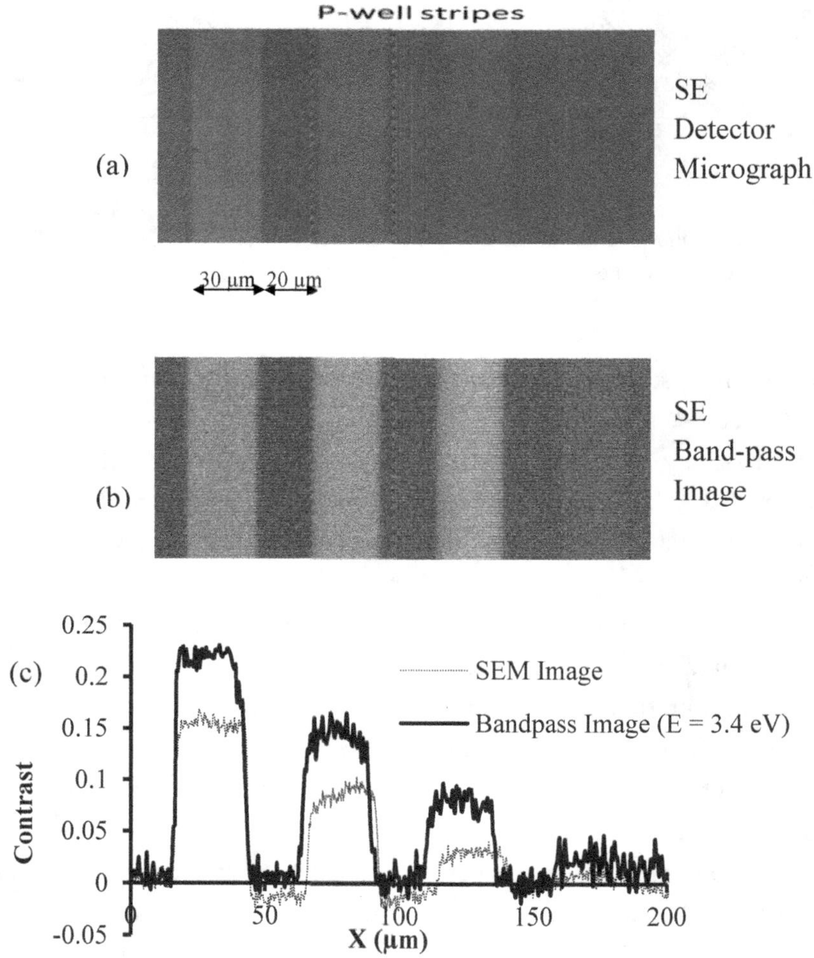

Fig. 4-6. Imaging p-doped stripes in the SEM taken with a 2 kV, 40 pA primary beam. (a) SE detector micrograph of the sample. (b) SE Band-pass image (E = 3.4 eV) of the sample. (c) Comparison of SE detector image contrast and SE band-pass contrast line profiles.

In the past, dopant contrast in the SEM was found to be significantly affected by parameters such as image scan speed and primary beam current due to effects such as electron beam recombination (Kazemian *et al.*, 2004). It is therefore important in the present context to investigate how the dopant band-pass contrast function $C_{BP}(E_j)$ varies with respect to these parameters.

The effect of varying the image frame time is shown in Fig. 4-7. Band-pass filtered images with different frame times were taken with a 2 kV, 40 pA primary beam at E = 3.4 eV and averaged along the vertical direction. The results in terms of band-pass contrast line-scans are shown in Fig. 4-7a. Although the band-pass contrast function increases as a function of frame time for all *p*-stripes, the way it increases seems to be strongly dependent on dopant density. At 30 s/frame, the band-pass contrast for all 4 stripes is distinct and its value steadily falls with decreasing stripe dopant concentration. This is similar to the case at 150 s/frame, but the difference in band-pass contrast between P18 and P19 is significantly reduced compared to the other stripes. At 300 s/frame, the difference in band-pass contrast between all three P17, P18 and P19 stripes becomes relatively small, and the overall trend with decreasing dopant density is reversed, that is, the band-pass contrast function is highest at P17 and lowest at P19.

Figure 4-7b plots changes in the average band-pass contrast of each stripe with increasing frame time. The graph provides more information about the contrast reversal effect which occurs between the P17, P18 and P19 stripes as the frame time is increased to 300 seconds. Although the band-pass contrast function increases, it increases for the wrong reason, the effective greater dose of the primary beam is recombining with holes in the *p*-stripe regions and erasing information about their initial differences in dopant density. It is interesting to note however, that the lowest density stripe P16 stripe rises much less sharply as a function of frame time, and this indicates that increasing the frame time might be beneficial for very low *p*-doped density regions. It indicates a possible use of slow scans to identify areas with low dopant concentration that would otherwise be difficult to observe and locate. More work with smaller dopant differences in the very low dopant density range is required in order to confirm that. In the light of the results shown in Fig. 4-7, the lowest frame time of 30 seconds is recommended.

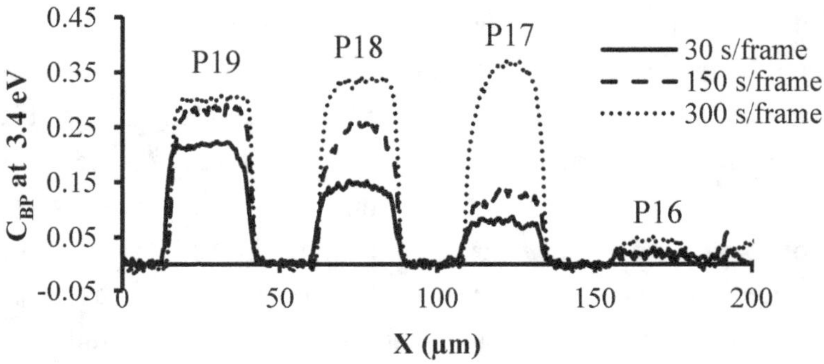

(a)

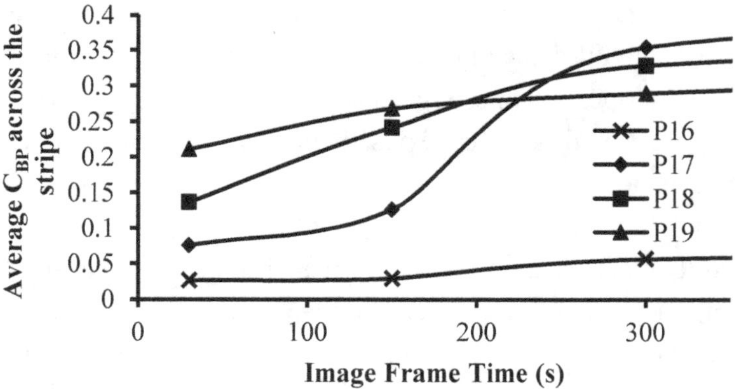

(b)

Fig. 4-7. Effects of scan rate variations on the 4 p-doped striped test sample. (a) Band-pass contrast (E = 3.4 eV) line-scans. (b) Average band-pass contrast function across each stripe.

The effects caused by increasing the primary beam current are shown in Fig. 4-8. Band-pass energy images for an SE energy of 3.4 eV for the primary beam currents of 40, 200 and 400 pA were taken with a 2 kV beam, and the line-scan information extracted from them is shown in Fig. 4-8a. When the beam current increases to 200 pA, the band-pass contrast function not only rises for all 4 stripes, but the profile shape over each stripe starts to become slanted, particularly for the P19 stripe. This effect becomes more pronounced at a beam current of 400 pA, where the band-pass contrast function drops sharply down from the left-hand side edge of each stripe, dipping down to negative values as it reaches the right-hand side edge. The image intensity at the right-hand side edges is lower than the *n*-substrate, resulting in negative band-pass contrast values, dipping down to –0.2. These results indicate that increasing the primary beam current has even worse effects than increasing the frame time, and the asymmetry of the line-scans across each *p*-dopant stripe suggests that other more complicated effects are occurring.

The more complicated effects caused by increasing the primary beam current are further evident in Fig. 4-8b, which plots changes in the average band-pass contrast function of each stripe with increasing beam current. The band-pass contrast function for P19, P18, and P17 seems to rise and go through a peak value at around 200 pA, and then falls down to levels below 0.1 at 400 pA. There seems to be another effect occurring beyond 200 pA which brings the band-pass contrast function value down. It is interesting to note that just as with the scan frame time variations, the lowest dopant density stripe, P16, has a significantly different type of variation than the other stripes. The P16 stripe continues to increase beyond the 200 pA value. The final result at 400 pA however, is that the band-pass contrast values for all stripes have similar values, and they therefore no longer carry any information about their initial stripe dopant density. The lowest primary beam current of 40 pA is therefore recommended.

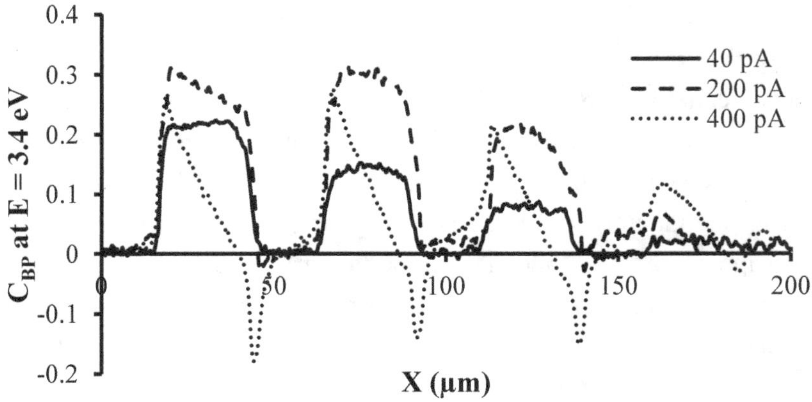

(a)

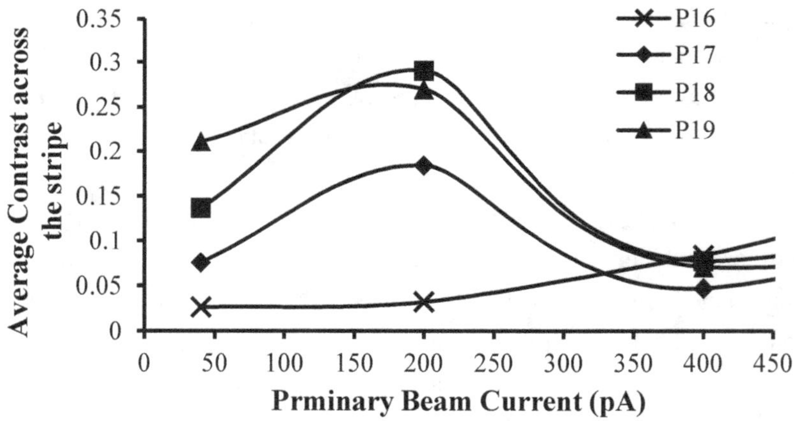

(b)

Fig. 4-8. Effects of primary beam current variations on the 4 p-doped striped test sample. (a) Band-pass contrast (E = 3.4 eV) line-scans. (b) Average band-pass contrast function across each stripe.

It should be noted that the results presented here do not provide any information about how the band-pass contrast function varies around the lowest primary beam current of 40 pA, since only three primary beam currents were used. This is also true for the frame scan time data; they do not provide enough information about how fast the band-pass contrast values change around the lowest scan time of 30 seconds. All that can be concluded is that both scan time and primary beam current should be kept as low as possible and not varied.

More information about the asymmetric skew effect on the band-pass contrast line-scan at high primary beam currents can be obtained by acquiring the SE energy spectral signal at different positions on the P19 stripe. SE energy spectral signals were obtained at a 1.27 nA primary beam current at positions closer to the stripe edges and compared to a spectral signal taken with a beam current of 40 pA, they are shown in Fig. 4-9a. There is a shift to the right in the1.27 nA off-centre spectra, compared to the spectral signal taken at 40 pA (at 40 pA the band-pass contrast function was constant across the width of the stripe). This indicates the possible presence of negative charging at the higher beam current, suggesting that the more complicated effects occurring might be coming from the build-up of negative charge.

The spectral signals from the two off-centre stripe positions at the higher beam current of 1.27 nA also show a strong dependence on SE energy. At SE energies lower than 4 eV, there is a stronger asymmetrical effect, while for SE energies above 4 eV, the asymmetric effect is relatively small. This is evident from the two band-pass energy images shown in Fig. 4-9b, one taken for a SE energy of 0.88 eV and the other taken at 4.22 eV. The band-pass image at SE energy 0.88 eV shows significant asymmetry, similar to the image for the SE energy of 3.4 eV, while at a SE energy of 4.22 eV, no significant asymmetry is visible. These results demonstrate the possibility of taking band-pass images at higher SE energies in order to mitigate charging effects at larger beam currents. However, the contrast levels of these stripes are not very different from one another, making these images only suitable for qualitative identification of p-doped areas and not quantitative dopant measurements.

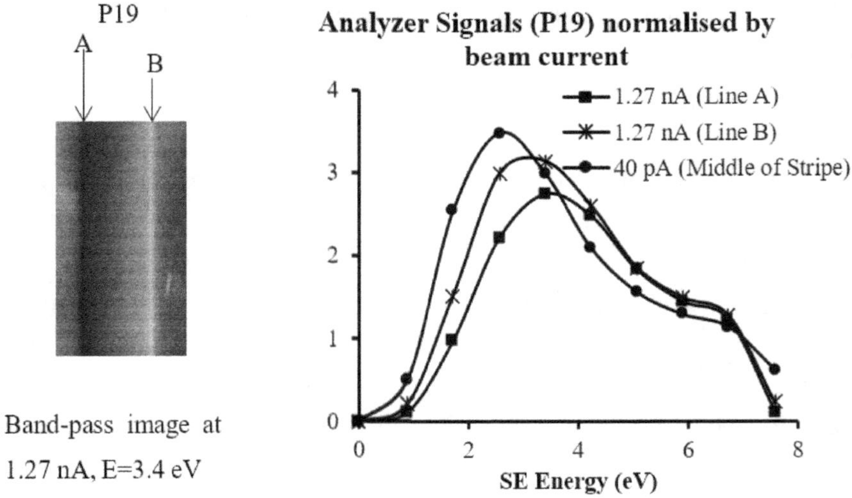

(a)

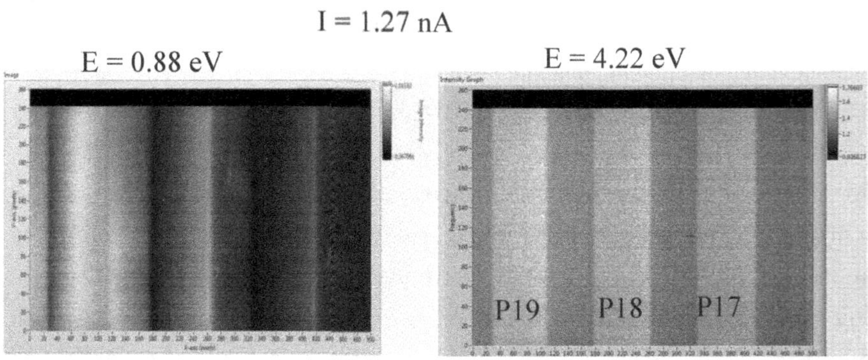

(b)

Fig. 4-9. Effects of higher beam current on SE energy spectral signals. (a) SE energy spectral signals obtained with 1.27 nA primary beam at two edges of the stripe A and B, compared to the signal obtained with 40 pA primary beam at the middle of the stripe. (b) SE band-pass images obtained with a 1.27 nA primary beam, at E = 0.88 eV (left) and E = 4.22 eV (right).

The conclusion that both the scan frame time and primary beam current should be as small as possible is not new, but the SE energy spectral approach does provide more information about the way high scan times and high primary beam currents diminish the dopant contrast effect. In practice, both these parameters can only be reduced to levels where shot noise from the primary beam is still manageable. For the doped stripe test sample, shot noise is reduced by averaging along the length of the stripe (242 pixels in the vertical direction). This is only possible because the dopant density along that direction is constant (1D layout), but for 2D dopant structures, the challenge of using low scan times and low primary beam currents while keeping shot noise to manageable levels will be more challenging.

Figures 4-10a–d show how the band-pass contrast function (evaluated in the middle of each stripe at a SE energy of 3.4 eV, and maximum average energy shift (at the edge of each stripe) vary linearly with the logarithm of the dopant concentration (N_A). These measurements were made from four *p*-stripe regions on the same *n*-substrate having the densities of 10^{16}, 10^{17}, 10^{18} and 10^{19} cm^{-3}. The error bars on these graphs were obtained by calculating the standard deviation of the band-pass function and average energy shift from pixels having the same dopant concentration (242 pixels that make up the vertical lines in the image), and therefore represent an estimate of the total noise per pixel, and will be referred to here as the total standard deviation (std) noise. The results show that the most sensitive dopant contrast cases (highest gradients), are obtained for a primary beam voltage of 2 kV on the 1.75 nm thin native oxide, and a primary beam voltage of 5 kV on the 3.91 nm thick native oxide. These results are implicit in the spectral signal information contained in Figs. 4-4a–d. The average error from linearity for the 2 kV, 1.75 nm case is 3.3% for the band-pass contrast function, and 3.6% for the average energy shift, while for the 5 kV 3.91 nm case, it is 3.2% and 6.4% respectively. There is clearly scattering of data present for the band-pass contrast function in the 2 kV, 3.91 nm case, and very low sensitivity for the 5 kV, 1.75 nm combination.

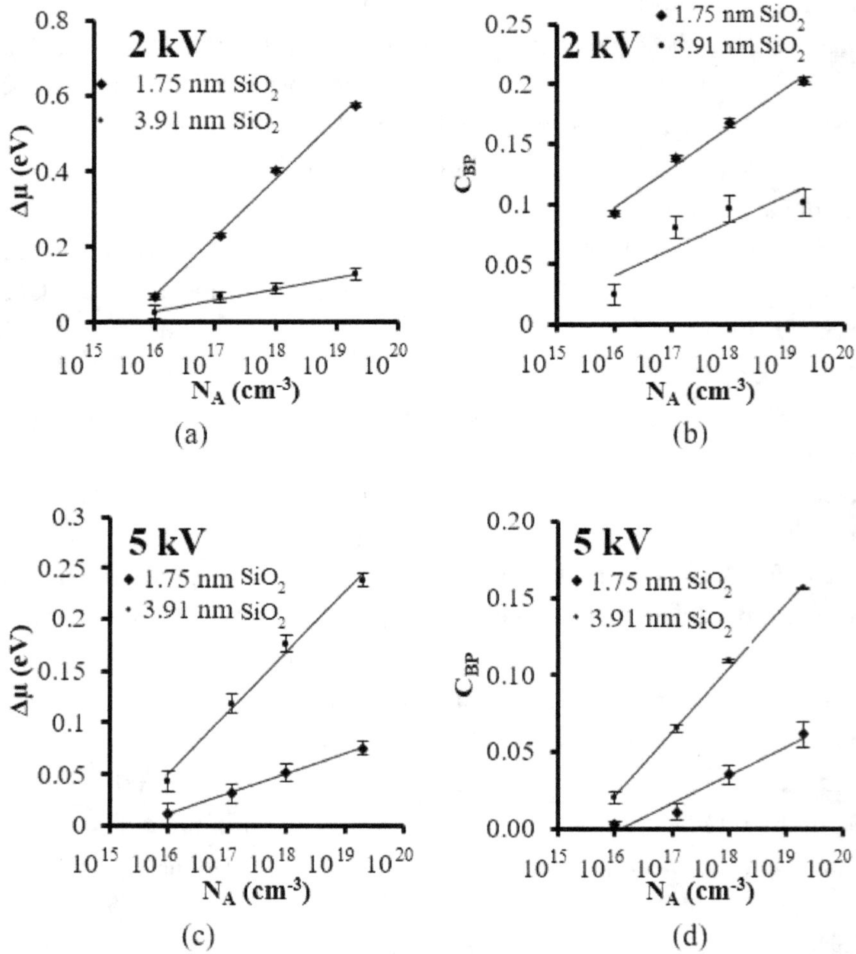

Fig. 4-10. Dopant concentration graphs of p-stripe density in the n-substrate test sample. The band-pass contrast function is evaluated at an energy of 3.4 eV in the middle of each stripe (point B). The average energy shift is taken at the stripe edge (point C). (a) Average energy shift at a primary beam voltage of 2 kV. (b) Band-pass contrast at a primary beam voltage of 2 kV. (c) Average energy shift at a primary beam voltage of 5 kV (d) Band-pass contrast at a primary beam voltage of 5 kV.

Table 4-1. Percentage experimental errors (a) for a native oxide thickness of 1.75 nm (b) for a native thickness of 3.91 nm. Combinations of lowest error highlighted.

%RMS Deviation		**1.75 nm SiO2**			
		Contrast		Δμ	
		5 kV	2kV	5 kV	2kV
Best fit line error	Average	57.2%	3.3%	2.3%	3.6%
	Range	1.0%–184.1%	2.0%–5.1%	0.7%–3.7%	1.7%–5.4%
Shot Noise	Average	49.8%	2.1%	35.8%	4.5%
	Range	13.3%–118.4%	1.6%–2.9%	9.2%–87.8%	1.0%–11.9%

(a)

%RMS Deviation		**3.91 nm SiO2**			
		Contrast		Δμ	
		5 kV	2kV	5 kV	2kV
Best fit line error	Average	3.2%	22.0%	6.4%	5.1%
	Range	1.6%–5.2%	10.9%–38.3%	2.7%–13.3%	0.9%–10.5%
Shot Noise	Average	5.9%	17.3%	9.5%	29.0%
	Range	0.5%–18.9%	11.1%–35.0%	2.7%–23.3%	12.9%–66.9%

(b)

A detailed list of deviation from a straight line fit together with the standard deviation noise errors are tabulated in Table 4-1. The best combinations of primary beam voltage with native oxide thicknesses are highlighted, reinforcing the observations already made in the SE energy spectral results shown in Figs. 4-4a–d: where the native oxide layer is thick, the strategy of using a higher beam voltage can produce reliable quantitative dopant contrast, and where the native oxide layer is thin, a lower primary beam voltage provides lower levels of percentage noise and higher dopant contrast sensitivity.

The results in Table 4-1 also indicate that overall, for the best primary beam voltage/native oxide thickness combinations (highlighted columns), the average percentage deviation from a straight line is either comparable or lower than the average standard deviation noise error. For the 2 kV primary beam voltage probing a 1.75 nm thick native oxide layer, these errors are typically lower than 5%, for the 5 kV beam probing the 3.91 nm thick native oxide, they are typically below 10%. The slightly higher values in the primary beam voltage of 5 kV with 3.91 nm thick native oxide layer are most likely caused by greater scattering in the thicker native oxide layer.

Noise analysis given in Fig. 4-11 shows that the better combinations of primary beam voltages and native oxide thicknesses are dominated by primary beam shot noise statistics (varying as $1/\sqrt{N}$), while for the other cases, there is evidence of the noise being increased by either native oxide scattering (the average error in the 2 kV thicker native oxide is 38%) or scattering occurring deeper within the doped silicon regions (the average error in the 5 kV thinner native oxide cases is 36%), causing it to depart from primary beam shot noise statistics. Both types of scattering are evident from the results shown in Fig. 4-11b and Fig. 4-11c, where the departure from ideal shot noise characteristics is greatest. This is the first-time that the effect of native oxide scattering on the information provided by different dopant contrast mechanisms has been quantitatively identified in terms of their noise characteristics, in the past, due to the much lower information content provided by SE detector brightness contrast, this was not possible, and complex contrast reversal effects were observed on samples with a thicker native oxide layer (El-Gomati *et al.*, 2005; Kazemian *et al.*, 2004; Mika and Frank, 2008).

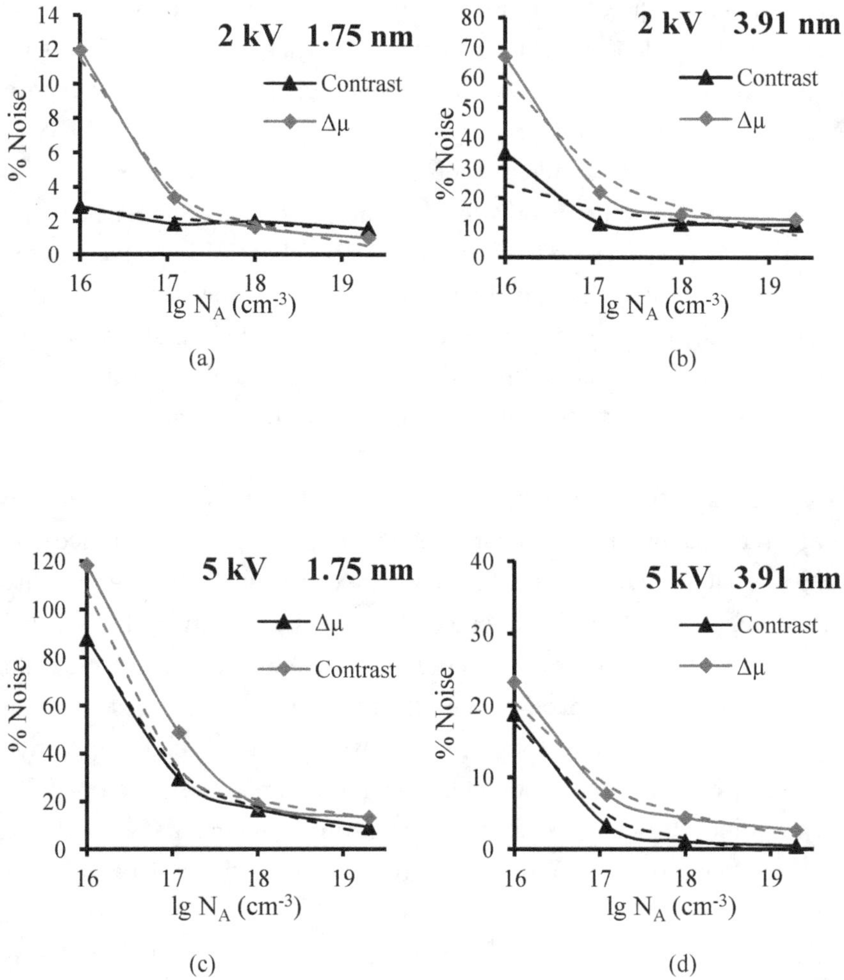

Fig. 4-11. Noise analysis of experimental results at 2 kV and 5 kV primary beam voltages, with 1.75 nm and 3.91 nm native oxide thicknesses. (a) 2 kV, 1.75 nm. (b) 2 kV, 3.91 nm. (c) 5 kV, 1.75 nm. (d) 5 kV, 3.91 nm.

4.4 *In-situ* estimate of native oxide thickness and surface band-bending energy

The results in Fig. 4-12 show how quantifying changes in the SE energy spectral signal shape can provide useful information about the sample. It gives *in-situ* (inside the SEM) estimates of the native oxide thickness by tracking how the spectral signal average energy (μ) and standard deviation (σ) change as the primary beam voltage goes from 2 to 5 kV, as shown in Fig. 4-12a. Different native oxide thicknesses were grown in the 1.5 to 4.5 nm range. Both the spectral signal's average energy and standard deviation vary up to around 40% from their initial values (at the 2 kV beam voltage) for the thicker native oxide thickness values. In principle, if the native oxide thickness of a doped semiconductor sample is not known, it can be estimated by tracking the change in average energy or standard deviation of the SE energy spectral signal.

SE energy spectroscopy can also estimate the surface band-bending energy (caused by trapped charge at the doped Si-SiO_2 interface) by fitting the Henke *et al.* (1979) analytical expression for the probability density function of intrinsic silicon (given by equation 1.2), using electron affinity as a free parameter. This procedure was carried out with reference to a fit first made to the experimental data obtained from an intrinsic silicon sample ($\chi = 4.05$ eV), shown in Fig. 4-12b. The effective change in electron affinity estimates from the fit are, 3.9 eV for a *p*-stripe interior point (10^{19} cm^{-3}), 0.15 eV above intrinsic silicon, and 4.15 eV for the *n*-substrate (10^{15} cm^{-3}), 0.1 eV below intrinsic silicon. This gives an overall surface band-bending energy of 0.25 eV going from the *n*-substrate to a point within the *p*-stripe region interior. This value is similar to previous estimates made either by reverse biasing a *pn*-junction (Elliott *et al.*, 2002), or by subtracting a signal energy shift from the theoretical in-built *pn*-junction potential (Kazemian *et al.*, 2007).

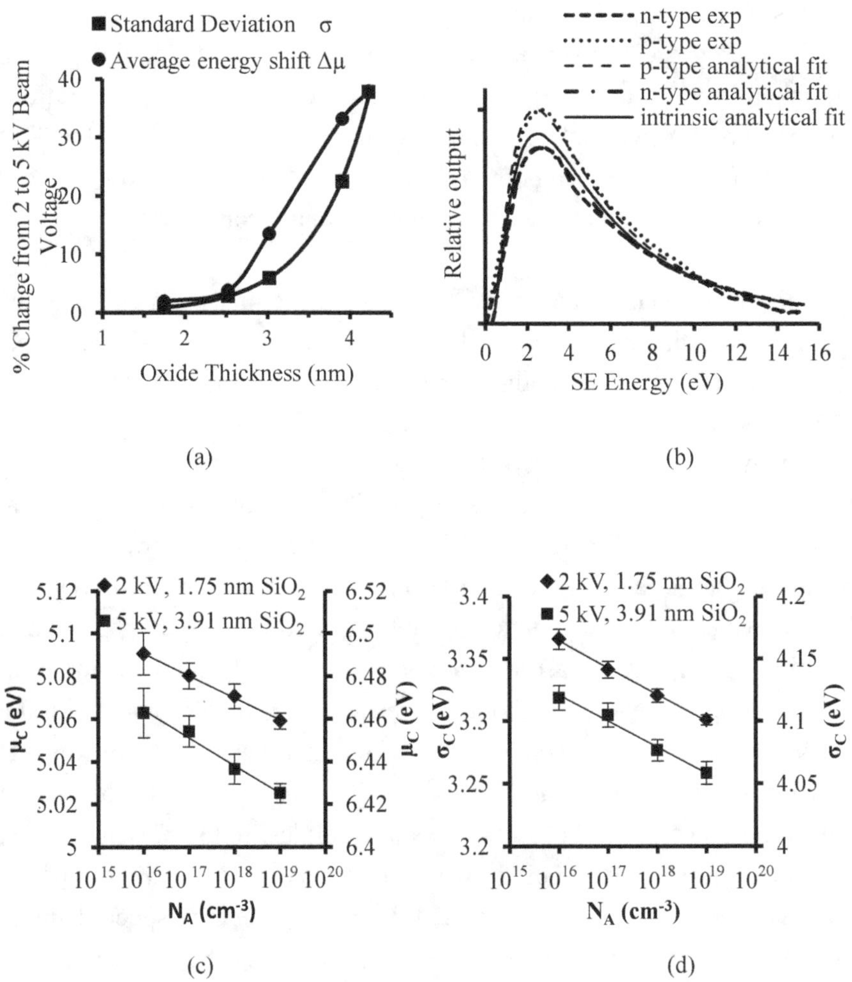

Fig. 4-12. SEM in-situ measurements from SE energy spectral shape changes on the *p*-stripe test specimen. (a) *In-situ* oxide thickness measurement. (b) *In-situ* surface band-bending voltage measurement. Electron affinity estimates from the fit are, 3.9 eV for the *p*-type, 4.15 eV for the *n*-type, 4.05 eV is used for the intrinsic silicon signal. (c) Spectral average energy for single dopant measurement (inside *p*-stripe regions). (d) Spectral standard deviation for single dopant measurement (inside *p*-stripe regions).

Figures 4-12c and 4-12d show a linear dependence between lg(N_A) and the SE energy spectral signal average energy (μ) and standard deviation (σ) taken from four *p*-stripes having different dopant concentrations (interior points), without reference to the *n*-substrate. This is the first time such results have been obtained, and they demonstrate how changes in SE energy shape can be used to quantify dopant concentration.

The discovery that quantitative dopant contrast measurements can be made by detecting small changes in the SE energy spectral signal shape, goes beyond the two main dopant contrast measurements of SE brightness contrast and S-curve energy shift that were reported previously in dopant contrast literature. Both the standard deviation and average energy (most probable energy) of the output SE energy spectral signal can be used. Since these results were produced by points outside the space-charge region (several microns beyond the *pn*-junction region), they indicate that a *pn*-junction is not required, and that in principle, quantitative dopant contrast measurements in the SEM can be made from a single dopant region, without reference to an underlying substrate. Experimental results in section 4.7 continue this theme and confirm that SE energy spectroscopy can be successfully used to quantify the dopant concentration of doped bulk silicon wafer samples. In the present context, it points towards there being two different types of ways by which dopant concentration can be extracted from the SE energy spectral signal. On way is to track the maximum average energy shift at a *pn* junction, the other, is to track changes in the SE spectral signal shape/height from points inside a single dopant region.

4.5 Contamination effects

In the preceding chapter, primary beam induced contamination effects on the SE energy spectral signal from metal samples were discussed. In this section, a similar investigation is carried out for *p*-doped and *n*-doped silicon samples. Figure 4-13a shows how the combined effects of beam induced contamination due to prolonged exposure times (×16) not only cause a fall in SE energy spectral signal peak height, but also reduce the height difference between *p* and *n* signals (dopant contrast). These results were produced for interior points on the 10^{19} cm^{-3} *p*-stripe (labeled here as P19) with a 1.75 nm thick native oxide at a primary beam voltage of 2 kV, where the 1× exposure time is 1.875 s. The results are similar to previous observations of contamination reducing dopant contrast derived from SE detector image brightness (Mika and Frank, 2008). However, in this case, a quantitative spectral analysis of these effects can be made, as shown in Fig. 4-13b, which indicates how the signal-to-noise of the band-pass contrast function varies with both SE energy and exposure time. The noise is calculated from the standard deviation over many pixels in the vertical direction, in the direction of the stripe (242 pixels). The maximum signal-to-noise value falls approximately by a factor of two when the exposure time increases by a factor of four, suggesting that it is inversely proportional to the square root of the exposure time. The optimal SE energy at which to evaluate the band-pass contrast function in order to maximize the signal-to-noise lies just beyond the SE peak height position, at 3.4 eV.

Prolonged exposure can be created either by lowering the scan speed or using multiple frames over the same area on the specimen, very similar results are obtained in both cases. The ability to quantify and analyze prolonged exposure effects, and derive the optimal energy for maximizing signal-to-noise, goes far beyond the complex SE detector image contrast reversal effects obtained previously (Mika and Frank, 2008; and El-Gomati *et al.*, 2005). Although, some degree of spectral analysis of the contamination effect was made using a high pass retarding field SE detector (Tsurumi and Hamada, 2013), its conclusions of reducing contamination by eliminating low energy SEs are not valid for band-pass contrast SE energy spectral signals, where the SE contribution for each point in the signal is detected over a narrow energy bandwidth (2.5 to 6% of the SE energy).

Figures 4-13c and 4-13d present results to show that the dopant contrast mechanism causing SE energy shifts is much less affected by contamination effects than the one that produces changes in SE signal height. Figure 4-13c indicates that 16 times greater exposure times only cause relatively small differences in the SE energy spectral signal trailing edge energy shift, while Fig. 4-13d indicates that the overall variation in the average energy shift is less than 5% from its initial value, compared to over 40% for the band-pass contrast function. It is interesting to note from Fig. 4-13d that a change in spectral signal standard deviation is even less affected by contamination effects, where deviation from its initial value is kept below 1%. This makes the prospect of extending the method of quantitative SEM SE energy spectroscopy to map 2D dopant concentration variations more feasible, since in that case, multiple frames over the same area on the specimen will be required to reduce shot noise effects.

One important difference between the effects of contamination for doped *p-n* junctions shown here and the metal samples used in the previous chapter is that normalizing to the background signal does not seem to compensate for the contamination effect in the *p-n* junction case. This is because the drop in SE spectral height caused by contamination seems to occur at quite different rates for *p*-doped regions, compared to *n*-doped regions.

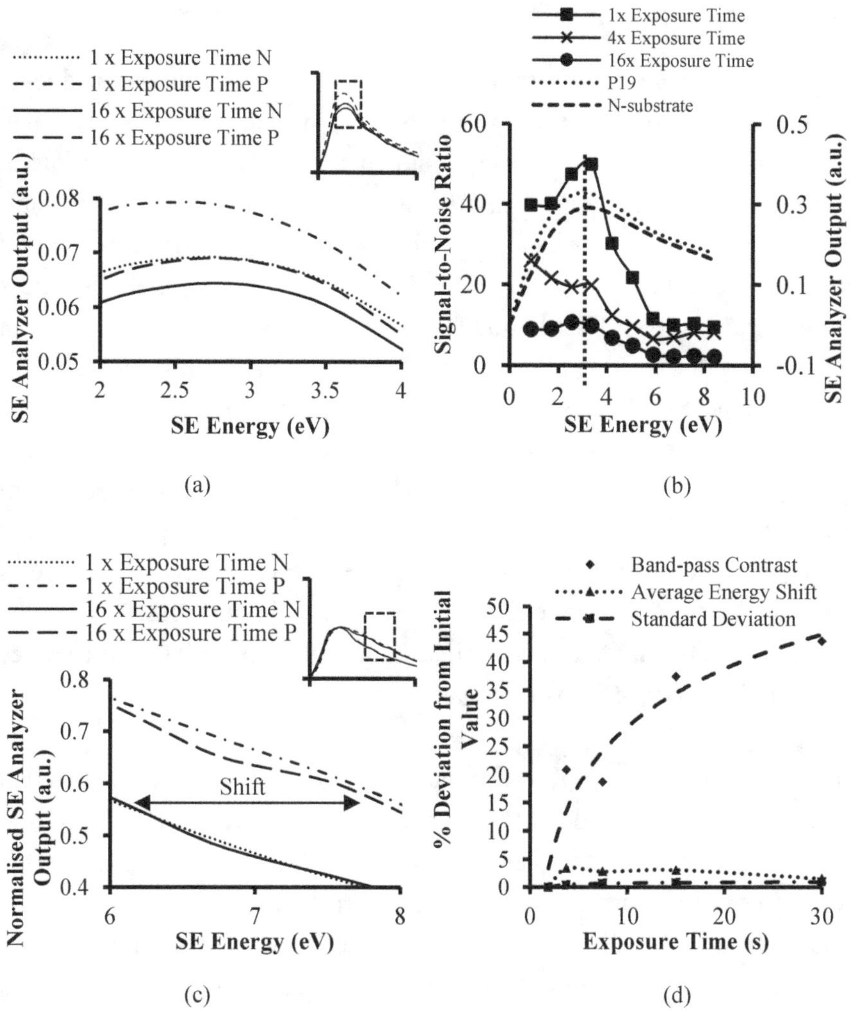

Fig. 4-13. Electron beam induced contamination effects. (a) Fall in SE energy spectral signal height for ×16 exposure time, where 1× exposure time is 1.875s. (b) Signal-to-noise of the band-pass contrast CBP as a function of SE energy and exposure time. (c) Energy shift at the trailing edge of the normalized SE energy spectral signal. (d) Percentage change in the band-pass contrast function CBP (E = 3.4 eV), average energy shift $\Delta\mu$, and standard deviation $\Delta\sigma$ as a function of exposure time.

4.6 Lateral dopant profiling across doped stripe edges

In this section, SE energy spectroscopy is used to obtain lateral dopant profiles of the doped stripes on the test sample, and its ability to map space charge regions in a SEM is shown for the first time. The lateral profiles of the band-pass contrast function and average energy shift across the left and right-hand edges of the 10^{19} cm^{-3} *p*-stripe (P19) at a primary beam voltage of 2 kV and 1.75 nm native oxide thickness sample are shown in Fig. 4-14a. The band-pass contrast profile rises to a maximum value inside the *p*-stripe region, while the average energy shift reaches a peak value within each edge (*pn*-junction region). There is obvious symmetry between the edge profiles, and there is a common region in which the band-pass contrast function and average energy shift rise up together in a similar way. The symmetry in lateral profiles across left-hand and right-hand edges suggests that patch field effects (transverse surface field effects) are minimal in this case, since the spectrometer is situated on the right-hand side. The relative difference between the two profiles is less than 4%.

More detailed lateral profile information (not shown here), for different *p*-stripe densities, 2 and 5 kV primary beam voltages, and 1.75 nm and 3.91 nm native oxide samples, supports the general observation that the best cases (most sensitive to dopant contrast) are for a primary beam voltage of 2 kV on the 1.75 nm thin native oxide, and a primary beam 5 kV on the 3.91 nm thick native oxide, consistent with the dopant concentration-lg(N_A) graph results shown in Fig. 4-10, and the spectral signal results shown in Figs. 4-4a–d. Greater scattering from the native oxide layer and scattering occurring from deeper within the sample give rise to asymmetries in the lateral profile, and cause profiles from different *p*-stripe concentrations to overlap with one another.

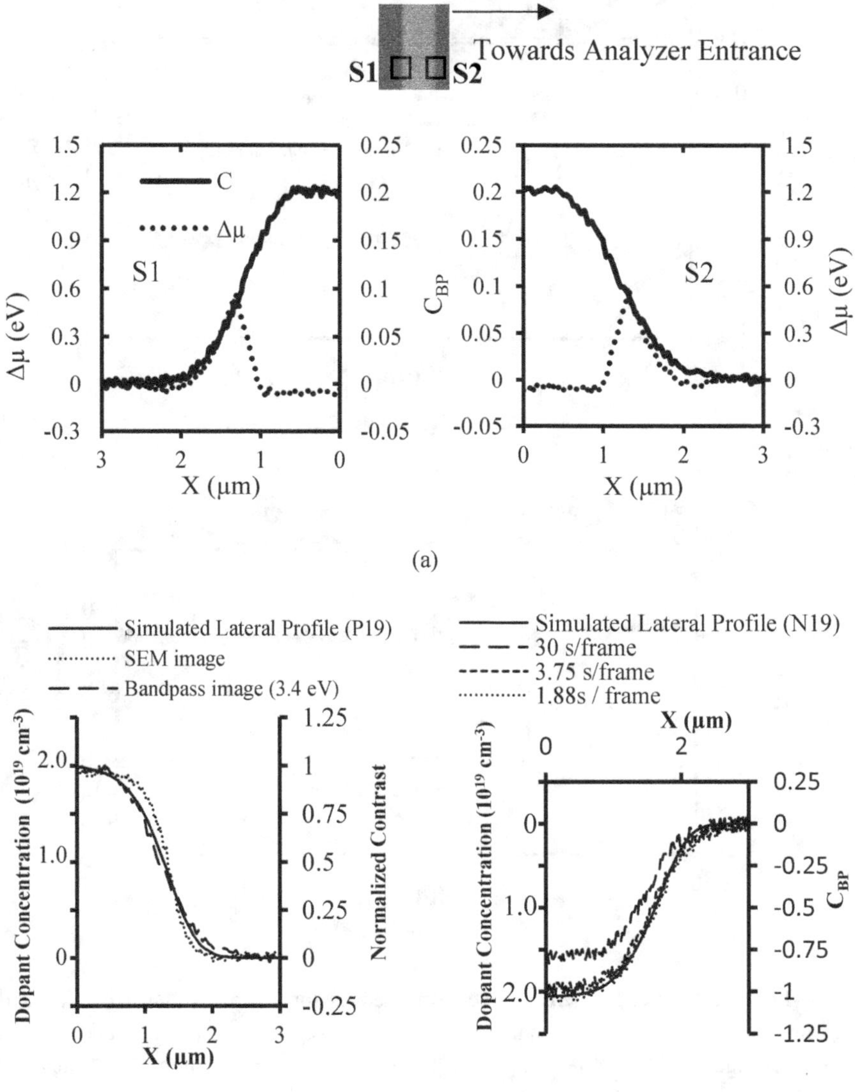

Towards Analyzer Entrance
S1
S2
C
Δμ
Δμ (eV)
C$_{BP}$
X (μm)
(a)
Simulated Lateral Profile (P19)
SEM image
Bandpass image (3.4 eV)
Dopant Concentration (10^{19} cm^{-3})
Normalized Contrast
(b)
Simulated Lateral Profile (N19)
30 s/frame
3.75 s/frame
1.88s / frame
(c)

(Fig. 4-14)

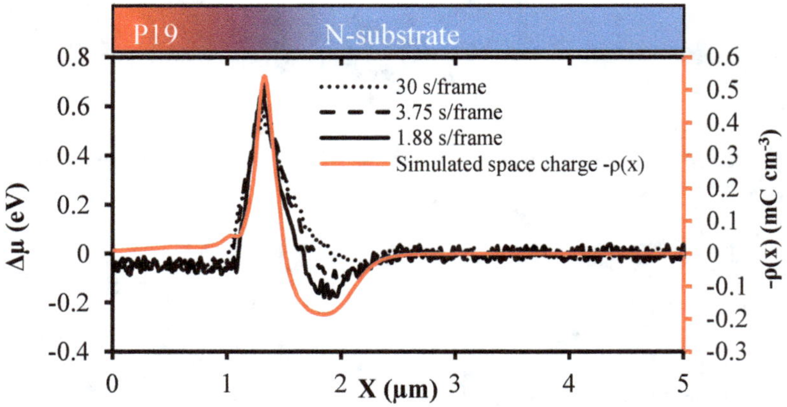

(d)

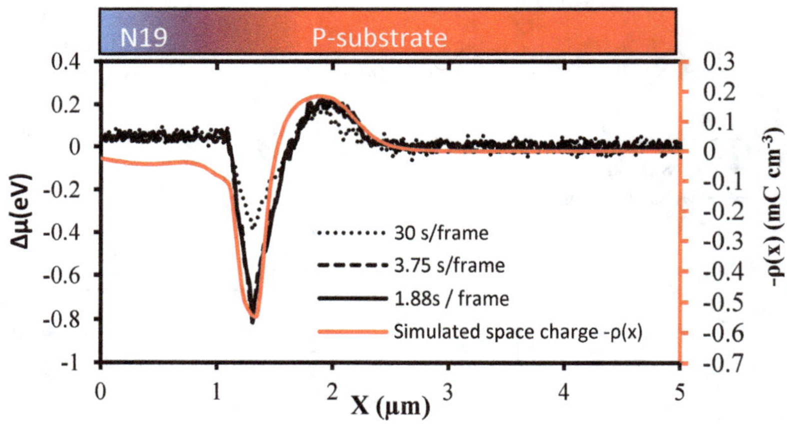

(e)

Fig. 4-14. Lateral profiling across doped stripe edges for a beam voltage of 2 kV and 1.75 nm native oxide thickness sample. (a) The band-pass contrast function and average energy change profiles. (b) Normalized band-pass contrast function, SEM detector contrast and dopant concentration TCAD simulation across S2 for the P19 edge. (c) Normalized contrast and dopant concentration for the N19 edge for different frame rates (number of frames adjusted to keep the same total dose). (d) Average energy shift and space-charge density from TCAD simulation for the P19 edge at different frame rates, same total dose. (e) Average energy shift and space-charge density from TCAD simulation for the N19 edge at different frame rates, same total dose.

Figures 4-14b and 4-14c indicate that there is good agreement between experimental band-pass contrast lateral profiles and dopant concentration profiles predicted by technology computer-aided design (TCAD) simulation for the 10^{19} cm^{-3} *p*-stripe (P19) and *n*-stripe (N19), see Appendix A3 for more details about the TCAD software simulation. A good fit was first obtained between TCAD simulations to the spreading resistance profiling (SRP) experimental depth profile data (Fig. A3-1 in Appendix A3), with an RMSD error of around 1.1%. The simulation was then subsequently used to generate the lateral profile with the same modeling parameters. The results in Fig. 4-14b show that the band-pass contrast function (evaluated at an SE energy of 3.4 eV) gives a better fit than the one generated from changes in SE detector brightness, giving an RMSD error of 4.9% and 10.5% respectively. The fact that SE detector brightness contrast correlates approximately to the dopant lateral profile had been reported earlier, where it was compared to SIMS lateral profile data (Elliott *et al.*, 2002). The TCAD simulated lateral profile for the N19 stripe edge with corresponding experimental band-pass contrast lateral profiles are given as a function of frame rate in Fig. 4-14c, while keeping the overall dose the same by varying the number of frames. Since the level of shot noise and contamination are therefore roughly the same for each frame rate (same total dose), the better agreement with the TCAD simulation profile for the faster frame rates therefore represents reduced primary beam electron recombination in a top positive depletion layer formed by dangling bonds at the Si-SiO_2 interface, providing direct experimental confirmation for the surface band-bending model originally proposed by Perovic *et al.* (1995).

Figures 4-14d and 4-14e show that there is good agreement between TCAD simulated space-charge density lateral profiles and the experimental average energy shift lateral profiles across P19 and N19 edge *pn*-junction regions. The match between experiment and simulation improves in the positive space-charge regions as the frame time decreases while keeping the overall dose the same, indicating that there is a recombination effect created by primary beam electrons. This is the first time SE energy shifts have been identified with the in-built space-charge lateral profile. In the past, they were linked to the *pn*-junction built-in voltage (Kazemian *et al.*, 2007).

The present experimental results indicate that the lateral variation of the electric potential within the region, deduced from the way it causes SE energy shift, is very different to the one predicted by simple one-dimensional *pn*-junction theory, or by TCAD simulation, which both calculate the potential to be a constant minimum negative value across the *p*-stripe interior region. Although the magnitude of the average energy shift varies linearly with the logarithm of dopant density, its lateral profile varies in accordance with the space-charge density profile, suggesting that the specimen surface has a metallic-like equipotential plane held at the same potential as the base plate potential, as shown in Fig. 4-15, forcing the surface potential in the *p*-stripe interior region to be zero. One possible explanation for this may come from the conductive plane being connected in some way to the presence of a beam induced carbon/graphite contamination layer, and the reason why its potential is tied to the base-plate potential might be due to surface currents and the conductive channel created by the primary beam through the specimen, supported by the fact that the measured specimen current is around 50% of the primary beam current. The primary electron beam energy is sufficient to liberate electrons from the valence band of silicon into the conduction band, and therefore creates a conductive channel to the base-plate (Srinivasan *et al.*, 2018b). The contamination layer forming an effective metal oxide semiconductor (MOS) structure has been proposed previously (El-Gomati *et al.*, 2004), and the present experimental results may help to further develop that model. Electrons being free to move on a top-metallic like plane may also explain why transverse surface field effects (patch fields) are suppressed, indicated by the symmetry of the lateral profiles shown in Fig. 4-14a. The presence of the positive surface states originating from dangling bonds at the Si-SiO_2 will also create an electric field screening effect (Chee *et al.*, 2011). Another important conclusion coming from these experimental results is that the positive depletion layer caused by dangling bonds acts independently from the space charge effect, effectively changing electron affinity, and making it possible to monitor dopant profiling by tracking SE spectral signal height changes. These dopant contrast mechanisms seem to act independently from one another. Further experiments aimed at investigating the nature of the top metallic-like ground-plane effect are required, and will be the subject of future work. In the present context, it is important to note that whatever its origin, this effect has made it possible, for the first time, to use a SEM to map the space-charge distribution inside a semiconductor sample.

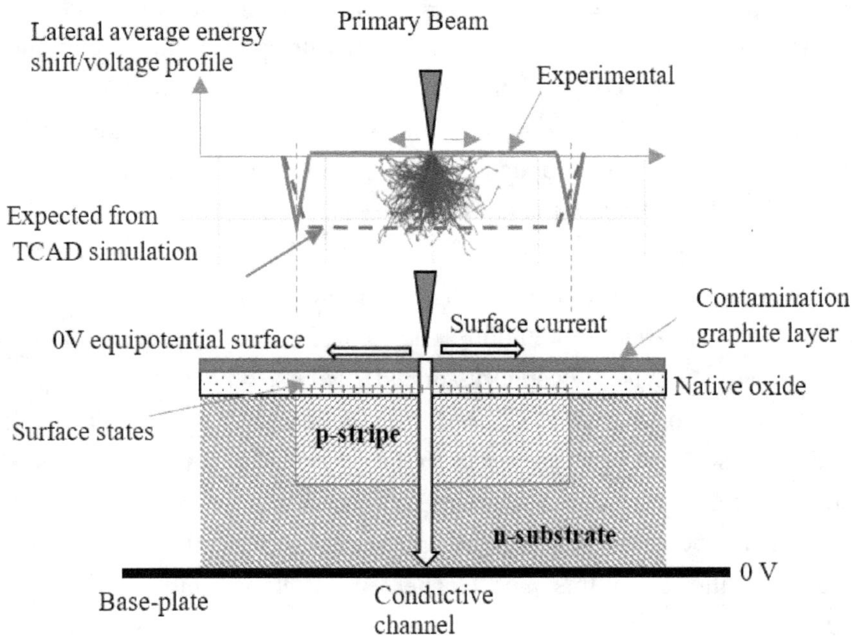

Fig. 4-15. Revised model for dopant contrast based upon the experimental average energy shift being zero for points above the *p*-stripe interior region.

4.7 Dopant concentration measurements of bulk doped silicon wafer samples

In section 4.4, certain SE spectral signal parameters such as standard deviation or average energy were found to depend linearly on the logarithm of dopant concentration, without the need to use a reference measurement to the substrate. In this section, the same approach is used to demonstrate how the local conductivity of bulk silicon wafers can be measured. At present, measurements of this kind are mainly obtained by the widely used four-point probe method (Valdes, 1954), and using the SEM for this purpose opens up new possibilities for semiconductor characterization, such as rapidly mapping local conductivity changes from point to point on doped silicon wafers. Alternatives to the four point probe method have already been sought, since direct contact with the wafer can be destructive and its reliability depends on the quality and repeatability of the contact between the probes and the sample surface. These non-contact techniques include direct current, radio frequency, microwave and time domain measurements (Krupka, 2013). The possibility of using a SEM is not usually considered.

The wafers used for the following experiments were prime grade Si wafers in the <100> orientation. They consisted of 4 *p*-type (Boron-doped) and 3 *n*-type (Phosphorus-doped) wafers, with different dopant concentration/conductivities. The dopant concentrations and the labels used for 7 such samples are listed in Table 4-2. The wafers were diced into 5×5 mm pieces and the pre-measurement sample preparation was identical to the other semiconductor doped samples: all samples were cleaned with an ultrasonic cleaner in acetone and isopropanol sequentially for 10 minutes each, before removing any oxide layers using the Oxford Plasma Pro Cobra 100 Deep RIE. The etched samples were then transported in a vacuum sealed box and inserted into the SEM.

Table 4-2. The dopant concentrations of 7 doped bulk wafer samples and their labels.

Label	Dopant Concentration
P1	2.5×10^{15} cm^{-3}
P2	2.9×10^{16} cm^{-3}
P3	8.0×10^{17} cm^{-3}
P4	1.9×10^{19} cm^{-3}
N1	9.2×10^{14} cm^{-3}
N2	1.1×10^{16} cm^{-3}
N3	1.2×10^{19} cm^{-3}

All SE energy spectral signals were obtained with a 2 kV and 40 pA primary beam, and a sample/inner cap electrode biasing of −10 V. Each spectral signal was acquired while scanning over an area of 5 by 5 μm and captured in 30s. Experimental spectral signals from the P1-P4 samples are shown in Fig. 4-16a. The SE energy spectral signal height at 4 eV, marked as E_P, was extracted and plot as a function of the logarithm of dopant concentration lg(N_A), as shown in Fig. 4-16b. The signal height at 4 eV showed no linear dependence with lg(N_A), giving an R-squared value of 0.0023. However, better results were obtained by normalizing the SE energy spectral signal height to its background signal (height at 0 eV). These signals are shown in Fig. 4-16c, and their signal heights at 4 eV were then extracted and plot as a function of lg(N_A), indicated in Fig. 4-16e. A strategy for further improving the linear dependence between SE spectral signal height and logarithm of dopant concentration came from aligning the SE energy spectral signals by their respective average energy values, shown in Fig. 4-16d, where the average energies of the spectral signals from P2, P3 and P4 were aligned to P1. The corresponding spectral signal height (at 4 eV) dependence with lg(N_A) after doing this is indicated in Fig. 4-16e. The R-squared value after normalizing to the signal background goes up to 0.9709, and after aligning to their average energy values, it goes up further to 0.9945.

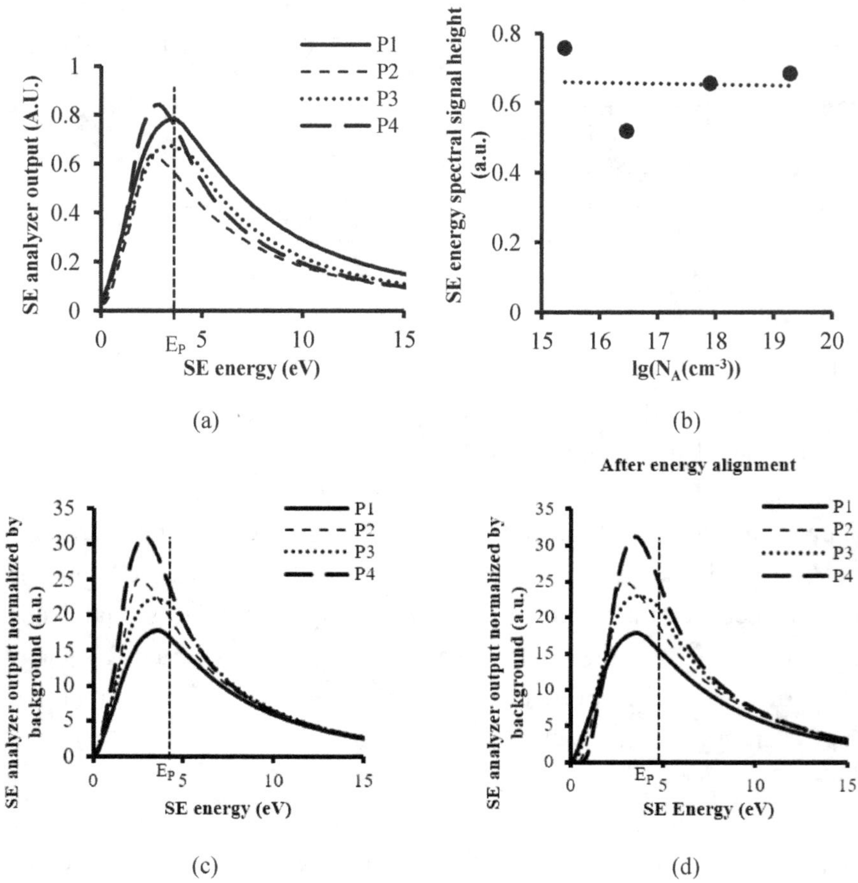
P1
P2
P3
P4
SE analyzer output (A.U.)
SE energy (eV)
E_P
(a)
SE energy spectral signal height (a.u.)
lg(N_A(cm^-3))
(b)
After energy alignment
SE analyzer output normalized by background (a.u.)
SE energy (eV)
(c)
SE Energy (eV)
(d)

(Fig. 4-16)

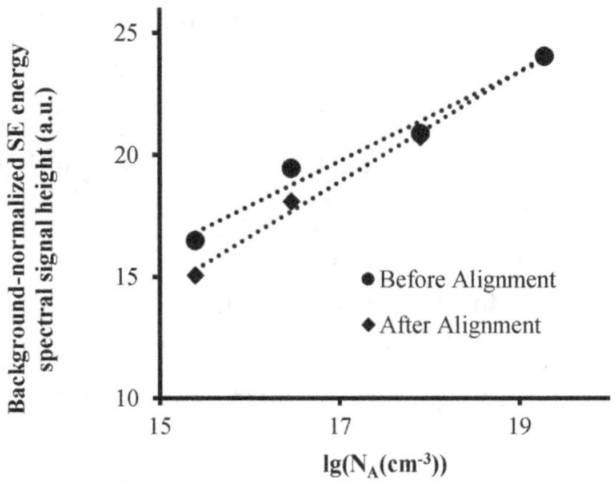

(e)

Fig. 4-16. SE energy spectral signals obtained from p-type Si bulk wafers. (a) SE energy spectral signals as measured in the SEM. EP is the SE energy at which the spectral signal height is recorded. (b) SE energy spectral signal height as a function of log of dopant concentration. (c) SE energy spectral signals normalized to the background. (d) SE energy spectral signals normalized to the background, where the average energies of the signals from P2, P3 and P4 are aligned to the average energy of the signal from P1. (e) SE energy spectral signal height normalized by background, with and without average energy alignment as a function of log of dopant concentration.

The two strategies of normalizing each SE spectral signal to its background signal and aligning SE energy spectral signals by their average energy values overcome two sources of error. The first is an error in spectral signal height caused by primary beam current/induced contamination, and the second is a spurious energy shift in the SE spectral signal position caused by sample back contact variations. These problems occur because each measurement requires a change of specimen, which inevitably cause changes in primary beam current, vacuum level, and sample back-contact conductivity to its holder. After compensating for these effects, a reliable linear relationship between signal height and log (N_A) is obtained, as shown by the data in Fig. 4-16e.

Experimental SE energy spectral signals were also obtained from the *n*-type bulk wafers with the use of background normalization and SE average energy alignment. The background-normalized SE energy spectral signal height from all 7 wafers (*p* and *n*-type) are plot as a function of lg(N) in Fig. 4-17a. The spectral signal height from the *n*-type wafers also varies linearly with lg(N), but differs from the *p*-type wafer specimens by decreasing with increasing dopant concentration. This trend is consistent with the experimental results obtained from the previous striped dopant test sample, where SE spectral signal height changes were tracked from within the interior region of each stripe.

The standard deviation σ was also extracted from the SE energy spectral signals of the bulk doped wafer samples, and its variation with the logarithm of the dopant concentration is shown in Fig. 4-17b. It was also found to vary linearly with lg(N), which is also consistent with the results obtained from the test doped stripe sample. The R-squared values of the linear fit for σ in *p*-type and *n*-type wafers are 0.9601 and 0.9712 respectively. However, as opposed to spectral signal height change, σ decreases with increasing lg(N) for both *p*-type and *n*-type wafer samples. This makes σ a more suitable parameter to measure local conductivity, since conductivity increases with dopant concentration for both *p*-type and *n*-type wafers. Using spectral signal standard deviation instead of SE energy peak height to track dopant concentration has the advantage of not requiring any normalization to the background signal or using any average energy alignment procedure. Figure 4-17c shows background normalized SE energy spectra for *p*-type and *n*-type bulk samples, which helps to explain the kind of dopant concentration variations depicted in Figs. 4-17a and 4-17b. The change in the spectral signal shape can be understood in terms of an effective change in electron affinity.

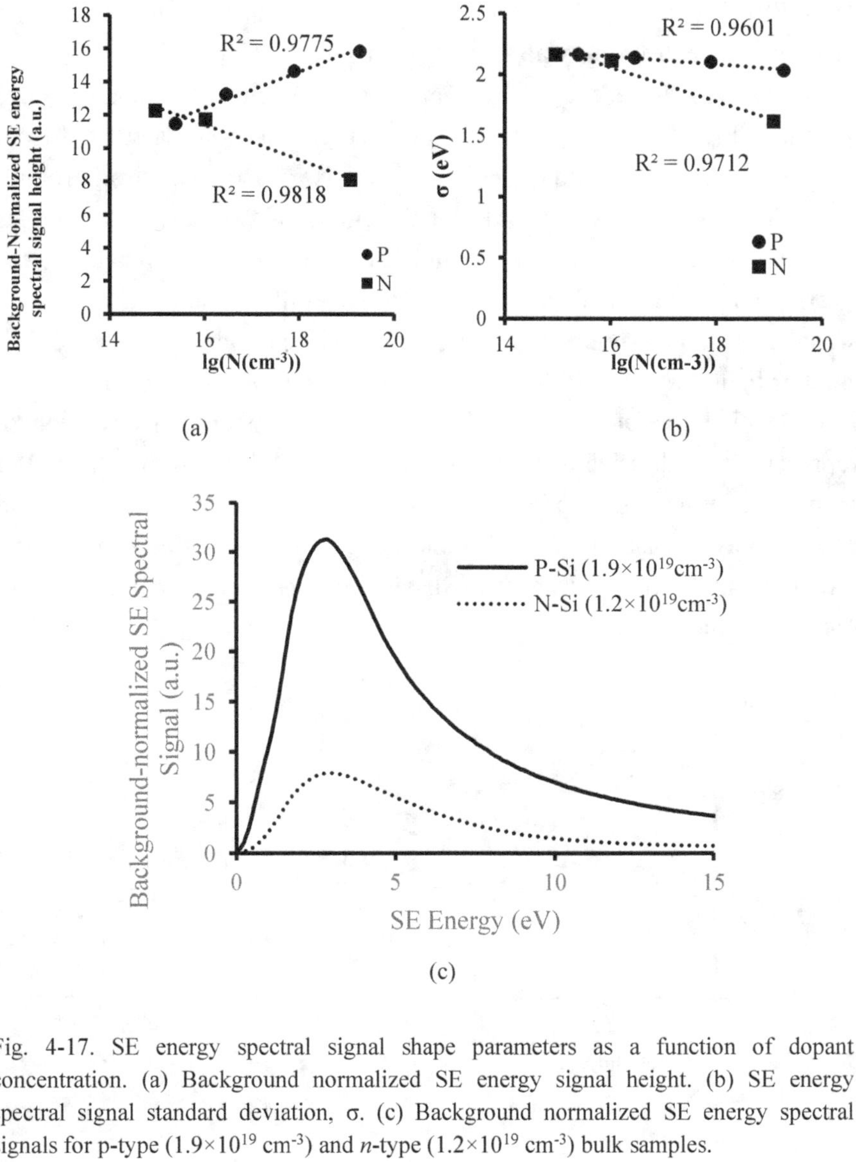

Fig. 4-17. SE energy spectral signal shape parameters as a function of dopant concentration. (a) Background normalized SE energy signal height. (b) SE energy spectral signal standard deviation, σ. (c) Background normalized SE energy spectral signals for p-type (1.9×10^{19} cm^{-3}) and *n*-type (1.2×10^{19} cm^{-3}) bulk samples.

Two different sections from the same wafer were used to assess the kind of variations caused by wafer dicing and sample preparation (cleaning), in addition to those caused by a change of specimen. The results are shown in Fig. 4-18. The error bars in these graphs were calculated from 5 spectra per point, indicating a shot noise error. The background-normalized SE energy spectral signal heights of two different samples from the same wafer showed no significant variation, as indicated by the overlapping shot noise error bars. The overall average difference in signal height measurement between the two samples is around 3.7%, smaller than the average shot noise error of 5% (size of the error bars). It is interesting to note that the shot noise for the standard deviation σ, represented by the size of the error bars in Fig. 4-18b, are much smaller than its average error of 2.1%. These results demonstrate that local conductivity measurements of bulk doped silicon wafer samples by tracking small changes in the SE energy spectral signal shape are repeatable and reliable.

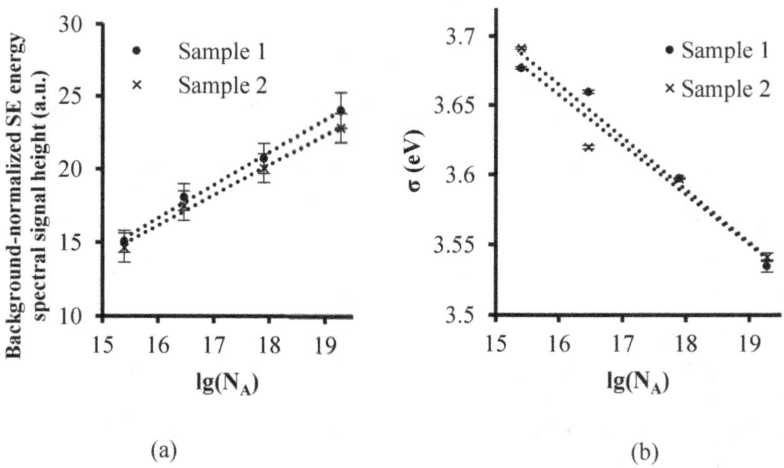

Fig. 4-18. SE energy spectral signal results obtained with two different samples from each doped wafer. (a) Background-normalized SE energy spectral signal height. (b) Signal standard deviation σ.

4.8 Solar cell dopant profiling

In this section, SE energy spectroscopy dopant contrast is applied to dopant depth profiling of a solar cell device. Figures 4-19a–b show schematic diagrams for the geometrical layout and preparation of a solar cell precursor sample provided by the Solar Energy Research Institute of Singapore (SERIS). The solar cell consists of a 200 μm *n*-type substrate ($N_D = 10^{16}$ cm^{-3}) and a 450 nm thick *p*-type ion-implanted layer. A 70 nm thick silicon nitride layer caps the solar cell surface. The sample was first immersed in 1% aqueous hydrofluoric acid (HF) solution to remove the silicon nitride anti-reflective coating present on the solar cell. The sample was then cleaved, and mechanically polished, with and without a metal coating (Fig. 4-19b). Metal coating was achieved by heating the wafer to 100°C and applying a layer of molten Field's metal before cooling. The experiments were conducted in the FEI NOVA SEM, with a 2kV, 40 pA beam primary beam, unless otherwise stated.

The purpose of metal coating is to suppress the appearance of jagged slanting surfaces next to the *p*-layer after cleaving, which can cause strong topographical effects in the SEM image. The solar cell was then orientated so that the depth direction (z) pointed in the horizontal direction (to the right), and the primary beam was scanned on the specimen in the z-y plane. The dopant concentration in the perpendicular transverse y-direction was constant.

SEM micrographs of its polished surface, with and without metal coating are shown in Fig. 4-19c. The micrograph of the sample without metal coating had a significant bright-band edge effect, making it difficult to identify the *p*-layer, while the micrograph of the sample with metal coating avoided the edge artifact and was able to provide better dopant contrast.

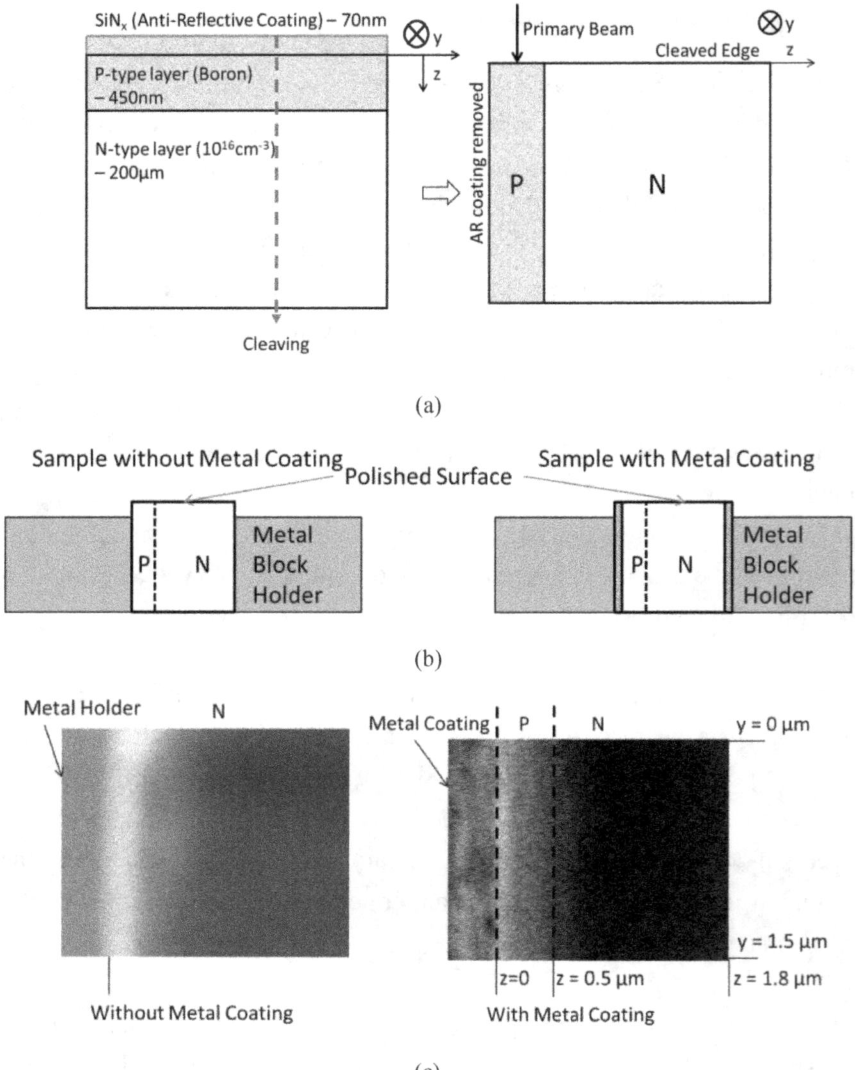

Fig. 4-19. Solar cell sample preparation before and after cleaving. (a) Sample schematic diagram before and after cleaving. (b) Schematic diagram with and without metal coating. (c) SEM micrograph with and without metal coating.

SE energy spectral data was acquired in spectroscopic imaging mode, where a series of band-pass images were taken, each at a different pass energy. Every band-pass energy filtered image was taken by moving to a fresh area along the *y*-direction (direction of constant dopant concentration), in order to reduce the effects of surface contamination. The specimen/inner cap electrode was biased to −10 V, and the pass energy ranged between 10 eV to 25 eV (corresponding to 0 to 15 eV SE energy), with a step size of 1.67 eV. Changes relative to the *n*-type substrate in the band-pass contrast function at 3 eV (close to the SE peak energy) and the average energy of the SE energy spectral signal were calculated for each pixel in the image. Line profiling data was then obtained by averaging over 242 pixels in the *y*-direction in order to reduce the effects of shot noise.

Experimental line distributions of both the band-pass contrast and average energy shift on the solar cell precursor sample are shown in Fig. 4-20a, and were found to closely follow ECV experimental depth dopant profile data. The RMSD (root mean square deviation) from the ECV profile is 0.66% for the band-pass contrast function, and 1.93% for the average energy shift over the depth range, z = 22 to 306 nm (labeled as the Linear Range). In this case, band-pass contrast and average energy shift have a similar profile, indicating that both space-charge density and *p*-dopant concentration vary in a similar way for a graded *pn*-junction having a side metal coating/holder connected to the region of maximum *p*-dopant concentration. This is quite a different situation to the previous dopant striped test sample. In this case, free electrons coming from the metal holder are able to create a maximum negative space-charge region where the *p*-type dopant concentration is a maximum. In the case of the previous dopant striped samples, the space-charge density fell to zero in the *p*-stripe interior regions due to the absence of any external source of free electrons, where the number of minority electron carriers are small.

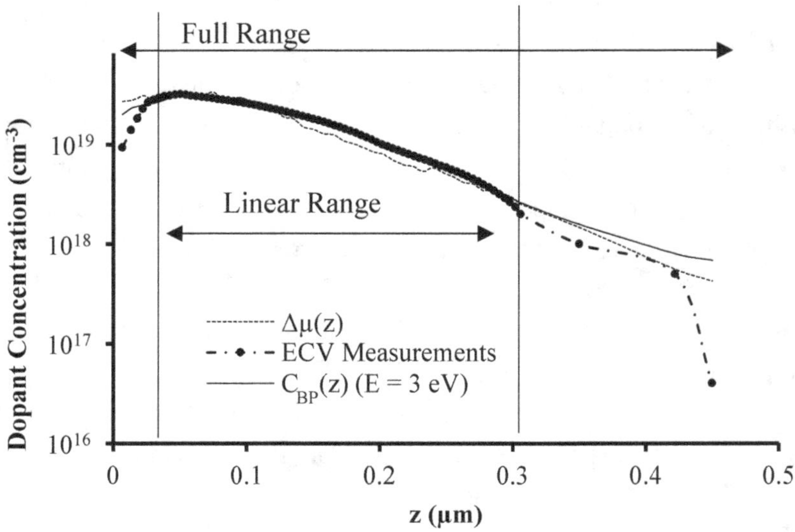

(a)

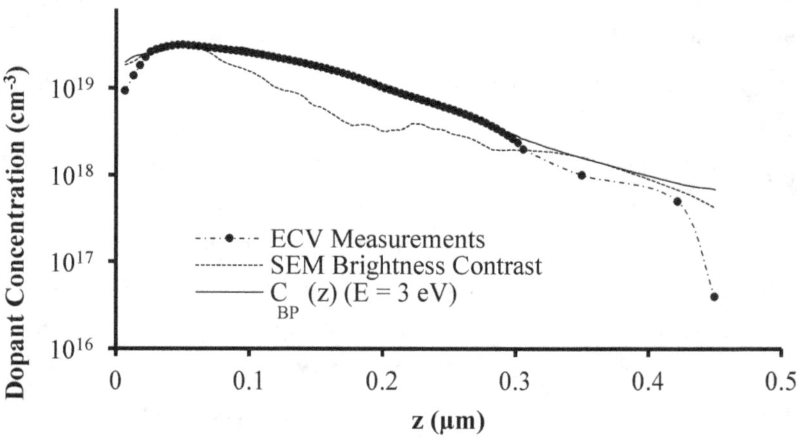

(b)

Fig. 4-20. Experimental SEM depth dopant profiles compared to ECV measurements. (a) Average energy and band-pass contrast profiles from the energy spectrometer. (b) Added data from brightness changes in the SE detector image.

Outside the Linear Range, the band-pass contrast and average energy shift profile remain relatively accurate, at around 5% RMSD. The only significant deviation to the ECV profile is at the edge of the *p*-doped layer (z = 450 nm). More information about this last data point from the ECV experiment is required before any conclusions about accuracy can be made. One possible explanation may come from the fact that this last ECV experimental data point lies within the Debye length, L_D, an effective screening length in a doped semiconductor defined by the following (Sah, 1991):

$$L_D = \sqrt{\frac{\epsilon k_B T}{q^2 N}} \tag{4.3}$$

where ϵ is the permittivity of the material (Si), k_B is the Boltzmann constant, T is the absolute temperature, q is the electron charge and N is the net dopant concentration. The dopant concentration measured by ECV at z = 450 nm is 1.2×10^{17} cm^{-3}, the net dopant concentration N is 1.1×10^{17} cm^{-3}, and taking into account the *n*-substrate, this gives a corresponding Debye length of 13 nm. This may explain the deviation mainly being isolated to z = 450 nm, and not present on the next data point at z = 422 nm, since the accuracy of the ECV technique is limited by the Debye length (Bock *et al.*, 2008).

Figure 4-20b displays the depth profile as extracted from brightness changes in the SE detector image. The match in this case to the ECV experimental profile data is significantly worse than the SE energy spectrometer band-pass contrast and average energy profiles, having an RMSD of 8.22%. These results demonstrate that with the right kind of specimen preparation, SEM dopant profiling using electron energy spectroscopy can give good agreement with a conventional dopant profiling method such as ECV on a real semiconductor device, like a solar cell, and that it gives better accuracy than the previous method of tracking brightness changes from a SE detector image. These results go beyond the kind of semi-quantitative SE detector image dopant profiling that was carried out on idealized test specimens in the past.

To understand how band-pass contrast and average energy dopant profiles are affected by recombination and contamination effects, parameters such as primary beam current and frame-scan time were varied. At high beam currents, the higher rate of carrier injection from the primary beam will increase electron-hole pair generation significantly in the semiconductor and affect measurement accuracy (Kazemian *et al.*, 2004), and there will also be higher levels of surface contamination. Figure 4-21 shows the SE band-pass contrast and average energy shift profiles at z = 50 nm for different primary beam currents. Both the SE band-pass contrast and the average energy shift decrease across the profile when the primary beam current is increased. At 108 pA, the profile RMSD error with respect to the reference profile at 40 pA is 6.7% for SE band-pass contrast, and 3.5% for average energy shift. At the much higher beam current of 330pA, the profile RMSD rose to 58.6% and 50.7% respectively. In practice, the acceptable range of the beam current was found to be approximately between 40 pA and 108 pA, which kept the profile RMSD of both the band-pass contrast and average energy profiles to within than a few percent.

It is instructive to compare the effects of increased total exposure time (at 30 s per frame) to those created by higher beam currents. Figure 4-22a shows the band-pass contrast profile for 0.5 mins (1 frame), 2 mins (4 frames) and 6 mins (12 frames). The band-pass contrast profile falls in height with exposure time. After 2 minutes of irradiation, the profile RMSD error from the original profile is 6.7%, while after 6 minutes the deviation increases to 22.5%. The increased exposure time effect on band-pass contrast profile is not as extreme as it is for the primary beam current: a primary beam current increase of 8.25 gives a profile RMSD of 58.6%, while the exposure time increasing by a factor of 12 gives a profile RMSD profile of 22.5%. However, it still causes unacceptably large changes in the band-pass contrast profile (more than a few percent).

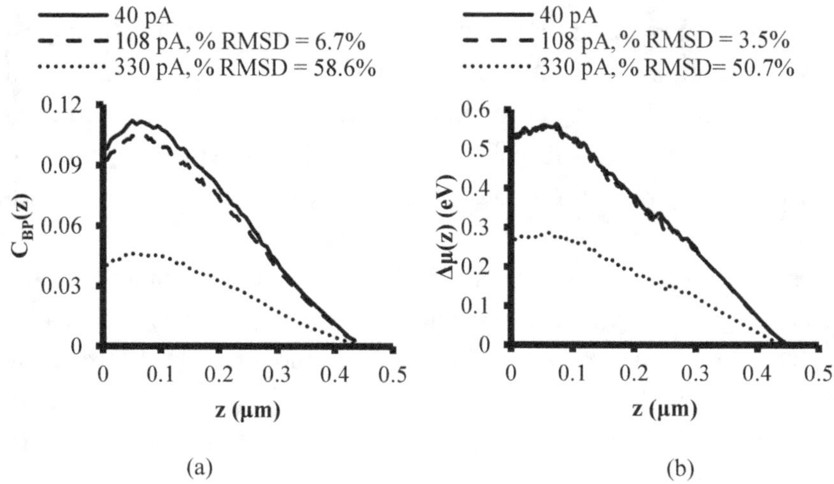

Fig. 4-21. Cross-sectioned depth profiling across the solar cell precursor sample obtained with different primary beam currents. (a) Band-pass contrast profile CBP(z). (b) Average energy shift profile Δμ(z).

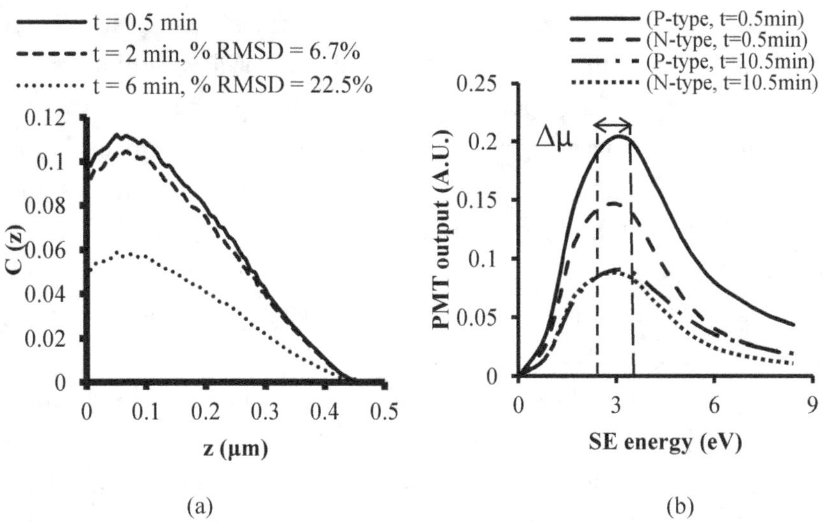

Fig. 4-22. Electron beam induced contamination effects as a function of frame scan time. (a) Band-pass contrast CBP(z) at different frame scan times t. (b) SE energy spectral signals obtained at *p*-type region (z = 50 nm) and *n*-type substrate at frame scan times t = 0.5 min and t = 10.5 min.

Another important aspect of the comparison between the effects of increased primary beam and exposure times is that it helps distinguish between recombination and surface contamination effects. Figure 4-22b shows how the SE energy spectral signal changes for increased exposure times. The spectral signal height falls significantly when the exposure time is increased by a factor of 21 (from 0.5 to 10.5 mins), which in turn, lowers the band-pass contrast function value. However, Fig. 4-22b also shows that the SE energy spectral signal shape remains largely unchanged, explaining why the corresponding average energy profile RMSD error is only 0.86% with respect to its initial profile. These results are consistent with the previous observations of surface contamination having a much greater effect on the SE energy spectral signal height, compared to its overall shape (average energy) for the stripe doped test samples (see section 4.5). In comparison, the average energy of the spectral signal is reduced significantly when there is an increase in primary beam current (Fig. 4-21b). The experimental results shown in Figs. 4-21 and 4-22 help to highlight the important role played by conductive paths of current down to the specimen base plate. These currents can exist both in the form of specimen surface currents, and a bulk current through the specimen. In this case, the maximum dopant concentration *p*-region area makes direct contact to a side metal holder through a metal coating. Increasing the number of scanned frames does not change the fact that there is still a 30 s time interval between frames, which gives time for charge to be transported to the side-metal holder, preventing any overall build-up of negative charge in the p-type region. This is why only surface contamination effects are observed by the SE spectral height being reduced, but the spectral signal's overall shape (and average energy) is largely unchanged. On the other hand, increasing the primary beam current directly increases the net amount of charge injected into the *p*-region, which gives rise to recombination, effectively reducing both the *p*-type dopant density and space-charge. This is why, in that case, both the spectral signal height and average energy shift are reduced.

4.9 Dopant mapping of intrinsic silicon irradiated by a gallium ion beam

Focused Ion Beam (FIB) instruments are widely used for micro-machining nanostructures. A focused ion beam, usually composed of Ga^+, is used for a variety of different tasks, such as milling, surface polishing, and deposition. As a natural part of performing these tasks, the ion beam is also used for imaging purposes, which means that many parts of the specimen surface are inadvertently irradiated by the beam, even in places where they do not undergo micro-machining. Depending on the primary ion beam voltage, current, and specimen, this may lead to unwanted side-effects. In silicon, it makes the surface amorphous and dopes it. At the lower dose levels used for focused ion beam imaging, these kinds of effects are not usually visible in the conventional SE detector image. This is therefore a natural application for SE energy spectroscopy, to see how much more information it can provide about the unwanted implanted ion distributions into a semiconductor specimen's surface while being imaged by a gallium ion beam.

A test pattern made of 5 μm square Ga^+ irradiation squares of different doses was defined on an intrinsic silicon bulk sample by a FEI Quanta 3D Focused Ion Beam (FIB)-SEM system, as shown in Fig. 4-23. The primary ion beam voltage was 5 kV. The beam scan frame rate was adjusted for each square in order to expose the silicon sample to different beam irradiation doses. A minimum dose of 334 μC cm^{-2} (D_0) was used, corresponding to an ion primary beam current of 0.1 nA and a scan time of 0.834s for an area of 25 μm^2. Figure 4-24a shows a beam-specimen interaction diagram generated from Monte Carlo simulation (SRIM program) for a 5kV Ga^+ ion striking the silicon sample surface, where Z indicates the distance from the specimen surface and X is the lateral distance. Figure 4-24b shows the predicted histogram of Ga^+ ions into the specimen as a function of depth. The average predicted depth is around 8.4 nm.

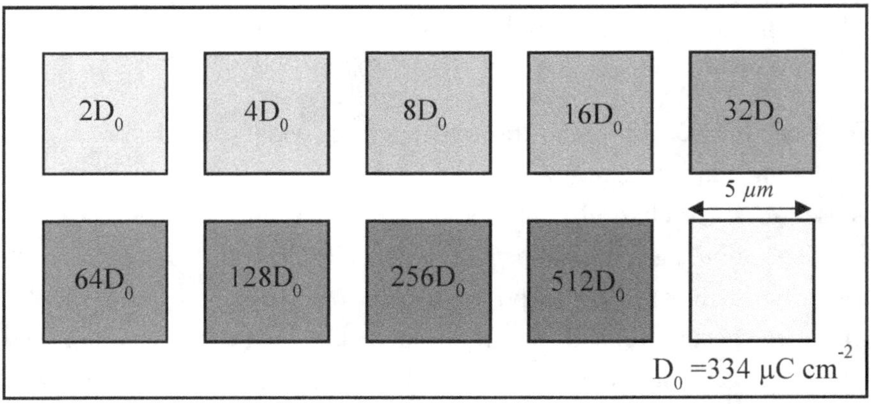

Fig. 4-23. Schematic diagram of Ga^+ beam irradiation areas/dose on an intrinsic bulk Si specimen. A 5 kV primary ion beam voltage was used.

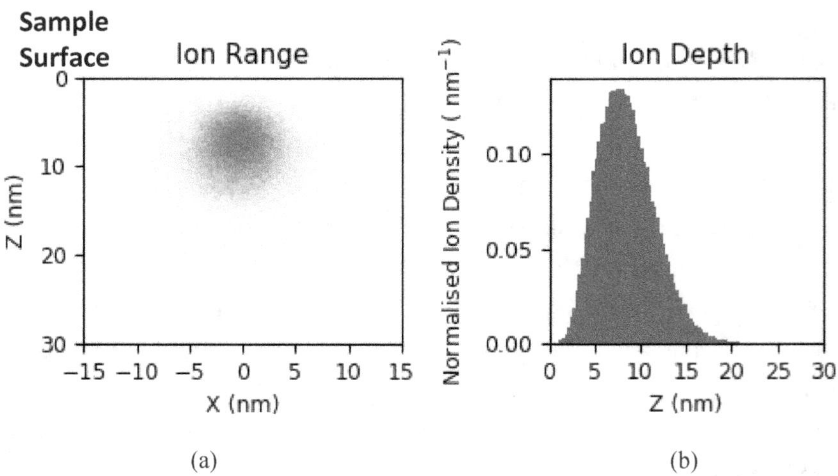

Fig. 4-24. Simulation from SRIM (and python plots) for Ga^+ ions (5 kV) in Silicon. (a) Beam interaction region, Z indicates the distance from the surface and X is lateral distance. (b) The predicted histogram of Ga^+ distribution as a function of depth. The average predicted depth is around 8.4 nm.

After ion beam exposure of the silicon sample, a 2 kV primary electron beam in the FEI Quanta 3D dual beam instrument was used together with the wide-angle toroidal energy analyzer attachment. The specimen/inner cap electrodes were biased to −10 V. Figure 4-25a shows the SE energy spectral signal obtained from the $8D_0$ dose square (2.67 mC/cm^2) in spot mode (over an area of a few μm^2 in the centre of the square) together with the one taken from the intrinsic silicon substrate, and Fig. 4-25b shows their corresponding normalized signals. The beam current used was 40 pA, and the spectral signal acquisition time was 0.3s per point for 100 data points along the spectral signal, giving a total time of 30s. These spectral signals have a similar form to those that were obtained on the bulk doped silicon samples (Fig. 4-16): as the level of doping goes up, the spectral signal height increases and its width narrows, causing its average energy and standard deviation to decrease. The cause for this comes from a surface band-bending effect, which gives rise to an effective change in the sample's electron affinity, demonstrated earlier in Figs. 4-12b–d. Since the escape depth of SEs from the silicon sample is approximately 2.7 nm (Lin and Joy, 2005), it is the net accumulation of positive charge over this depth that causes a change in SE energy spectral signal shape. The escape depth is about three times smaller than the average depth (8.4 nm) of the implanted ion beam as predicted by the Monte Carlo simulations shown in Fig. 4-24b.

Figure 4-26 shows quantitative dopant contrast data that has been extracted from SE energy spectral signals taken from the ion beam irradiated square areas shown in Fig. 4-23. Figure 4-26a plots the change in band-pass contrast function ΔC_{BP} at a pass energy of 2.66 eV (peak height position) as a function of dose (Log base 10), relative to intrinsic silicon, and Fig. 4-26b plots the corresponding spectral signal standard deviation.

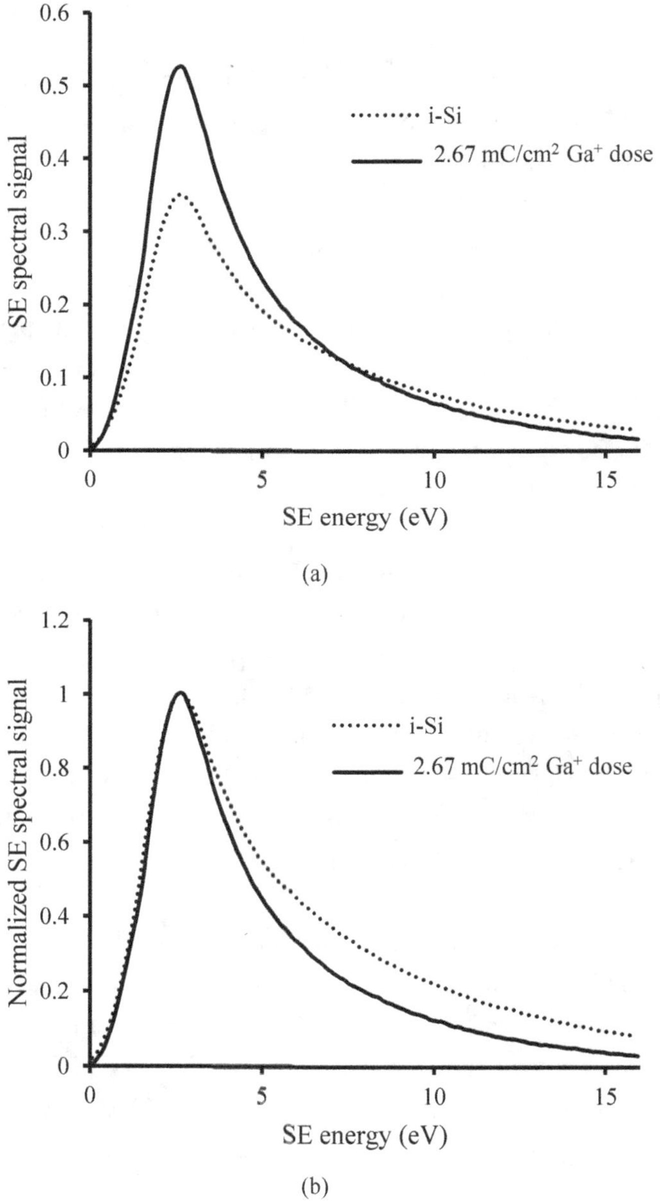

Fig. 4-25. SE energy spectral signals from a 2 kV electron beam probing a Ga^+ irradiated square area on a bulk intrinsic silicon sample ($8D_0$ dose square). (a) Output spectral signal. (b) Normalized SE energy spectral signal.

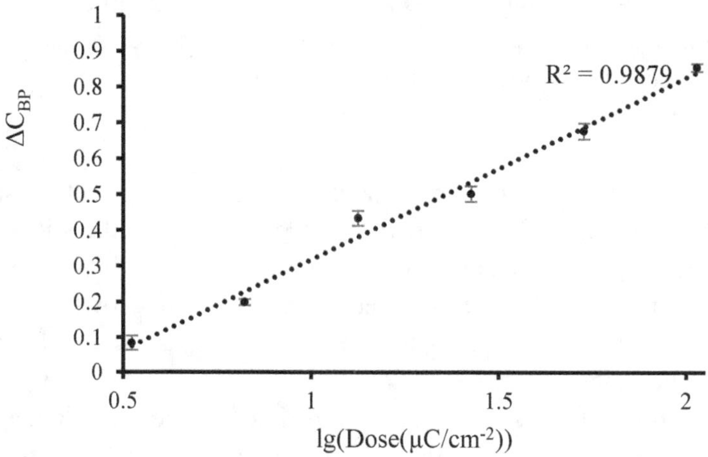

(a)

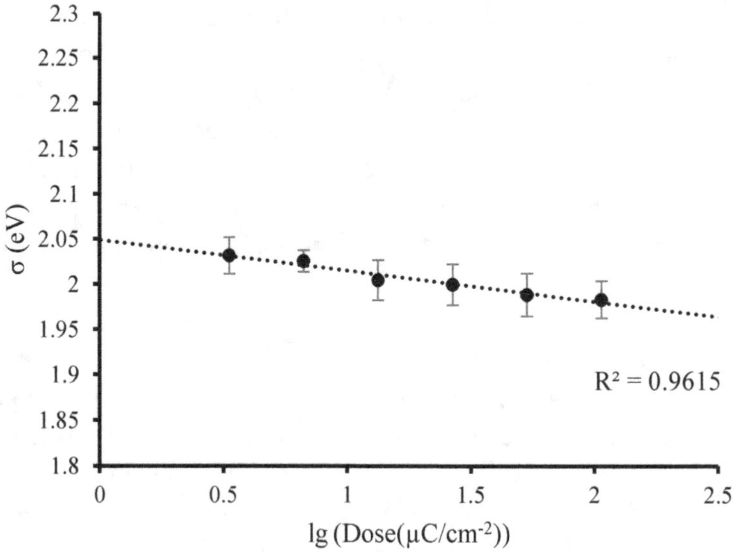

(b)

Fig. 4-26. Dopant concentration plots as a function of dose for a 2 kV electron beam probing Ga^+ irradiated square areas on a bulk intrinsic silicon sample. (a) ΔC_{BP}, band-pass contrast function at a pass energy of 2.66 eV relative to intrinsic Si. (b) σ, standard deviation of the SE energy spectral signal.

The quantitative dopant contrast variations shown in Fig. 4-26, like the previous quantitative data for doped bulk silicon samples, are well approximated by straight line variations, and therefore demonstrate that SE energy spectroscopy can be used to quantify the doping effect of a Ga^+ ion beam irradiated intrinsic silicon sample.

Figure 4-27 shows how the dopant contrast mapping information on the Ga^+ beam patterned irradiated silicon sample obtained by the wide-angle toroidal electron energy attachment compares with both the conventional SE detector image and a EDS map acquired from a windowless Oxford ultimate extreme system. Both the SE and band-pass spectral images were averaged over ten frames, with 16 s per frame, each having a total acquisition time of 160s. The acquisition time for the EDS image was 4 minutes (240s). The lower doped Ga^+ irradiated square patterns are more clearly visible in the band-pass function image, compared to either the SE detector image or the EDS map, indicating that there is a higher dopant contrast sensitivity advantage of SE energy spectroscopy over conventional SEM techniques. Note that there is a heightened contrast effect coming from the edges of the most irradiated square pattern, in both the SE detector and band-pass function images, which is absent in the EDS map. This most likely comes from a topographical effect, a height discontinuity caused by the Ga^+ beam sputtering the specimen's surface.

Another test irradiated pattern is shown in Fig. 4-28. In this case, the focused Ga^+ ion beam was used to define ring shaped patterns. The outer diameter is 5 μm, while the inner diameter of the three top rings from left to right are, 4.5 μm, 4 μm and 3.5 μm, giving corresponding ring widths of 0.25 μm, 0.5 μm, and 0.75 μm. A fixed concentration of 2.67 mC/cm^{-2} (8 D_0) was used. The band-pass contrast image (for one and ten frame averaging), shown in Figs. 4-28c and 4-28d, are compared with the standard SE detector image from the electron and Ga^+ primary beams in Figs. 4-28a and 4-28b. The band-pass images are able to provide much clearer definition of the ring pattern than the SE detector images.

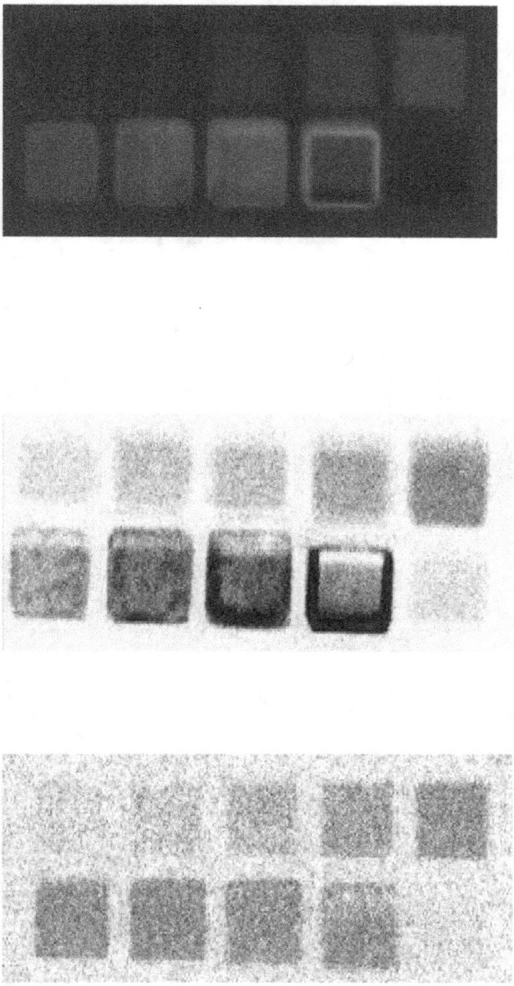

SE detector image

(a)

Spectral band-pass contrast (C_{BP}) image at 2.66 eV

(b)

EDS Map

(c)

Fig. 4-27. Dopant contrast imaging of Ga^+ beam irradiated areas on a bulk intrinsic Si sample with a 2 kV primary beam voltage. (a) Conventional SE detector image (averaged over 10 frames). (b) Band-pass contrast imaging for a pass energy of 2.66 eV (averaged over 10 frames). (c) EDS map from a windowless Oxford ultimate extreme system.

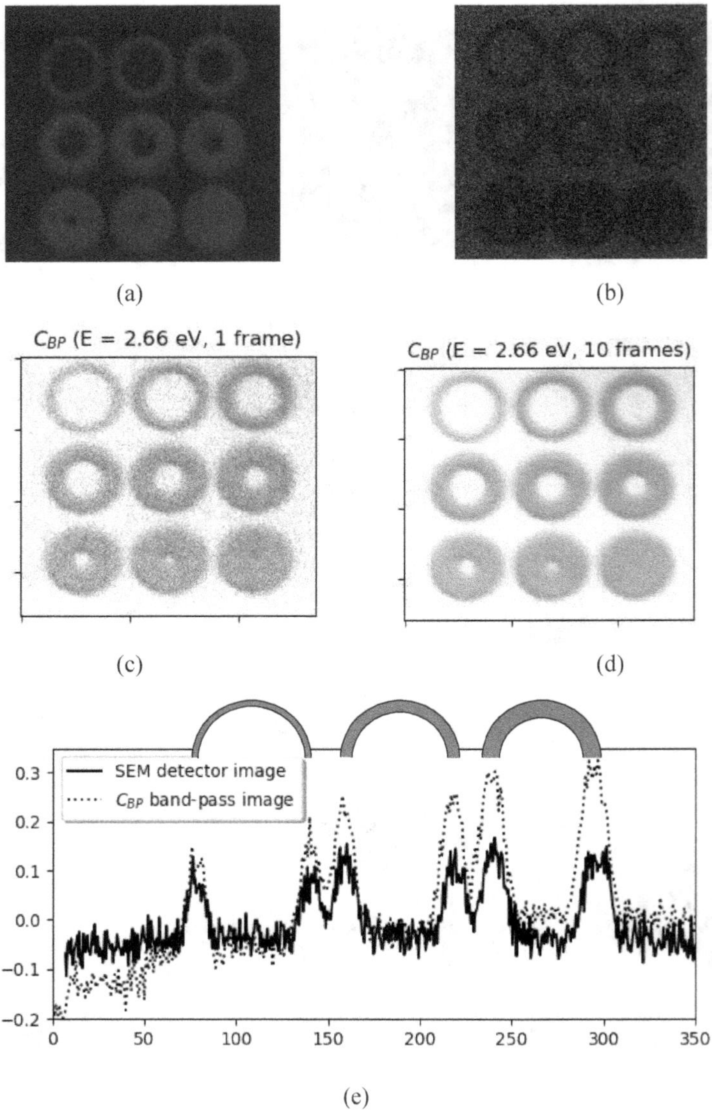

Fig. 4-28. Dopant contrast mapping of Ga^+ beam ring shaped irradiated areas. (a) Conventional e-beam SE detector image. (b) Ga^+ beam generated SE detector image. (c) Band-pass contrast images at 1 frame (pass energy of 2.66 eV). (d) Band-pass contrast images at 10 frames (pass energy of 2.66 eV). (e) Horizontal line-scans through top three ring centers.

Figure 4-28e shows horizontal SE detector line-scans from the electron beam and band-pass contrast averaged images over ten frames taken through the centres of the top three ring patterns. The band-pass contrast line–scan signal is around 3 times higher than the SE detector line-scan signal for the thicker ring (0.75 μm), but only marginally higher for the thinner ring (0.25 μm). The lower factor of improvement in the case of the thinner ring can be explained by the relatively large pixel size used in these images, around 0.082 μm.

The experimental results shown in Figs. 4-27 and 4-28 demonstrate that SE energy spectroscopy provides a way of enhancing the image contrast of FIB irradiated patterns on an intrinsic silicon sample inside a dual-beam FIB/SEM instrument. Its enhanced contrast is around a factor of three times better than the contrast of the conventional SE detector image. Overall, SE energy spectroscopy can both better quantify and visualize the amount of inadvertent irradiation that a silicon sample is exposed to inside a FIB/SEM dual beam instrument. The work reported here has obvious applications for monitoring +He primary beam ion irradiation of specimens inside the Helium Ion Microscope (Wirtz *et al.*, 2019).

Chapter 5

Probing and Mapping Charge Distributions

The probing and mapping of charge distributions in the SEM is an important application for SE energy spectroscopy. Primary beam induced charging can happen in a variety of different ways, but most commonly it occurs when observing specimens that contain non-conductive materials. It happens when a high voltage focused primary electron beam (> 5 kV) strikes an insulator, causing it to charge negatively, which subsequently deflects and defocuses the beam, making it very difficult to form a topographic image of the specimen's surface. The specimen surface can charge up to negative voltages that are comparable in magnitude to the primary beam voltage, and often leads to electrical breakdown.

One common strategy to avoid negative charging with high primary beam voltages is to coat the specimen with a metallic layer (usually gold), which is grounded. This succeeds in shielding the adverse field effects of the buried charge back on to the primary beam and at the same time provides a conductive path to ground that avoids the unstable build-up of charge. The method is widely used for observing biological, medical, and insulator specimens (Pawley, 2008), and is usually adequate for simple imaging applications. However, it does not prevent charge from being deposited into non-conductive regions, since the primary beam penetrates the top metal layer (usually only a few to tens of nanometers thick), nor does it provide any way of monitoring how much charge has been deposited into the specimen. Since the SE detector image is mainly dominated by SEs generated from within the top metal layer, charging contrast effects originating from below the specimen surface will be difficult to recognize or interpret. This is because there is no reliable way of separating charging artifacts from surface topography in the SE detector image. Charging artifacts not only depend on the type of specimen being observed, but also depend on operational parameters such as primary beam voltage/current and scan/exposure time.

Another common strategy for avoiding charging of non-conductive specimens is to vary the primary beam voltage and attain an approximate charge balance condition, a stable operating point where the specimen as a whole does not accumulate any net positive or negative charge. This is made possible because as the primary beam energy is changed, there are two energies for which there is a total scattered electron yield of unity, known as the crossover energies E_1 and E_2 (Wuhrer and Moran, 2016), where $E_2 > E_1$. They both typically lie in the 0.5 to 4 keV energy range. Figure 5-1 shows a schematic diagram of the total electron yield from the specimen as a function of primary beam landing energy. Above unity, the specimen charges positively, below unity, it charges negatively.

The second crossover energy, E_2, is the stable energy to which the primary beam voltage is usually tuned in order to attain an approximate overall charge balance condition, and its value depends on the instrument being used and the material being probed. The problem here is that although an overall charge balance condition can often be obtained, there is very little information available about the local charge distribution present on the specimen surface. The SEM SE detector image only gives broad qualitative contrast changes in which it is difficult to separate local charging effects from surface topography variations. This is another instance where the more quantitative information available through SE energy spectroscopy can help. By using SE energy spectral information, charging can be distinguished from surface topography, and local areas of accumulated charge on the specimen surface can be mapped out separately.

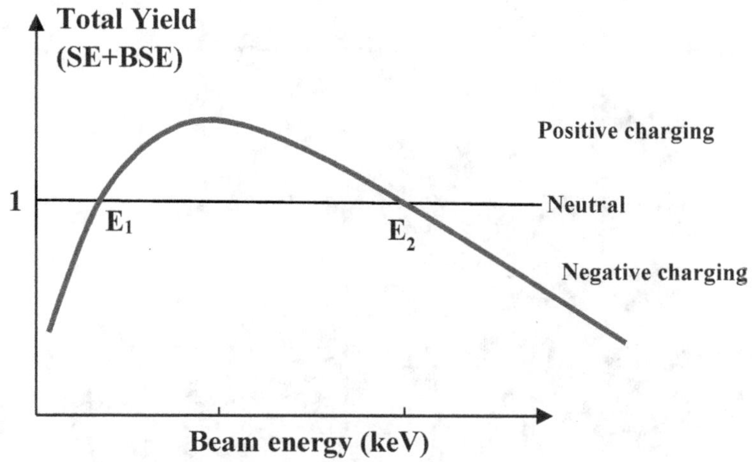

Fig. 5-1. Total electron yield as a function of primary beam landing energy. The crossover energies E_1 and E_2 represent an overall charge balance condition.

Belhaj *et al.* (2000) used SE energy spectroscopy to quantify extreme levels of negative charging of insulator samples in the SEM. They irradiated both spherical and flat Al_2O_3 insulators by a 20 kV primary electron beam in a Philips 505 SEM equipped with a standard Everhart-Thornley SE detector, and one of their SE images is shown in Fig. 5-2. It displays the appearance of various image artifacts superimposed upon a distorted SE image of the specimen surface. Belhaj *et al.* showed that if the primary beam voltage was then subsequently reduced to 10 kV, an image of the specimen chamber interior was obtained. They concluded that this mirror mode of operation had been caused by the specimen charging negatively, where the specimen voltage magnitude exceeded 10 kV. The fact that some of the anomalous contrast features in the 20 kV primary beam voltage SE image (Fig. 5-2), resembled some of the same topographical features that appeared in the 10 kV mirror image (not shown), provided evidence that a significant number of detected electrons in the 20 kV case came from tertiary electrons generated from the specimen chamber walls/column. This is consistent with the hypothesis that the specimen voltage was reaching large negative values, and thereby accelerating SEs to high energies (> 10 keV).

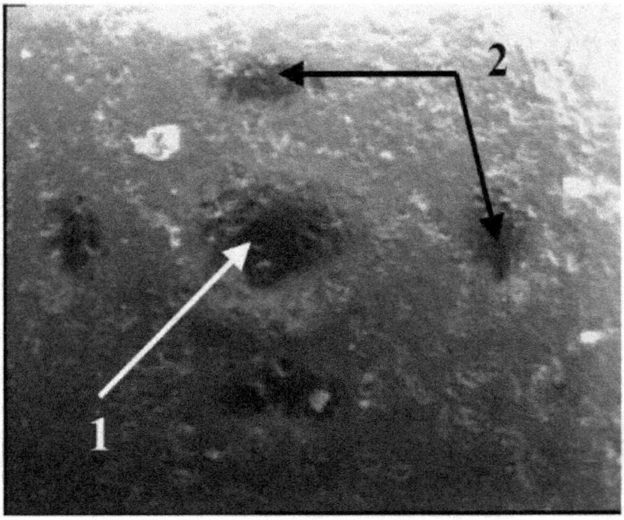

Fig. 5-2. Observation of a flat Al_2O_3 insulator sample at a primary beam voltage of 20 kV inside a SEM by Belhaj *et al.* (2000). The dark spot in the middle corresponds to the distorted view of the last output aperture of the SEM gun (1). The screw holes are also observed (2). Copyright 2000 by Hindawi. Reprinted with permission.

In order to acquire more quantitative information, beyond what was available in the SE detector image, Belhaj *et al.* used the small-angle toroidal band-pass energy analyzer attachment designed by Rau and Robinson (1996) to obtain energy spectral information. A schematic diagram of this energy analyzer attachment is shown in Fig. 5-3a. This analyzer has for the most part, been applied to BSE energy spectroscopy applications, and was used by Belhaj *et al.* in this instance to capture the energy spectrum of both SEs and BSEs under highly negative specimen charging conditions. The entrance polar angular spread and central polar angle of this analyzer design are relatively small, 20° ± 1°, compared to say that of the wide-angle toroidal energy analyzer (45° ± 8°). An array of 20 *pn* junction silicon diode detectors was used in the azimuthal angular direction at the analyzer exit plane. The experimental energy spectrum obtained by Belhaj *et al.* (2000) showing negative charging together with a schematic representation of an uncharged sample obtained by the small-angle toroidal energy analyzer attachment, is shown in Fig. 5-3b.

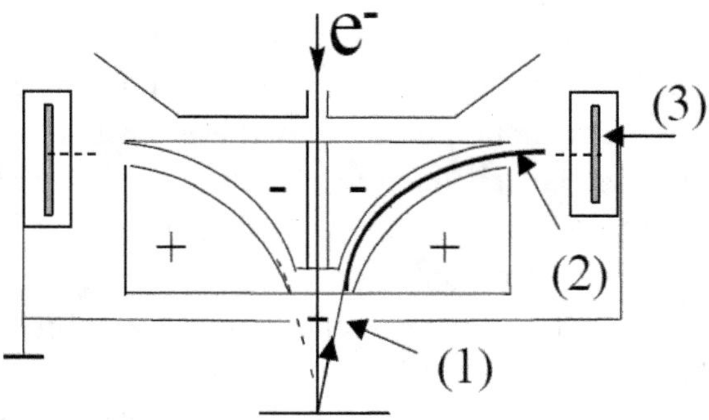

(a)

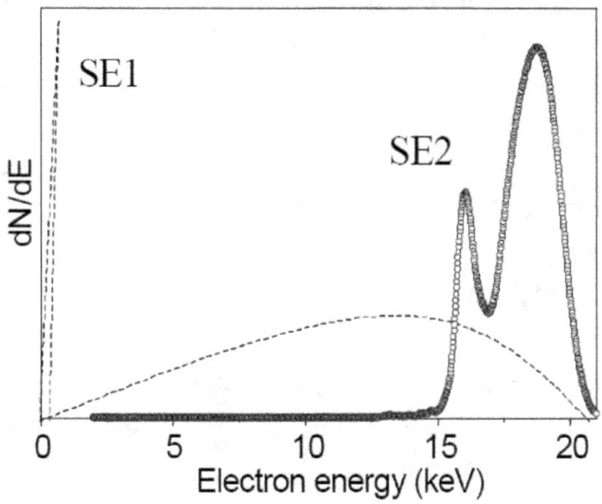

(b)

Fig. 5-3. Use of the small-angle toroidal band-pass energy analyzer attachment designed by Rau and Robinson (1996) to study charging of an Al_2O_3 insulator sample at a primary beam voltage of 20 kV inside a SEM by Belhaj *et al.* (2000). (a) Schematic illustration hollow cone of emitted electrons at polar angle range of 20° ± 1° entering at the input annular diaphragm (1), traveling between two toroidal electrodes at the potentials, +V and –V, going on to output annular gap (2), and detector (3). (b) SE energy spectrum. Dashed line indicates charge-free condition (schematic drawing), and solid line represents an experimentally measured spectrum. Copyright 2000 by Hindawi. Reprinted with permission.

Figure 5-3b shows that there is a shift of the secondary electron peak from its uncharged initial value of a few eV, to around 16 keV. The large shift in SE energy peak position confirms the runaway unstable nature of negative charging of insulator specimens when irradiated by high voltage electron beams. A charge balance condition is only reached once the landing energy is comparable to the second crossover E_2 energy, which in this case, requires the specimen to charge to approximately −16 kV. These levels of negative voltage create a strong electric field distribution between the specimen and its zero volt surroundings in the SEM specimen chamber, effectively preventing the possibility of obtaining any meaningful information about the specimen from the SE detector image. The small-angle band-pass SE energy toroidal attachment in this extreme charging condition is able to provide an estimate of how much negative charging is occurring. In principle, a much more reliable SE image could of have been obtained using the SEs that pass through the spectrometer at or around the energy of 16 keV, but this was not reported by Belhaj *et al.* (2000).

The work carried out by Belhaj *et al.* (2000) demonstrates how SE energy spectroscopy can be used to quantify extreme levels of negative charging inside the SEM, where a high voltage primary beam directly irradiates a non-conductive specimen. It demonstrates that SE energy spectroscopy not only provides a way to quantify the amount of charging that is occurring, but it also helps to better understand and interpret puzzling artifacts that appear in the SE detector image. In these types of situations, a large negative potential on the specimen accelerates SEs to have relatively high energies as they pass through the analyzer, therefore, this kind of SE energy spectroscopy is very similar to BSE energy spectroscopy. This is why the small-angle toroidal energy analyzer attachment used by Belhaj *et al.* (2000), although having many limitations, such as a relatively small entrance polar/azimuthal angular range, or a restricted detection energy range (> 2 kV), was still able to acquire useful SE energy spectral information for this application.

5.1 Monitoring localized charging in metal-insulator-semiconductor (MIS) samples

Srinivasan *et al.* (2018) used the wide-angle toroidal energy analyzer attachment to monitor controlled local negative charging occurring in a metal-coated insulator specimen. Their specimen, consisted of a test metal-insulator-silicon (MIS) structure, where a top gold layer (thickness between 10–30 nm) was formed on to a silicon nitride 200 nm thick insulator layer, which had been deposited on a 500 μm thick silicon substrate, as shown in Fig. 5-4a. The specimen was placed on to a grounded metal base plate, and there was the option for grounding or leaving the top metal layer floating. All experiments were conducted in a FEI XL30 FEG SEM.

In a metal-insulator-semiconductor (MIS) specimen with a floating top metal layer and grounded base-plate, as shown Fig. 5-4b, there will be a certain amount of buried charge deposited into the insulator, but most of the primary beam will be able to go through the thin insulator layer (200 nm thick) and reach the semiconductor substrate. Once there, it will excite electrons into the conduction band and produce a sizeable specimen current, providing the possibility to create controlled stable charging conditions. A first set of experiments in the SEM was carried out to see if any charging was observable in the SE detector image by irradiating the specimen with a 10 kV primary beam in spot mode. Subsequent SE images were taken with the standard Everhart–Thornley detector around the irradiated spot, with a 6 s/frame slow scan over an area of 3 × 2.6 μm. The irradiation time and beam current in spot mode were systematically increased, and the thickness of the gold layer, d_{AU}, was varied. When the top layer was grounded, no charging was observed in the SE image. However, for a floating top metal layer, under the right conditions, a bright area in the image was observed, as shown in Fig. 5-5. This indication of buried charge appeared to saturate with time, and after the primary beam was switched off, the bright charged area in the image disappeared gradually, taking around 15 min to disappear completely.

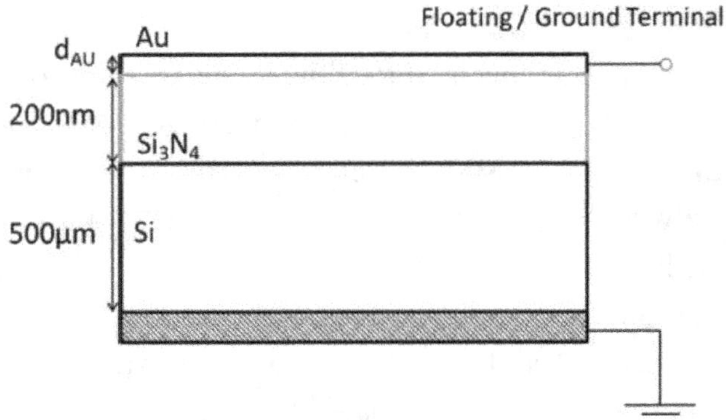

(a)

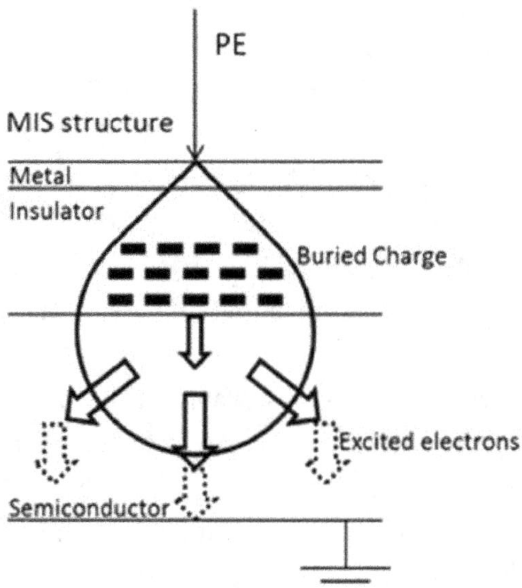

(b)

Fig. 5-4. The metal-insulator-silicon (MIS) sample characteristics. (a) Sample geometrical layout. (b) Primary beam induced scattering and charging.

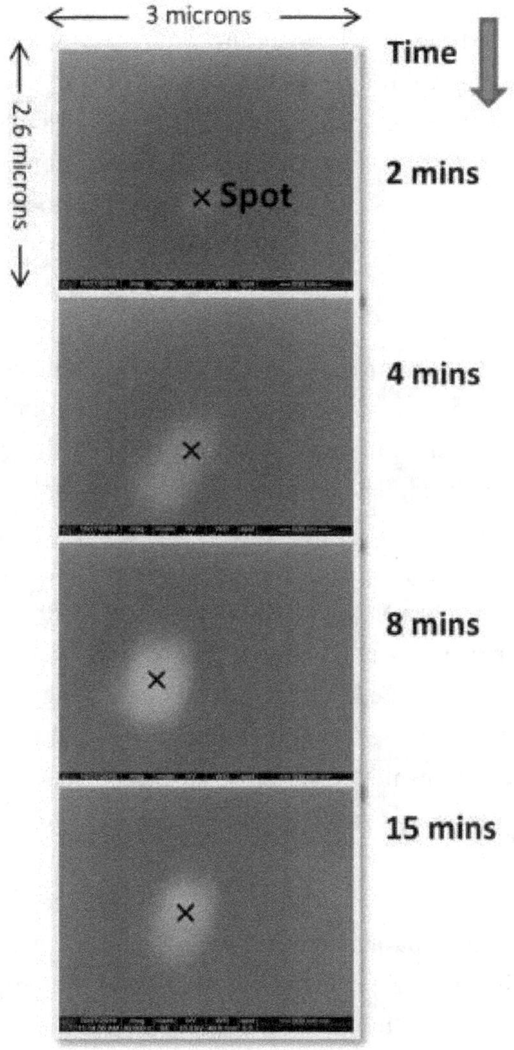

Fig. 5-5. Build-up of buried charge in the secondary electron image with exposure time for the test metal-insulator-semiconductor specimen inside a scanning electron microscope with a primary beam voltage of 10 kV in spot mode. The cross symbol represents the position of the primary beam scan area.

After obtaining the preliminary SE detector images of controlled charging in the MIS test specimen shown Fig. 5-5, more quantitative results were acquired through the use of the wide-angle toroidal energy analyzer attachment. The specimen and inner cap electrode were biased to –10 V in order to maximize its output signal-to-noise characteristics, and the 10 kV primary beam was kept in spot mode. 72 points per spectral signal were taken over a range of 0–20 eV with a dwell time of 0.125 s per point.

Figure 5-6 shows experimental energy analyzer signals from the MIS specimen for a variety of different conditions. They indicate that under specific experimental conditions, a second peak in the energy analyzer signal can be observed. The first peak in these signals is located to be around 2.8 eV, and the second peak appears around 6 eV. While surface contamination effects are present in these experimental results, they do not change the SE energy analyzer signal shape; they only cause relatively minor changes in SE signal height with irradiation time. This can be clearly demonstrated by electrically connecting the top gold layer to the same voltage that is biasing the specimen substrate (−10 V), which then effectively screens the presence of the buried charge and produces a SE energy analyzer signal that is similar to a bulk gold sample, as shown in Fig. 5-6a.

The results in Fig. 5-6a also show that although contamination effects cause the SE signal height to fall with irradiation time, most of the drop occurs within the first 5 minutes, falling by around 7%, beyond that, the SE signal appears to be stable. The floating gold SE analyzer signal was acquired just after irradiation (zero time), its increased height and second peak is shown alongside the contamination effects in Fig. 5-6a, indicating that contamination effects in this case play a relatively minor role.

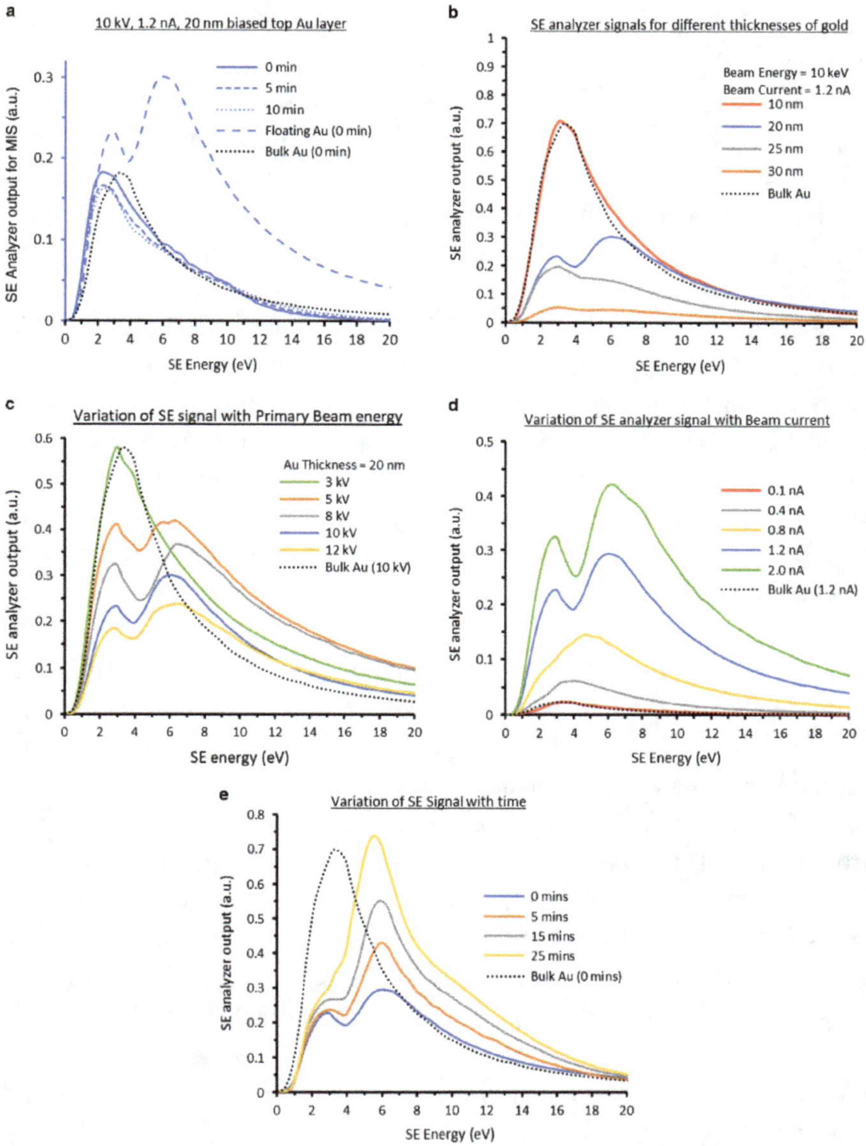

Fig. 5-6. Experimental energy analyzer signals for the metal-insulator-semiconductor (MIS) specimen (Srinivasan *et al.*, 2018). Bulk Au signal normalized accordingly for comparison. (a) Contamination effects illustrated by biasing the top Au layer (screening the buried charge); (b) as a function of top gold layer thickness; (c) as a function of primary beam voltage; (d) as a function of primary beam current; and (e) as a function of irradiation time.

Figure 5-6b indicates that at a 10 kV beam voltage, there is an optimum thickness of the top gold layer at which a second peak appears (at 20 nm), while Fig. 5-6c demonstrates that the second peak only starts appearing for beam voltages above 3 kV (for the 20nm thickness top gold layer). Figure 5-6d indicates that the appearance of the second peak is also dependent on the primary beam current being above a certain value (1.2 nA for a beam voltage of 10 kV and 20 nm gold layer thickness). The effect of the second peak, although present, is too small to be visible below a beam current of 0.8 nA. Figure 5-6e shows how the second peak rises progressively with primary beam irradiation time. The analyzer signal in this case was directly acquired after constant irradiation intervals (5 minutes in primary beam spot mode). The 5 minute interval irradiation time was significantly larger than the energy analyzer signal acquisition time (12 s). The position of the second peak in Fig. 5-6e remains approximately at 6 eV as its height grows with irradiation time.

Srinivasan *et al.* (2018) performed Monte Carlo simulations in order to better understand the experimental results shown in Fig. 5-6. They indicated that SE1s, the SEs generated directly by the primary beam, escape from an area approximately equal to the primary beam probe area (5 nm beam diameter), while SE2s, the SEs generated from BSEs, escape from a much larger area, one that has a diameter of tens of nanometers. The simulations also indicated that the primary beam scatters over an area of tens of nano-meters as it crosses the insulator interface (located at 20 nm below the specimen surface), suggesting that the transverse dimensions of the buried charge will also be approximately this size.

Further Monte Carlo simulations were used to analyze the fraction of electrons stopped as a function of depth. These results predicted that the fraction of stopped electrons increases steadily inside the insulator, and is higher for a thicker gold top layer. This is due to the fact that the primary beam interacts more with the specimen as it slows down and loses energy. The amount of buried charge inside the insulator is expected to rise in accordance with the fraction of number of electrons stopped inside the insulator. On the other hand, SEs that escape from the specimen's surface typically come from a depth of around 0.5 nm, the escape depth for gold (Lin and Joy, 2005), and a thicker top metal layer also results in a greater screening effect of the buried charge. This may explain why there is, in practice, an optimal thickness for creating buried charge, as indicated by the experimental results shown in Fig. 5-6b.

The dependence of buried charge SE contrast as a function of primary beam voltage as shown in Fig. 5-6c is straightforward to explain. Below a certain primary voltage, 3 kV in this case, a relatively small fraction of the primary beam electrons reach the insulator layer, and it therefore does not generate a significant amount of buried charge (indicated by the absence of a second peak in the energy analyzer signal). As the primary beam voltage increases, more of the primary beam electrons penetrate deeper into the sample and the SE yield therefore also reduces.

Srinivasan *et al.* (2018) presented a simple model to explain how buried charge can create contrast in the SE energy analyzer signal. A buried charge layer in the insulator will tend to increase the local density of free electrons at the specimen surface at and around the primary beam position through electrostatic induction, as illustrated in the schematic diagram shown in Fig. 5-7a. Assuming that the deposited charge extends to several tens of nanometers in the transverse direction, the increase in the local density of surface electrons will also extend over tens of nanometers across the specimen surface, and there will then be a preferential rise in the number of SE2s compared with SE1s, since the SE2s emerge from a much larger surface area than SE1s (schematically shown in Fig. 5-7b). In addition, as already noted in section 2.2, when the specimen/inner cap electrodes are biased, the lower energy part of the SE energy spectrum is dominated by a defocussing effect and the presence of a low energy background signal coming from internal scattering on the inner cap electrode and analyzer exit aperture. This may help to explain why there is hardly no specimen charging contrast effects that are observable in the first SE peak, but occur mainly in terms of the formation of a second higher energy peak.

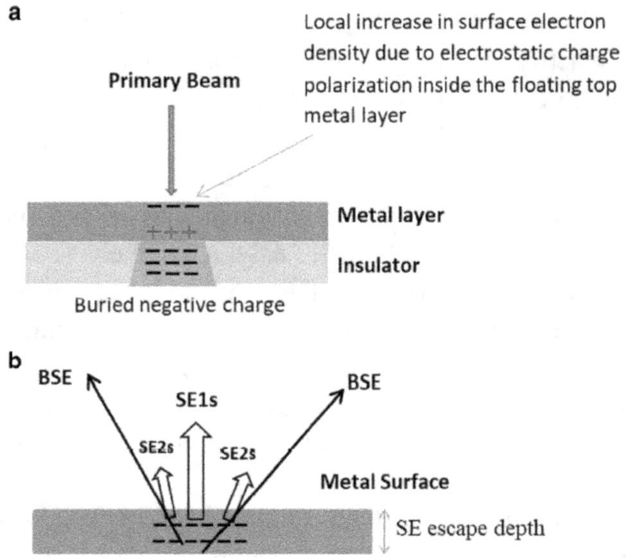

Fig. 5-7. Simple model for the secondary electron (SE) buried charge contrast. (a) Increase in electron surface local density due to electrostatic induction. (b) Relative enhancement of SE2s due to their higher surface area of emission.

Another useful application for SE energy spectroscopy is to study the kind of local charge distribution created in a MIS specimen when observing it under approximate charge balance conditions, where there is no visible charging observable in the SEM SE detector image. The test sample chosen for this study is a specially prepared MIS structure, where a top metal pattern exposes underlying square areas of a silicon nitrite layer, 300 nm thick, prepared on silicon substrate. These samples were fabricated in the ISI research institute, Brno, Czech Republic. The top metal layer consists of a 20 nm gold layer on 2–3 nm thick titanium, while the silicon substrate has a thickness of 525 μm. SEM micrographs imaging this test MIS structure at different primary beam voltages are shown in Fig. 5-8. The width of the larger tracks is 10 μm each, while the width of the smaller tracks is 1 μm each.

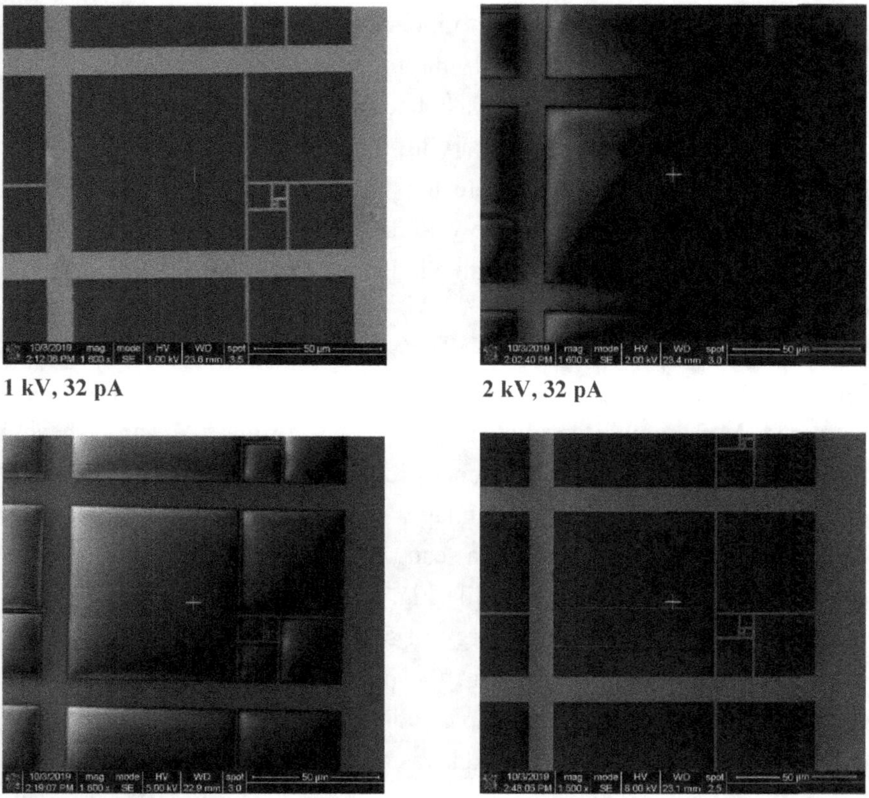

1 kV, 32 pA **2 kV, 32 pA**

5 kV, 56 pA **8 kV, 35 pA**

Fig. 5-8. SEM Images of a patterned MIS test sample at different primary beam voltages. Magnification of 1600x and scan time of 12s per frame. The width of the larger tracks is 10 μm each, while the width of the smaller tracks is 1 μm each.

The images in Fig. 5-8 show that there is no visible charging for the primary beam voltages of 1 kV and 8 kV, while negative charging occurs for the primary beam voltages of 2 and 5 kV. The 8 kV image displays a lower level of brightness and topographic contrast compared to the 1 kV image. This is consistent with the MIS results obtained earlier (see Fig. 5-6c), where maximum charging was caused by the primary beam voltage being in a certain range. In this instance, the explanation for there being no charging for the 8 kV primary beam voltage is that the beam energy is too high in order for it to interact with the silicon nitride layer. A primary beam voltage of 1 kV therefore represents the approximate charge balance condition. This was checked by alternating between low and high magnifications, and observing that there was no visible change in the brightness of the SE detector image. At other primary beam voltages, alternating between low and high magnifications produced distinct brightness changes, an indication of non-stable charge conditions.

Figure 5-9 shows a comparison of the SE detector image of the test patterned MIS sample irradiated by a 1 kV primary electron beam, alongside experimental results obtained from the wide-angle toroidal energy analyzer attachment. The images shown here were averaged from 16 other images, each taken with a scan rate of 12s per frame. Figure 5-9a shows the SE detector image, while Fig. 5-9b is the corresponding band-pass contrast image for a SE energy of 5.9 eV, and Fig. 5-9c depicts SE energy spectral signals taken from the points A, B, C, D and E on the specimen, as marked on the band-pass contrast image of Fig. 5-9b. The SE energy spectral signal for each pixel was acquired in image mode, where a series of images were taken, each with a different analyzer deflection plate voltage (ranging from −6 to −13 V in 1 V steps). Between each deflection plate voltage setting, after acquiring 16 images, the specimen was scanned at higher scan speeds in low magnification mode for 30s. This procedure was found to be an effective way of resetting the build-up of charge that had occurred while acquiring each set of 16 images. In each image, 500 × 242 pixels were used.

The results shown in Fig. 5-9c indicate that charging in the silicon nitrite region causes the SE energy spectral signal to increase in height and shift to the right. These effects combine to produce bright regions in the band-pass contrast image (Fig. 5-9b). There was no visible sign of charging occurring in the SE detector image (Fig. 5-9a).

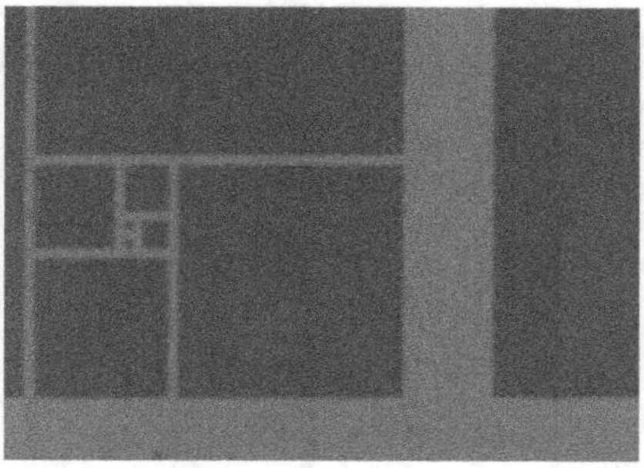

(a)

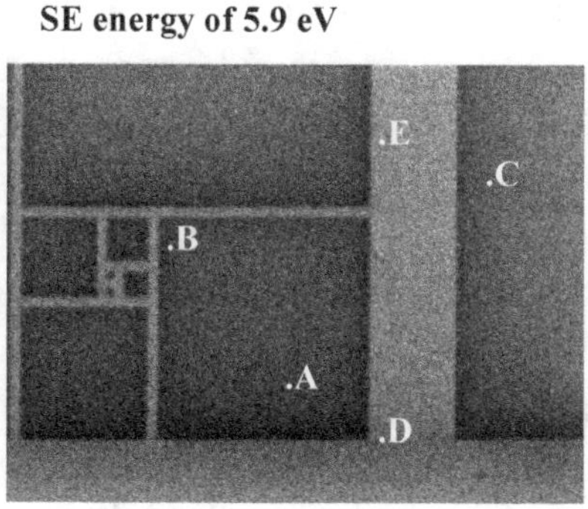

(b)

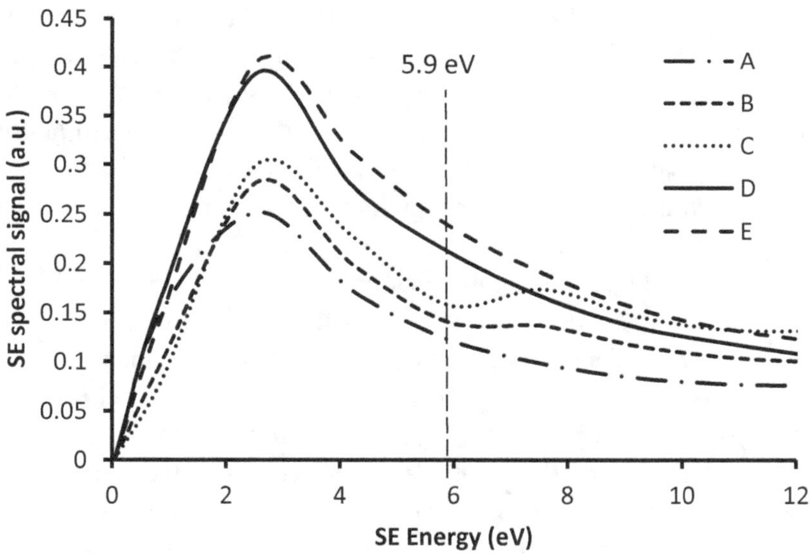

(c)

Fig. 5-9. SE energy spectral imaging with the wide-angle toroidal energy analyzer attachment for a patterned MIS test sample at a 1 kV primary beam voltage. (a) Conventional SE detector image (no visible charging). (b) Band-pass filtered energy image taken at a SE energy of 5.9 eV. Different locations denoted by points A, B, C, D, and E. (c) Experimental SE energy spectral signals for the different points, A, B, C, D and E on the sample (shown in (b)).

The experimental results shown in Fig. 5-9 show that although a charge-free condition is indicated in the conventional SE detector image, SE energy spectroscopy is able to demonstrate that some degree of local charging is actually occurring. Figure 5-10 shows how this small amount of charging can be further quantified and mapped by tracking changes in the spectral signal's average energy μ. Figures 5-10a and 5-10b show histograms of the average energy μ for gold and silicon nitrite respectively, extracted from the SE energy spectral signals from all pixels in the image. Overall, the average energy values for silicon nitrite regions are higher than those from the top layer gold regions. This provides a way for mapping the charge distribution in the silicon nitrite region and distinguishing it from gold regions by color coding, as shown in Figs. 5-10c and 5-10d. The color coding is adjusted so that green to white corresponds approximately to a μ range of the gold (7.75 to 8.25 eV), while white to blue corresponds to the μ range of silicon nitrite (8.8 to 9.3 eV). The intermediate white region for μ is extended in order to clearly separate charging effects from differences created in the SE energy spectral signal average energy μ caused by material contrast.

The color coded local charge distribution maps shown in Figs. 5-10c and 5-10d indicate that the larger areas of silicon nitrite appear to charge more negatively (darker blue), which is expected, since those areas are more exposed to the primary beam. Silicon nitrite areas close to the metal appear whiter, indicating that their proximity to the top layer metal results in less charging. This is also expected, since the much greater mobility of electrons in the metal has a screening effect that will tend to neutralize charge build-up. There is also clearly a directional effect present at the edges of the metal tracks, where one side appears dark blue, and the other side appears dark green. This is most probably caused by the fact that the analyzer is located to one side of the specimen. However, the most important aspect of these results is that they show how local charge distributions created by LVSEM observation can be monitored and mapped through SE energy spectroscopy, even when no charging is visible in the conventional SE detector image.

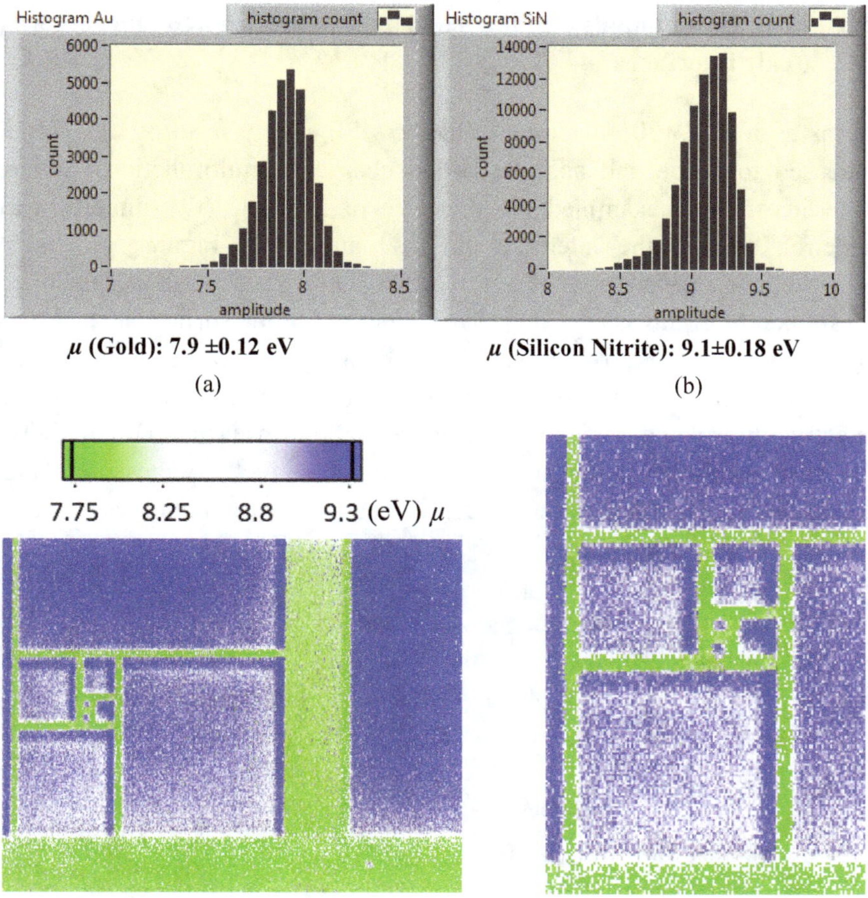

Fig. 5-10. Charging distribution extracted from SE energy spectral image data in terms of average energy μ per pixel. (a) Histogram of μ (gold). (b) Histogram μ (silicon nitrite). (c) Color coded charge distribution map plot from average energy μ extracted from the SE energy spectral signal for each pixel. Each pixel is first sorted by SE spectral signal height (threshold value) and then its μ scaled appropriately so gold occupies a lower range (green) and silicon nitrite occupies a higher range (blue). (d) Magnified image of an area in (c).

5.2 Probing and analyzing buried charge at multifunctional oxide interfaces

Srinivasan *et al.* (2014) demonstrated how SE energy spectroscopy could be used to probe and analyze buried charge at multifunctional oxide interfaces. They examined the presence of a 2DEG (two-dimensional electron gas) at the interface of LAO and STO. Figure 5-11 is a schematic representation of the LAO/STO hetero-interface specimen examined by Srinivasan *et al.* (2014). One half of the surface area of this sample was exposed to a collimated high energy proton beam, while the other half was shielded from it. The exposed and unexposed regions were electrically isolated, which was confirmed by the absence of conductivity between the regions when subjected to direct conductivity measurements using Al wire bonding to the interface. The proton irradiation causes the embedded interface to become insulating, while the other half, not exposed to the high energy proton irradiation, remains conductive. The conductivity of the interface is controlled by adjusting the fluence of the proton beam. However, for the irradiated samples investigated, the interface was made completely insulating. More details about this specimen are given by Mathew *et al.* (2013).

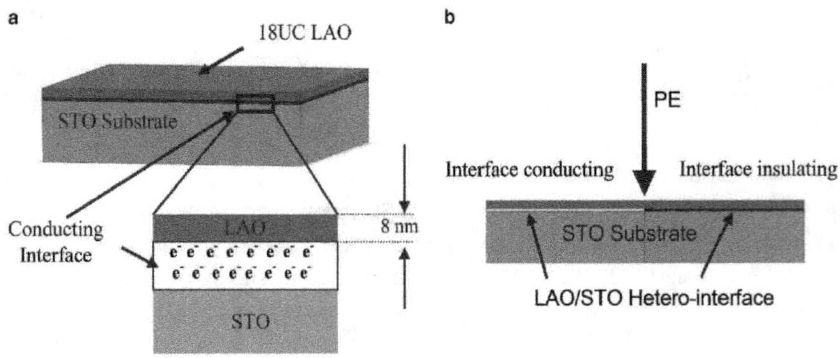

Fig. 5-11. Schematic representation of the specimen. (a) Representation of the 2DEG formed at the interface of crystalline LAO and crystalline STO substrate. (b) Conducting and insulating interface regions side by side on the same sample (the interface was made insulating by proton irradiation). 2DEG, two-dimensional electron gas; LAO, $LaAlO_3$; STO, $SrTiO_3$.

The wide-angle toroidal energy analyzer attachment was used with both the specimen and inner cap electrode biased to −10 V. The deflection voltage was ramped in steps of 200 mV and each point had a primary beam dwell time of ~150 ms. The SE analyzer signal was collected from regions that were exposed to the primary beam for the first time in order to minimize the effects of contamination. All experiments were carried out inside a Philips XL30 ESEM-FEG with primary beam energies ranging from 2 to 5 keV. The deflection voltage was varied from 6 to 25 V, corresponding to SE energies from 0 to 31.75 eV, and analyzer pass energies of 10 to 41.75 eV.

SE analyzer spectral signals were obtained separately for the conducting LAO/STO hetero-interface specimen and an uncoated STO substrate, as shown in Fig. 5-12. The SE analyzer signal obtained from the STO substrate contains a single peak, as expected. However, the SE analyzer signal taken from the surface directly above the conducting LAO/STO hetero-interface shows the appearance of a second high amplitude SE peak. The first peak is similar to the one in the uncoated STO spectral signal. The second peak occurs at a higher deflection voltage (SE energy) in comparison to the first SE peak and has an amplitude that is almost twice as high, indicating the presence of a high SE yield region.

SE analyzer signals were then obtained from surfaces directly above both the conducting and insulating LAO/STO parts of the hetero-interface, as shown in Fig. 5-13. Both these SE analyzer signals have first and second peaks, however, while their first peaks have similar amplitudes and energy positions, their second peaks differ significantly from one another. The second peak amplitude of the conducting interface spectral signal is around 2.5 times greater than the corresponding amplitude for the insulating interface spectral signal. On the other hand, the energy position of the second peak in the insulating interface is around 1.5 times larger than the second peak energy position for the conducting interface spectral signal.

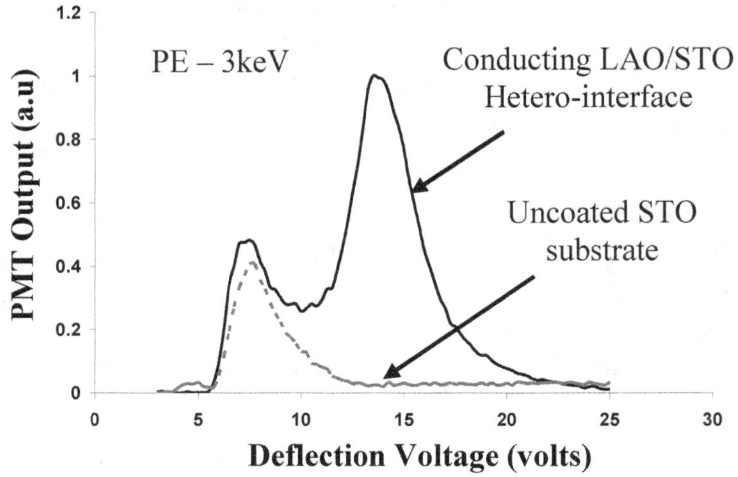

Fig. 5-12. Experimental SE analyzer signals obtained from an uncoated STO substrate (shown in dotted line) and from the LAO/STO heterointerface with conducting interface (solid line). A primary beam acceleration voltage of 3KV was used.

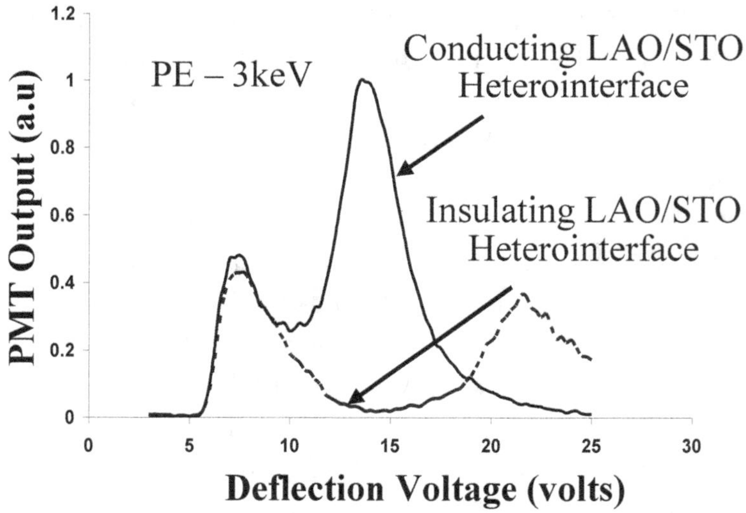

Fig. 5-13. Experimental SE analyzer signals obtained from the LAO/STO heterointerface with conducting interface (solid line) and the insulating LAO/STO heterointerface (dotted line). A primary beam acceleration voltage of 3 kV was used.

The experimental results shown in Fig. 5-13 demonstrate how SE energy spectroscopy is able to not only quantify the amount of charge trapped at the LAO/STO hetero-interface, but also capture the energy with which it is bound. Trapped charges are emitted due to interactions with the primary beam and backscattered electrons (BSEs), and come out with energies higher than the other SE electrons from the cascade. In the case of the SE spectral signal obtained from the specimen when the interface becomes insulating compared to its initial conductive state, the second SE peak is much reduced in height and shifts further to the right, indicating that there are now less trapped electrons at the interface and that they are more difficult to remove.

The high contrast between the second peak of the SE analyzer signal obtained from the conducting and insulating parts of the LAO/STO interface can only be attributed to a change in the electrical properties of the interface, because the physical structure of the specimen (film thickness, material density) remains unaltered by the high energy proton irradiation. These results suggest that SE energy spectroscopy in the SEM can provide a contactless way of monitoring charge present at the interface of thin film oxides.

The primary beam energy was found to be an important parameter in maximizing contrast in the SE analyzer signals on the LAO/STO hetero-interface sample. Figure 5-14 shows SE analyzer signals obtained from the surface of the specimen above the conducting LAO/STO interface at various primary beam energies. The SE analyzer signal has a high dependence on the primary beam energy; the distinct high amplitude second peak occurs only at a particular value of primary beam energy, 3 keV in this case. At the primary beam energies of 2 and 5 keV, the second peak is not present. At 4 keV, the second peak does occur in the SE analyzer signal, but it has an amplitude that is reduced to around 75% of the 3 keV signal value, and its peak shifts to the right.

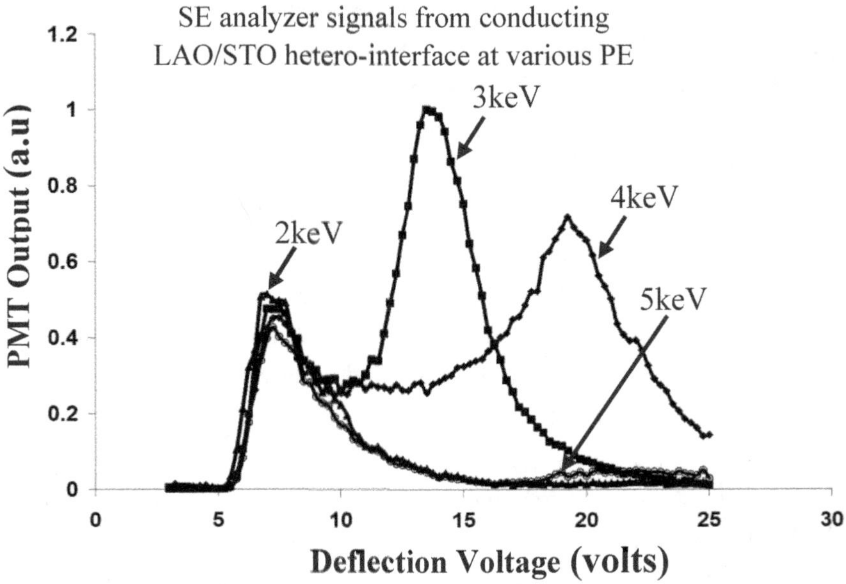

Fig. 5-14. Experimental secondary electron signals obtained from LAO/STO hetero-interface at various primary beam energies. The signals are obtained at primary beam electron energies of 2 keV, 3 keV, 4 keV and 5 keV.

Srinivasan *et al.* (2014) used Monte Carlo simulation of primary beam electron trajectories to help understand the dependence of the SE energy spectral signal with primary beam energy. The simulation results indicated that the interaction volume of the primary beam increases as the primary beam energy increases, as expected. The percentage energy loss of primary beam electrons were plot as a function of depth from the specimen surface, as shown in Fig. 5-15, and indicated that for a primary beam energy loss of 10–50%, only the 3 keV energy loss curve crossed the interface (depth 8 nm from the surface). The 2 keV curve lay within the interface depth, while the 5 keV curve lay well beyond it. These simulations predict that a primary beam energy of 3 keV interacts the most with the interface, confirming the results obtained by experiment (shown in Fig. 5-14). In the case of the 2 keV beam, the majority of the primary beam interaction occurs within the top LAO layer, for higher energies of the primary beam (4 and 5 keV), most of the interaction of the primary beam with the sample occurs within the bulk substrate. In general, the optimal value of primary beam energy required to perform SE energy spectral analysis of the trapped charge at the interface is dependent on the thickness of the deposited thin film (LAO in this case), and is approximately proportional to it.

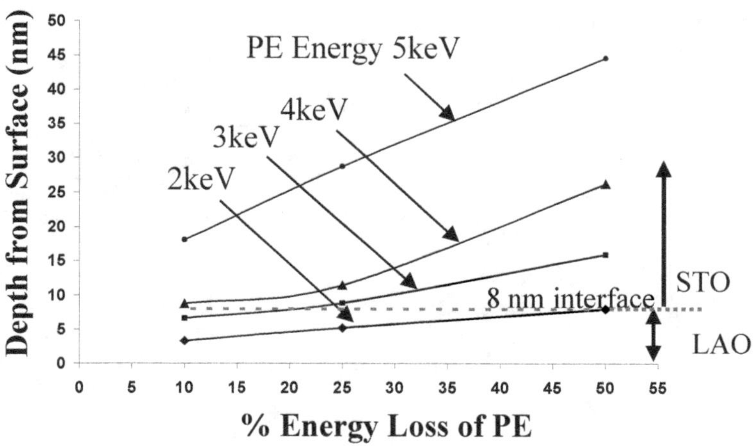

Fig. 5-15. Plot of the percentage energy loss of primary beam electrons against the depth from the surface of the LAO/STO specimen derived from Monte Carlo simulation. The dotted line indicates the LAO/STO interface at a depth of 8nm from the surface.

Overall, Srinivasan *et al.* (2014) demonstrated how SE energy spectroscopy can quantify information about charge being present at the interface between multifunctional oxide interfaces, in a way that goes beyond observing brightness changes in a SE detector image. The appearance of trapped charge typically appears as a second peak in the output SE energy spectral signal.

Chapter 6

Conclusions and Future Work

6.1 Conclusions

The experimental spectral results presented in this book suggest that SE energy spectroscopy has the potential to become a useful analytical technique in the SEM for mapping materials science properties on the nano-scale. They demonstrate that it is possible to separate, identify and quantify a wide variety of different contrast mechanisms in the SEM by tracking small changes in the SE energy spectrum shape and capturing its fine spectral features. Although there are traces of these same contrast mechanisms present in the conventional SE detector image, they are more difficult to interpret and quantify. This is because the SE detector provides a bulk signal in which contrast mechanisms containing information about the sample and its interaction with the primary electron beam combine together in complex ways. By capturing the SE energy spectrum in addition to the SE image, more information can be used to help separate them, and better quantify them.

In some cases, the SE energy spectral signal contains information that is not visible in the conventional SE detector image. Small changes in SE kinetic energy produced by inelastic collisions with valence band electrons or caused by local charging effects of non-conducting specimens lie in this category. This makes it possible for SE energy spectroscopy to capture bulk density of states information and map out local charge distributions in LVSEM. In the case of elemental metal regions on the sample, there is a linear relationship between SE spectral signal peak height/FWHM and atomic number. This, when combined with fine spectral features/shape contrast information from other parts of the SE energy spectral signal, has the potential for carrying out quantitative material identification/analysis for LVSEM. In comparison, the SE yield of conventional SE detectors does not seem to have any quantifiable relationship with specimen atomic number. In the subject of quantitative dopant contrast, SE energy spectroscopy solves longstanding problems that have previously made the technique unreliable, such as the occurrence of complex contrast reversal effects in the SE detector image caused by the primary beam voltage or the scan rate changing. The SE experimental spectral results presented in chapter 4 of this book show how surface band-bending, buried space-charge, and native oxide scattering can all be separately identified and quantified, and how that in turn, leads to the opportunity of using the SEM to make new kinds of *in-situ* measurements, such as monitoring the thickness of native silicon dioxide layers, or mapping the surface conductivity of bulk silicon wafer samples.

An important conclusion emerging from the research work reported in this book is that in order to achieve SE energy spectroscopy inside the SEM, the first and foremost challenge to overcome is to develop an appropriate electron energy spectrometer for it. The ability to track small shape changes in the SE energy spectral signal requires the use of a high precision energy analyzer, one that has been specifically designed to integrate into the objective lens optics/specimen stage layout of a SEM, and one that can provide a high output signal-to-noise ratio. There is a need to make such electron energy analyzers available for SEMs, either as an add-on accessory or as a built-in feature to a SEM electron detection system. Another important conclusion emerging from the work carried out here is that fine spectral features and small changes in the SE energy spectral shape can be tracked and meaningfully quantified in routine SEM specimen chamber HV operating conditions, despite the presence of beam induced contamination effects. This is very different to the other electron energy spectroscopy techniques normally used in surface science studies, such as AES and XPS, where UHV conditions are mandatory.

6.2 Future SE energy spectral experiments

The discovery that accurate bulk density of states information can be extracted from fine spectral features appearing in the SE energy distribution for some metal samples at low primary beam voltages (0.5 to 2 kV) is one of the most important findings of the research work reported in this book. However, much more investigation is required to better understand it. It is tempting to go directly on to investigate the possibility of carrying out angle-resolved secondary electron emission spectroscopy (ARSEES) in the SEM in order to capture electronic band structure information. This would not require too much more instrumental modification to the present experimental setup. Tilting the specimen through an angle of 0 to 45° coupled with specimen rotational movement can generate data for a polar angular range of −45° to +45° about the specimen normal. In the azimuthal angular direction, there is already a large angular range of detection in the present toroidal analyzer prototype attachment (100°) and in principle, a variety of different ways to use parallel detection to capture SE azimuthal angular distribution can be used. Meanwhile, there are still many more useful spectral results that need to be obtained with the present experimental setup as it currently stands.

The high degree of correlation between experimental SE energy spectral bulk valence band DOS data and density functional theory (DFT) for some of the metal specimens presented in chapter 3 suggests that the contribution from surface states is relatively small. This would make it different from angle-resolved photon emission spectroscopy (ARPES), a technique that gathers its information mainly from the specimen surface, usually from depths that lie within 0.5 nm, limiting it to operate in UHV conditions with pressures that are typically below 5×10^{-11} Torr (Lv *et al.*, 2019). In fact, one of the main limitations cited for ARPES is its difficulty in providing bulk density of states information. The inherent ability of SE energy spectroscopy to acquire bulk density of states information and be spatially-resolved (on the nano-scale), may enable it to take on a complementary role to ARPES, and find its own useful set of applications. The fact that it does not need any specimen preparation and operates in HV conditions are of course, already significant advantages, but a comparative study on a wide set of samples with ARPES will help further clarify the role played by surface states, and to better understand its relative weaknesses and strengths.

Within the subject of quantitative material contrast in the SEM, more detailed comparison of SE energy spectral results with EDS is required. Since EDS information normally comes from regions that are several microns deep below the specimen surface, as opposed to SE generation, which occurs in the nanometer depth range, SE energy spectroscopy will most likely always complement EDS, rather than compete with it. This means that even in situations where EDS is routinely used and well established, SE energy spectroscopy may be able to provide additional information, particularly when specimens have nanometer thick top layers or nano-particles on their surface.

The potential for SE energy spectroscopy to become a quantitative material analysis tool for LVSEM applications, particularly at primary beam voltages around 1 kV, is important for a variety of different reasons. The inability of EDS to operate at such low primary beam voltages is a significant limitation. LVSEMs have many well-known advantages over conventional SEMs, such as higher signal yields, a smaller beam/specimen interaction volume, greater surface information, and the possibility of minimizing charging effects while inspecting non-conductive specimens (Joy and Joy, 1996; and Wuhrer and Moran, 2016). These advantages, coupled with certain SEM instrumentation improvements over the last few decades, such as the better brightness/lower energy spread of field emission sources, and higher sensitivities of electron detectors, has led to significant improvements in SEM image spatial resolution and output signal-to-noise. At present,

LVSEMs have a spatial resolution of a few *nm* at primary beam voltages as low as 1 kV, and are used to examine a wide variety of biological and non-conductive specimens on the nano-scale (Erdman and Bell, 2013). The possibility of superimposing quantitative material information on to LVSEM images through SE energy spectroscopy has therefore a wide area of application.

LVSEM SE energy spectroscopy using the present wide-angle toroidal energy analyzer attachment design is not compatible with through-the-lens SEM detection systems. This is because it operates as an add-on attachment mounted on the specimen stage inside conventional SEM columns, in the field-free region between the specimen and lower objective lens pole-piece. The present prototype analyzer attachment allows for a working distance of 15 mm. Through-the-lens SEMs immerse the specimen in a magnetic field and/or apply a strong electric field above it in order to slow the primary beam down to low landing energies. One way of improving the image resolution for LVSEM applications with the present wide-angle toroidal energy analyzer attachment is to install it in a conventional SEM column (below the objective lens operation) that has some additional way of reducing the chromatic aberration effect. This might for instance be achieved through the use of a primary beam energy monochromator. Alternatively, it could also be achieved through the use of a chromatic aberration corrector unit.

The general finding that different parts of the SE energy spectral signal convey different types of information, with more surface/valence band information coming from its rising edge, and more indirect backscattered information coming from its peak height/trailing edge shape, needs to be tested on a greater range of specimens and primary beam operating conditions. The experimental results demonstrating that there is a linear dependence of peak height or FWHM of the SE spectral signal with the atomic number of metal samples provides an important first step. More investigation on how to use the change in the rising edge overall shape and the fine spectral features embedded within it to enhance quantitative material contrast needs to be carried out. Although this information cannot provide finger-print material identification like EDS, it may come quite close to it for many applications.

More work is required in order to better understand the spectral peaks that appear in the density of states SE energy spectral experimental data. At the end of chapter 3, they were compared to previously reported DFT theoretical predictions for the W, Cu and Pt specimens from different sources, and to an XPS experimentally derived one for the Au specimen. Greater precision in the comparison needs to be made, where one common theoretical framework is used to systematically generate DFT distributions for all samples. This is required in order to determine which DFT model/simulation best fits the experimental data. More experiments are also needed in order to understand the role played by beam-induced contamination. Greater reduction of the surface contamination effect might be obtained by applying some form of local specimen heating, perhaps by joule heating. This will allow for faster spectral acquisition rates and a greater amount of signal averaging. Also, negative biasing of the specimen and inner cap electrode can be tried. These methods may reduce the need to collect SE energy spectra from different areas of the specimen in order to provide adequate signal averaging, since they will increase output signal levels by over an order of magnitude from a single scan area. If successful, they will also make 2D nanaoscale mapping of DOS information possible.

On the subject of material identification, more investigation is needed to understand why there seems to be enhanced material contrast when the specimen and inner cap electrode are biased to negative voltages (such as −10 V). The experimental results in the context of distinguishing between materials having close atomic numbers (platinum and gold) in chapter 3, seem to indicate that sample/inner cap electrode negative biasing is an effective strategy, more understanding about the reasons behind this is required. Possible reasons might be that sample/inner cap electrode biasing shifts the SE energy peak into a lower energy range, and captures SEs that have a different angular emission range. The whole subject of whether material contrast is enhanced by increasing or changing the captured SE angular emission range is an important topic for future work.

In the subject of dopant contrast, more experiments are required to understand why the specimen surface appears to behave like a conductive equipotential plane, taking the same electric potential as the base-plate potential in regions that have no buried space-charge. The most likely reason for the surface potential being pinned down in this way comes from there being multiple conductive pathways to ground (to the bottom base plate), both through the specimen and across its surface. Experiments which can measure the specimen bulk current and surface current separately as a function of primary beam current for a variety of different semiconductor specimens may therefore help in better understanding this effect. For solar cell samples, it may also be beneficial to carry out SE energy spectroscopy experiments in tandem with the EBIC (Electron Beam Induced Current) technique (Reimer, 1998b).

Within the topic of mapping space-charge and dopant concentration by SE energy spectroscopy, a wider range of *pn* junction geometries need to be investigated. The contribution of shot noise for the one-dimensional test stripe samples used in this work was made relatively small by averaging the spectral signal along the length of the stripe, however, for a genuinely two dimensional *pn* junction layout, that strategy cannot be used. For a fully two-dimensional geometry, a series of images at different pass energies needs to be acquired, and additional image averaging will be required to reduce shot noise to an acceptable level, which will inevitably increase the overall exposure time of the specimen to the primary beam, further increasing contamination effects. For the purposes of acquiring space-charge information, the shape of the SE energy spectral signal is much less affected by the build-up of contamination than the SE spectral height, however, an increase in recombination effects also needs to be avoided. How to keep shot noise, contamination, and recombination effects simultaneously down to manageable levels for fully two-dimensional *pn* junction samples is therefore a subject for future work.

On the more general topic of semiconductor characterization, SE energy spectroscopy has several future possible applications. The method of monitoring the change in shape of the SE energy spectral signal between two different primary beam voltages as a way of determining the top oxide layer thickness *in-situ* needs to be tested on a greater variety of specimens. The experimental results in chapter 4 demonstrated that this approach can in principle be used for tracking thickness variations of a silicon native oxide layer, but further work is required to transform it into a general oxide thickness mapping technique that can be applied to determine thickness variations of a grown top silicon dioxide layer, such as a covering passivation layer. For thicker top oxide layers, more complicated effects are expected to occur, such as primary beam induced charging. In addition, it would also be useful to probe top oxide layers on other materials (other than silicon) and other thin top insulator layers.

One of the most promising applications of SE energy spectroscopy for the subject of semiconductor characterization emerging from the experimental results presented in chapter 4 is its ability to measure the local dopant concentration of bulk semiconductor wafer samples. These results demonstrate that dopant concentration can be measured through the surface band-bending effect alone, without the need to make any reference measurement to an underlying substrate. At present, the four-point probe method (Valdes, 1954) is widely used to make conductivity measurements of bulk wafer specimens, but using the SEM opens up the possibility of mapping local conductivity variations across the surface of a large wafer sample. In order to test out this idea, the method needs to be implemented in SEMs that can accommodate full wafer size specimens. More work is also required in order to compare the accuracy and reliability of the SE energy spectral approach to experimental results obtained by the standard four point probe method on the same specimen.

The SE energy spectrum generated by a Ga^+ focused beam was acquired for a range of different metals, and the charge deposited on the surface of an intrinsic silicon sample when inspecting it by a focused Ga^+ beam was mapped out *in-situ* by using SE energy spectroscopy inside a FIB-SEM dual beam system. A similar study can be made for the Helium ion beam microscope (HIM), which may help to better understand its imaging contrast mechanisms and the way its primary helium ion beam interacts with the specimen surface while acquiring its topographical image. Although the HIM has significant advantages over the SEM, such as a smaller interaction volume, a higher depth of field, and less charging of non-conductive specimens at high primary beam voltages, it also has the inherent action of sputtering the specimen surface, which is undesirable for many applications (Postek *et al.*, 2011). SE energy spectroscopy inside the HIM can provide more information about how a helium ion primary beam interacts with different specimens. Acquiring the SE energy spectrum in a HIM instrument has already been proposed for the purposes of identifying different material phases, and the need to use SE energy analyzers that are specifically designed for the HIM has already been acknowledged (Wirtz *et al.*, 2019).

In the subject of using SE energy spectroscopy to map out surface charge distributions when examining non-conducting samples, a greater variety of specimens is required, beyond the test MIS samples reported in chapter 5. SE energy spectroscopy has the ability to capture local charge build-up on non-conductive specimens when observed in low primary beam landing energy charge balance conditions. In high precision applications such as Critical Dimension SEM metrology (CD-SEM), this type of information may provide a way of correcting for measurement errors caused by local charging (Arat *et al.*, 2019).

6.3 Future instrument improvements

There are a number of analyzer attachment design improvements and developments that can be recommended for the future. The possibility of incorporating some form of multi-channel detection array in the azimuthal angular direction in order to create angle-resolved secondary electron spectroscopy (ARSEES) in the SEM has already been mentioned. Since electrons exit the analyzer converging towards the rotational axis of symmetry, a simple electrostatic zoom lens can be used to project the angular distribution on to a flat 2D scintillator/electron multiplier CCD camera detector positioned on the other side of the rotational axis. In this way, the output azimuthal angular distribution can be captured as an image. It should be noted that the use of parallel azimuthal angle detection by an electric electron toroidal energy analyzer design has previously been used for angle-resolved gas phase ultraviolet photoemission studies (Leckey and Riley, 1985). This was later developed into an improved parallel toroidal electron energy analyzer ARPES system for acquiring electronic band-structure information (Leckey and Riley, 1988; and Broekman *et al.*, 2005). Although not exactly the same electric toroidal electron energy analyzer design as used here, a similar parallel azimuthal angle detection approach can be used for achieving ARSEES in the SEM.

There is also the obvious challenge of how to incorporate SE energy spectroscopy into high resolution in-lens SEM columns, the kind that are usually used for LVSEM. Some previous proposals to achieve this have already been reported, and were outlined in section 1.3 of chapter 1. They provide a starting point for future SEM column redesign. Figure 6-1 shows a schematic diagram of one possible way to integrate a multi-channel time-of-flight electron energy spectrometer (Khursheed, 1992) into a SEM column. A fully magnetic immersion objective lens is integrated into the walls of the specimen chamber, and a drift tube region is located in a separate chamber above it. Scattered electrons are collected by a micro-channel plate (MCP) detector located at the top of the drift tube region, and the primary electron beam is pulsed by a beam blanker positioned in the upper part of the SEM. A monochromator unit is located either close to the beam blanker unit, or integrated to be a part of it. The monochromator provides the possibility of reducing the primary beam energy spread when acquiring the SE energy spectrum, since specimen biasing in order to lower the primary beam landing energy also leads to a degradation of SE transit time/energy information. One advantage of this column design over many previous through-the-lens column designs is that it does not involve deflecting SEs off-axis. SEs spiral up through the lens bore and naturally spread out as they traverse the drift tube region to be detected on a rotationally symmetric MCP detector. This also gives it a high transport efficiency for capturing SEs and BSEs (both well over 50%).

In the normal imaging mode of operation, during which the SE energy spectrum is not captured, it is possible to bias the specimen negatively, allowing for the capture of high resolution images at low landing energies, like in other widely used LVSEM columns. This also, conveniently shortens the overall SE transit time, and time gating can be used to separate the BSE signal from the SE one. Then, taking advantage of the fact that the primary beam is inherently pulsed, the routine imaging mode can also become a dose controlled one, where the ratio of the pulse period T, to its width Δt, can be varied by altering the blanking signal repetition rate (Fig. 6-1b). This mode of operation will be useful for a variety of different applications where precise control of primary beam irradiation time is required, such as preventing the build-up of charge or contamination at the specimen surface, or avoiding damage to certain surface sensitive samples.

In the SE spectral acquisition mode of operation, the repetition period T will typically be one to two orders of magnitude larger than the pulse width, Δt (Fig. 6-1c), giving time for the MCP detector to acquire the output time dispersion signal, with T varying from say 50 to 200 ns. In-lens deflection coils may be required for large field of view applications, but would be switched off while acquiring the SE energy spectrum.

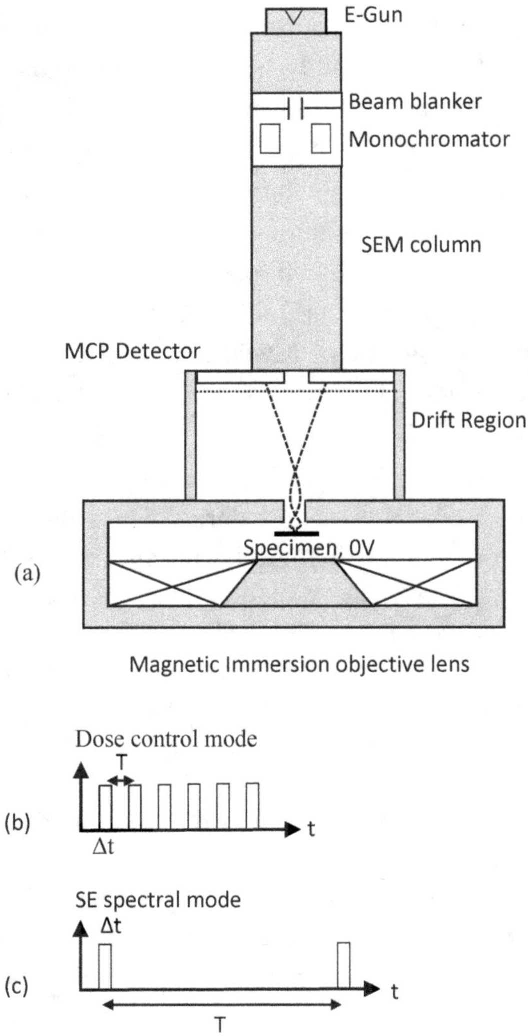

Fig. 6-1. SE time-of-flight spectrometer inside a magnetic immersion objective lens column. Scattered electron trajectory paths represented by dashed lines. (a) Schematic drawing of column layout. (b) Blanking pulse for dose control mode, Δt is the pulse duration, T is the repetition period. (c) Blanking pulse for SE spectral signal acquisition.

Figure 6-2 shows a schematic diagram of the kind of SEM column design needed for a position sensitive multi-channel SE Wien Filter deflection spectrometer (Khursheed and Karuppiah, 2001). This column has a monochromator and fully immersion magnetic objective lens, like the time-of-flight spectrometer column, but in addition, it has two Wien filter deflectors. Wien filter 2, located above lens bore, deflects SEs off-axis and disperses them according to their energies ($E_2 > E_1$), while leaving the primary beam unaffected. Wien filter 1, located in the upper part of the column, is excited in the opposite sense to Wien filter 2, and together, they minimize primary beam chromatic and geometric aberrations. The strength of excitation on both Wien filter units will be relatively weak, since the primary beam energy is several orders of magnitude greater than the kinetic energies of the SEs.

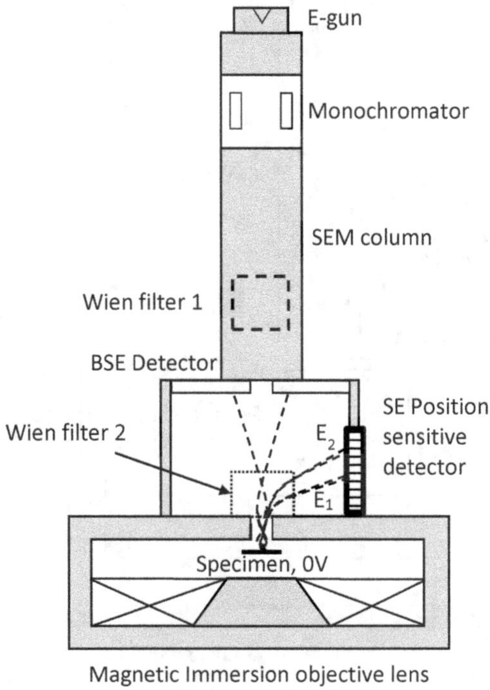

Fig. 6-2. Position sensitive multi-channel SE energy Wien filter deflection spectrometer integrated into a magnetic immersion objective lens column. Scattered electron trajectory paths represented by the dashed lines. SE energies, $E_2 > E_1$.

Figure 6-3 shows the schematic drawing of a SEM column designed to incorporate a Wien filter/spherical deflector for LVSEM applications, where the primary beam is retarded down to low landing energies by biasing the specimen negatively (Kienle and Plies, 2004). There are also two Wien filters used in this case, however, since the SEs travel up the column with energies close to the primary beam energy, a spherical deflector is used to assist in the deflection of SEs, in order to prevent the aberrations on the primary beam from becoming too large. An electrostatic zoom lens helps magnify the SE projected image on to a position sensitive scintillator/CCD camera detector. In this column design, depending on the energy difference between SEs and BSEs, it may be possible to collect BSEs and SEs simultaneously, otherwise, SE and BSE energy spectra are captured separately, one after the other.

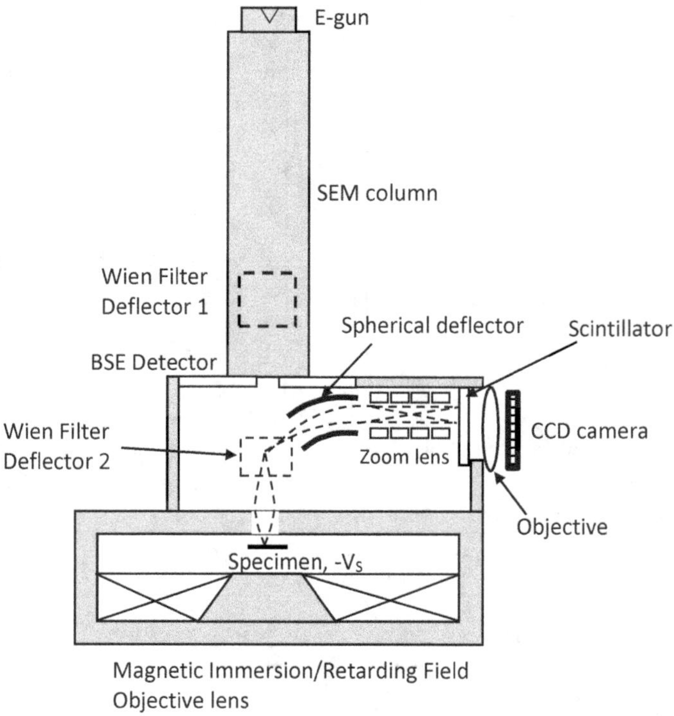

Fig. 6-3. Parallel energy Wien filter/spherical deflector spectrometer integrated into a LVSEM column. Scattered electron trajectory paths represented by dashed lines.

Figure 6-4 shows the schematic diagram for a round magnetic beam separator SE energy spectrometer SEM column design (Khursheed and Osterberg, 2006). This type of column/spectrometer is also suitable for LVSEM, where the specimen is biased to a negative voltage in order to retard the primary beam down to low landing energies. In this case, the primary electron beam column is orientated horizontally (to the left), and directs a converging beam of electrons into the center of a round beam magnetic sector which is able deflect the primary beam by 90° without incurring large aberrations. The SEs, which subsequently travel back up through the objective lens bore, are focused into the center of the beam separator with the help of a transfer lens and deflected to the right. A post magnetic deflector focuses SEs and BSEs on to a position sensitive detector. Two separate post-deflectors can also be used to better separate detection of SEs and BSEs. Photons can be conveniently detected above the beam separator in this design.

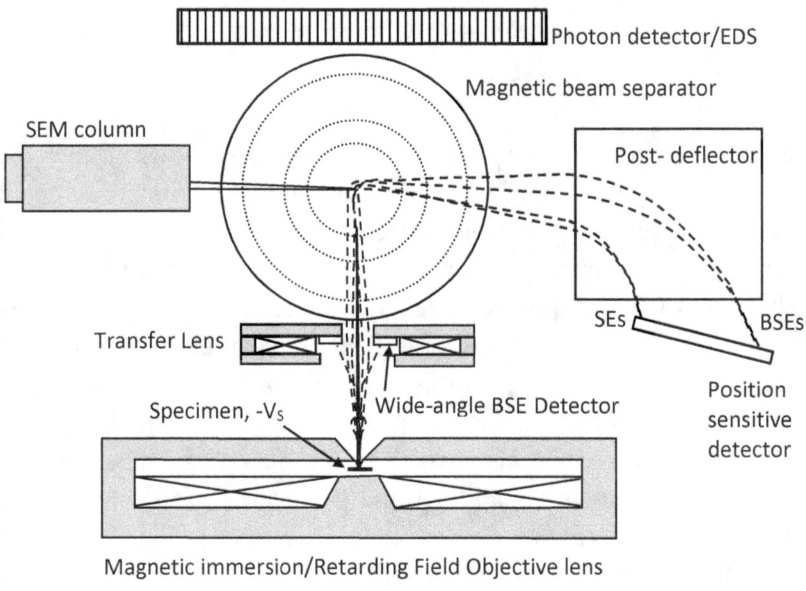

Fig. 6-4. Schematic layout of a LVSEM column incorporating a magnetic beam separator electron energy spectrometer. Primary beam electron ray paths represented by solid lines, scattered electron ray paths represented by dashed lines.

All the SEM column in-lens SE energy multi-channel designs shown in Figs. 6-1 to 6-4 provide for much better separation between SEs and BSEs than any of the previous through-the-lens SE detection systems that were proposed for SE energy filtering (see section 1.5, chapter 1). A common problem for the Hitachi S-4800, Hitachi Regulus 8220, Magellan 400 FEG, Helios Nanolab, and FEI XL30 S-FEG SEM through-the-lens SE detectors was that a spurious background signal generated from high energy electron scattering on grids/electrodes was detected, in addition to the main signal. In the case of the in-lens spectrometer column designs shown in Figs. 6-1 to 6-4, there is much less mixing of higher energy electrons with the lower energy ones since scattered electrons are separated in terms of their energies. Another general point is that since all the in-lens spectrometer column designs presented here involve multi-channel energy detection, in principle, the overall data-acquisition time for acquiring SE spectroscopic image information should be approximately comparable to conventional SEM image acquisition times.

The in-lens SE energy spectral SEM column designs shown in Figs. 6-1 to 6-4 represent broad concepts only, and important practical details relating to their scanning/detection systems still need to be worked out. It will probably not be possible to achieve a high field of view by using in-lens deflectors while acquiring SE energy spectral information, because that would both deflect SEs as well as primary beam electrons. It should also be noted that the column designs shown in Figs 6-3 and 6-4 can operate in a low landing primary energy mode (LVSEM) by biasing the specimen negatively while also acquiring the SE energy spectrum. The columns designs shown in Figs. 6-1 and 6-2 can only operate in a low landing primary energy mode for imaging purposes, and not while capturing the SE energy spectrum, which is why a monochromator unit is proposed for them.

Apart from dedicated SE spectral column designs, another area for further instrumental development is to make separate attachments dedicated to capturing SE emission polar and azimuthal angular distributions, which work in parallel to the SE energy analyzer attachment. SE emission angular distribution information, together with the SE energy spectrum, will most likely be helpful for better quantifying a variety of different contrast mechanisms in the SEM. The simplest approach to achieve this would be to use miniature detector arrays in the SEM chamber.

The full range energy spectral results presented in chapter 3 highlight the advantages of being able to acquire higher energy spectral data together with SE energy spectral information. In general, two separate toroidal sector energy analyzer units can operate in parallel, each covering a separate azimuthal angular range, each operating at different pass energies. In this way, a higher energy portion of the scattered energy spectrum can be obtained while the SE energy spectrum is being captured. There are advantages in being able to capture high energy fine spectral features together with the low energy SE spectrum. UHV electron beam energy spectral systems have for instance, been able to correlate surface/bulk plasmon peaks in the reflection electron energy loss spectrum (REELS) to various features and shape changes in the SE energy spectrum (Werner *et al.*, 2008; Ding *et al.*, 2008; and Da *et al.*, 2019). In order to make this viable for pass energies above 3 keV, the present wide-angle toroidal energy analyzer attachment size needs to be increased in order to avoid electrical spark over problems. However, a larger analyzer attachment has the undesirable effect of increasing the working distance, which results in poorer image resolution. Redesign of the present wide-angle toroidal energy analyzer attachment is therefore required in order to both operate at high pass energies (above 3 kV) and also at the same time, have low working distances (less than say 15 mm).

One way to enlarge the analyzer size is to integrate it into the specimen chamber outer walls, as shown in Fig. 6-5. This design strategy has several advantages. Firstly, depending on the size of the lower lens pole-piece, it can further reduce the working distance, typically to less than

5 mm. Secondly, it will be big enough to avoid any electrical spark over problems for pass energies up to 30 kV (the maximum primary beam voltage normally used in SEMs). Another advantage is that a larger analyzer unit avoids the use of impractically small exit apertures when high energy resolution is required.

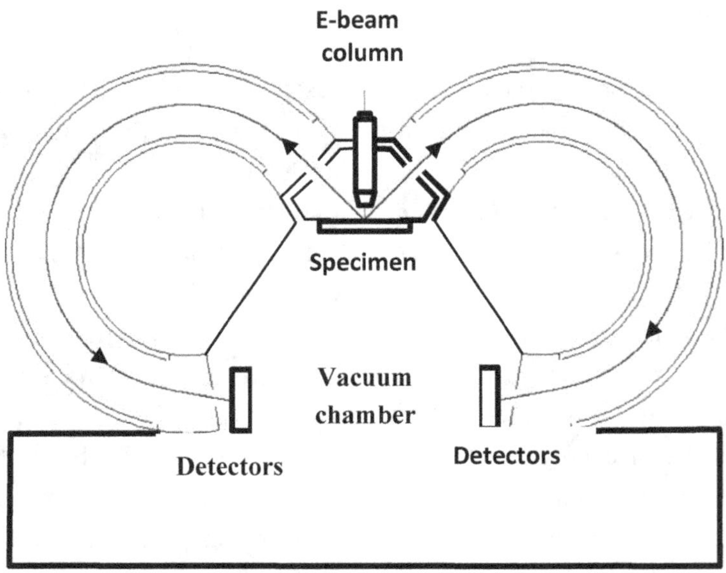

Fig. 6-5. Enlarged wide-angle toroidal energy analyzer integrated into the specimen chamber outer walls.

Capturing bulk or surface plasmon peaks for REELS in the SEM will require both an improvement in analyzer energy resolution as well as reduction of beam induced contamination on the specimen surface. This is because the width of energy detection varies linearly with pass energy, so a better relative energy resolution is required in order to resolve spectral peaks located at larger energies. A better relative energy resolution always comes at the price of lowering output signal-to-noise, and this necessitates the use of longer data acquisition times (more averaging) to keep shot noise effects down to manageable levels, which in turn, makes the output spectral signal more vulnerable to beam induced contamination. The strategy of heating the specimen holder might be a simple way of achieving this.

One way to improve the wide-angle toroidal energy analyzer attachment performance, without having to enlarge its size, would be to redesign it so that the pass energy of scattered electrons travelling through it can be controlled. In order to achieve this while keeping the specimen at ground potential, extra pre-retardation or pre-acceleration conical electrodes need to be incorporated at the analyzer entrance, and the conical electrodes surrounding the specimen in the present analyzer design need to be removed. The purpose of these extra conical electrodes at the analyzer entrance is to accelerate or decelerate incoming electrons while, at the same time, applying a collimating action. This collimating action effectively controls the virtual source position, ensuring that the analyzer's focusing action is similar to the one in the original design. Although the exit focal position is slightly extended, the analyzer is still predicted to have second-order focusing capability. A second exit aperture unit can be fitted on to the analyzer when it is operated in the high/low pass energy mode.

Figure 6-6 shows simulated trajectory paths through a modified wide-angle toroidal energy design which uses conical shaped electrodes to act as a retarding field pre-focusing lens. The upper and lower sections of the conical electrodes need to be biased differently in order to compensate for asymmetric electric field effects. These simulations were performed using Lorentz 2D by Integrated Engineering Software. The inner and outer plate radii of this toroidal design is the same as the original design, 14 and 22 mm respectively (Khursheed and Hoang, 2008). The first conical electrode is located close to the outer analyzer zero volt casing, and has upper and lower voltages, V_{1U} and V_{1L} respectively. These voltages are set negatively in order to retard the incoming electron beam energy to be close to the intended analyzer pass energy. The second pair of conical electrodes, V_{2U} and V_{2L}, provides a collimating action before the electron beam enters the analyzer. The third conical electrode, set to voltage V_3, is both connected to an inner casing surrounding the analyzer deflection plates and the inner deflector plate, while the outer deflector plate, set at V_D, is adjusted separately to scan the pass energy. The pass energy of the analyzer is given by (E_S- eV_3), where E_S is the emission energy of the scattered electrons, and *e* is the electric charge of an electron.

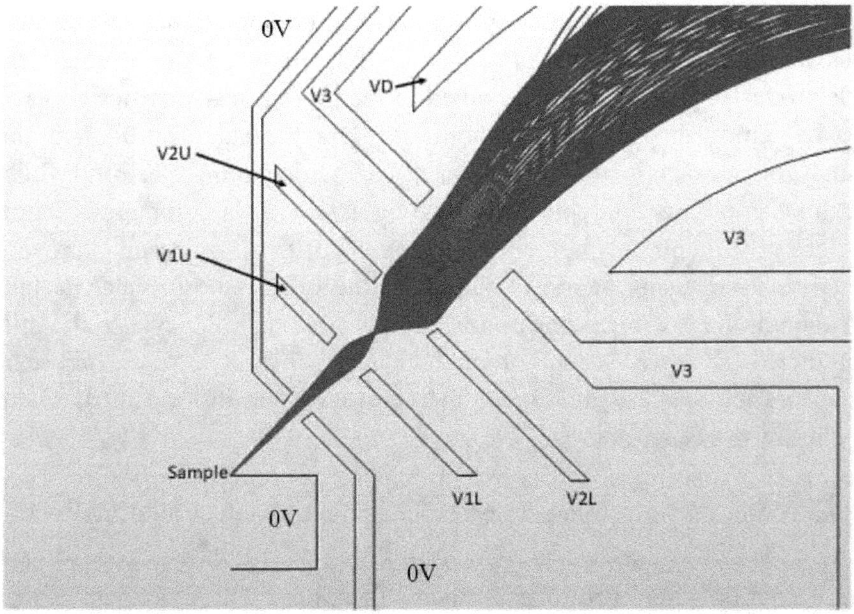

Fig. 6-6. Simulated electron trajectory paths from the sample through a series of conical electrodes that form a point of focus as they retard down the electron beam energy and subsequently apply a collimating action to keep the apparent source position constant.

For the analyzer entrance electrode design shown in Fig. 6-6, trajectory paths are traced for 2 keV scattered electrons leaving a 0 V sample with polar emission angles ranging from 43 to 47°. These electrons are first retarded down by V_{1U} and V_{1L} which are set to −1990 volts. They are then collimated by V_{2U} which is biased to −1997 V and V_{2L}, which is biased to −2000 V. They then go on to travel through the analyzer with

V_3 set to −1990 V and V_D set to −1996 V in order to maintain a pass energy of 10 eV through the analyzer.

Figure 6-7a shows electron trajectory paths through the analyzer (at a pass energy of 10 eV), illustrating how they are brought to a focus at the analyzer exit. Figure 6-7b shows a close-up diagram of the exit point of focus, which lies just beyond the present analyzer exit. A second aperture unit is positioned at the new exit point of focus.

Figure 6-8 depicts the simulated exit focal beam width as a function of emission angular spread, confirming that the analyzer is predicted to have a second-order focusing action (geometric aberrations having a cubic dependence with entrance angular spread), similar to the original analyzer without pre-retardation. Note that the trajectory paths travel close to the analyzer deflection plates, indicating that the assumed polar angular spread of 43 to 47° is close to the maximum allowable entrance spread for a pass energy of 10 eV. An entrance aperture, not shown here, can be fitted on to the 0V analyzer casing in order to suitably limit the entrance angular spread.

If a 200 μm wide exit aperture is used, the energy spread in the exit rays due to linear energy dispersion is calculated to be 0.26 eV and only the electrons having an entrance polar angular spread of −15 to 22 mrad around the central ray contribute to the detected signal. The beam width due to geometric aberration is calculated to be 84 μm, assuming that it is equal to twice the *rms* value of the total beam width distribution shown in Fig. 6-8 (between −15 to 22 mrad). This exit beam width translates into a 0.11 eV geometric aberration limited energy spread, significantly smaller than the 0.26 eV energy spread produced by linear energy dispersion. When both these energy spreads are added in quadrature, the total energy resolution is estimated to be around 0.28 eV. Therefore, for the exit aperture width of 200 μm, the energy resolution is predicted to be mainly determined by exit aperture size, and not limited by geometrical aberrations.

(a)

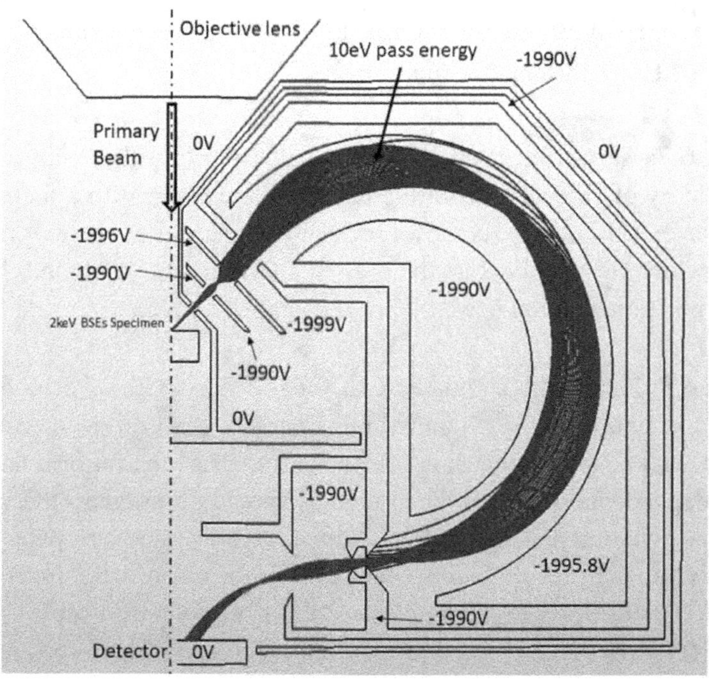

Toroidal energy analyser

Final Aperture

(b)

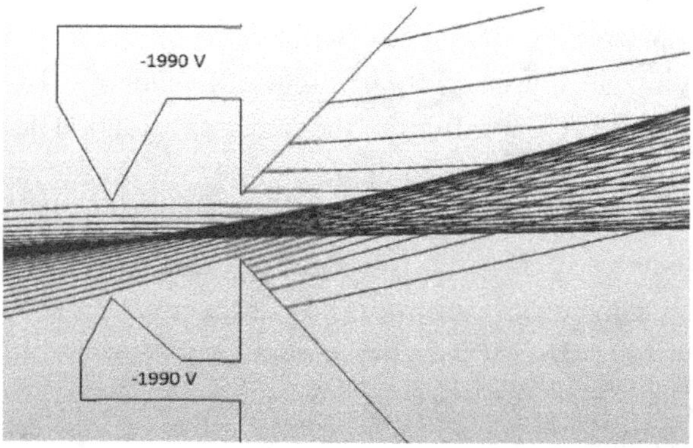

Fig. 6-7. Direct ray tracing simulation of 2000 eV electrons slowed down to a pass energy of 10 eV through the retarding field wide-angle toroidal analyzer design. (a) From the specimen to the detector (b) Around the exit aperture.

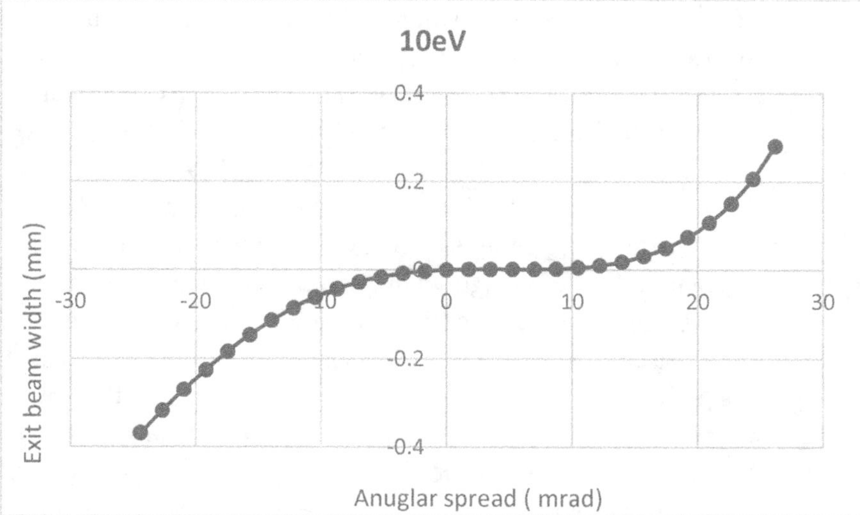

Fig. 6-8. Simulated exit focal beam width as a function of emission angular spread.

Fractions of an eV energy resolution for 2 keV incoming scattered electrons will most likely be more than sufficient for LVSEM REELS applications. Just how much beam induced contamination needs to be reduced in order to obtain meaningful information about the specimen is something that needs further investigation. One notable point however about the simulation results shown in Figs. 6-6 to 6-8 is that the optics of the wide-angle toroidal energy analyzer operating in retardation mode is predicted to be significantly better than the hemispherical deflector analyzer (HDA). The HDA does not typically perform well beyond retardation factors of 10 (factor of energy reduction), fringe fields at the analyzer entrance limits its performance (Sise *et al.*, 2016). In comparison, these simulation results predict that the wide-angle toroidal energy analyzer should operate comfortably with a retardation factor of 200 (2 keV down to 10 eV). The present toroidal analyzer attachment also has two other significant advantages. Its large angular collection in the azimuthal direction gives it over two orders of magnitude better signal-to-noise performance than the HDA (1-2° compared to 360°), and its relatively small size means that it can achieve much smaller working distances (< 15 mm). These potential advantages not only make it relevant to electron energy spectroscopy in the SEM, but also make it suitable for use in surface science UHV spectral systems. It therefore opens up several promising directions for future work.

For the case of accelerating SEs through the analyzer, in order to increase their kinetic energies by say 10 eV, 10 volts is added to all the analyzer internal electrode voltages. This will produce a similar spectral result to the present analyzer design when the specimen/inner cap electrodes are biased to −10 volts. However, the use of the pre-focusing conical lens system provides the flexibility of refocusing the lower energy electrons on to the exit aperture plane by additionally adjusting V_{2U} and V_{2L}, the voltages on the collimating electrodes. This design is therefore a significant improvement over the present one which uses two conical cap electrodes around the specimen.

To conclude this brief summary of future redesign possibilities for the wide-angle toroidal energy analyzer, it is interesting to note that the analyzer is also predicted to have a parallel energy mode of operation for retarding high energy electrons down to low pass energies. Figure 6-9 shows how this might be achieved for 2 keV energy scattered electrons being retarded down to a 10 eV pass energy. A multi-slit aperture is used at the exit plane, and the exiting electron beam is both accelerated and deflected upwards by two plates (0 V upper plate and −500 V lower plate). This causes the energy dispersion within the exit electron beam to be magnified on to a vertical detection plane at or close to the analyzer's rotational axis. Figure 6-9 depicts ray tracing for an electron energy spread ranging from −0.2 to 0.2 eV in 0.1 eV intervals for an emission polar angular range between 43 to 47°. An approximate energy resolution of around 0.05 eV can be estimated from the separation of the rays landing on the detector plane in Fig. 6-9b. The multiple slit widths on the exit aperture plate were chosen arbitrarily and are not all of the same size. They are used here only for illustrative purposes, in order to show how a high degree of multi-channel energy separation might be achieved.

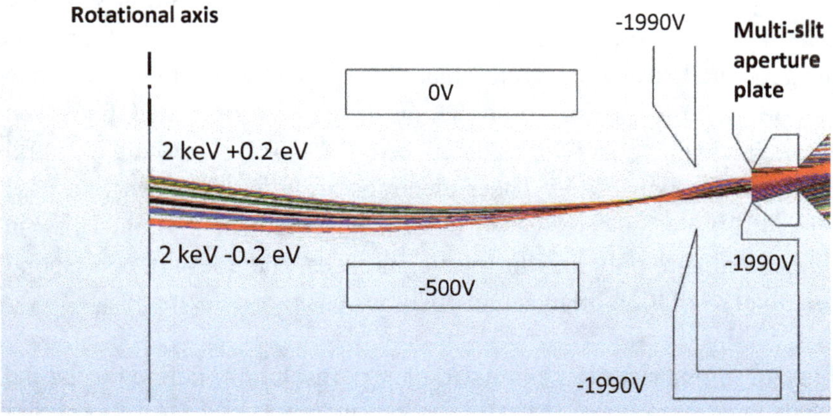

(a)

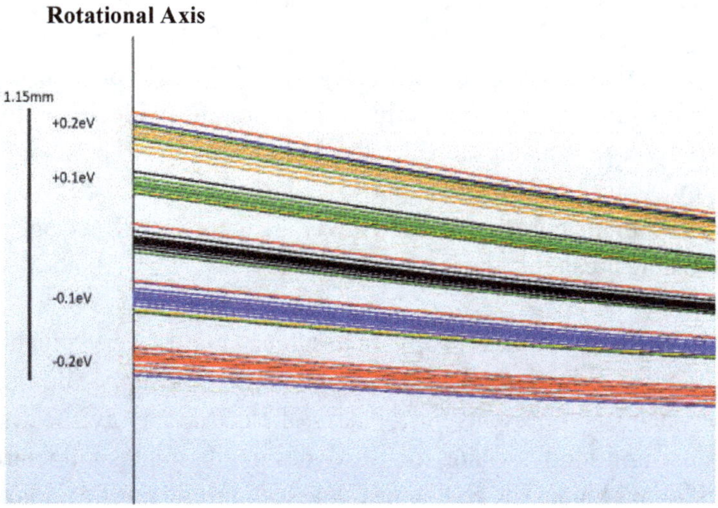

(b)

Fig. 6-9. Direct ray tracing simulation of a parallel energy retarding field wide-angle toroidal analyzer design. A 2 keV beam with a ± 0.2 eV energy spread is retarded down to a pass energy of 10 eV. (a) From a multi-slit exit aperture to the detection plane. (b) Region close to the detection plane.

A better estimate of the analyzer resolution on the detector plane can be obtained by simulating a wide-range of energies and polar angles, while at the same time, modelling the detector plane to be a position sensitive pixel array. The pixel size is referred to here as bin width. A histogram is then generated from electrons that land on the position sensitive detection plane. Figure 6-10 shows the histogram simulated for seven energies having an energy spread between -0.4 to 0.2 eV around a central emission energy of 2 keV. These electrons are retarded down to a pass energy of 10 eV and are recorded for a bin width of 3.3μm. This bin width is used here only for illustrative purposes, and does not correspond to the pixel resolution of an actual position sensitive detector.

Table 6-1 summarizes the simulated energy resolution, defined to be the full half-width maximum (FWHM) of the spectral signal for different bin widths on the detector plane. These simulation results predict that the energy resolution in most cases, will be well below 0.05 eV. The focusing action is clearly better in the −0.4 to −0.1 eV range, where the energy resolution is predicted to be below 0.02 eV. Whether it is really practical to make a parallel energy version of the wide-angle toroidal energy analyzer as an add-on SEM attachment remains to be seen, it is of course, much easier to make as part of a larger dedicated SEM spectral system. The main purpose here however, is to present it as one of the many possible areas for future work.

Overall, SE energy spectroscopy is a subject that is very much in its infancy, and it brings a fresh set of challenges and opportunities. For material scientists it presents a new class of electron energy spectral data to generate and study, while for SEM designers and manufacturers, it provides new kinds of challenges and opportunities for combining electron energy spectroscopy with scanning electron microscopy.

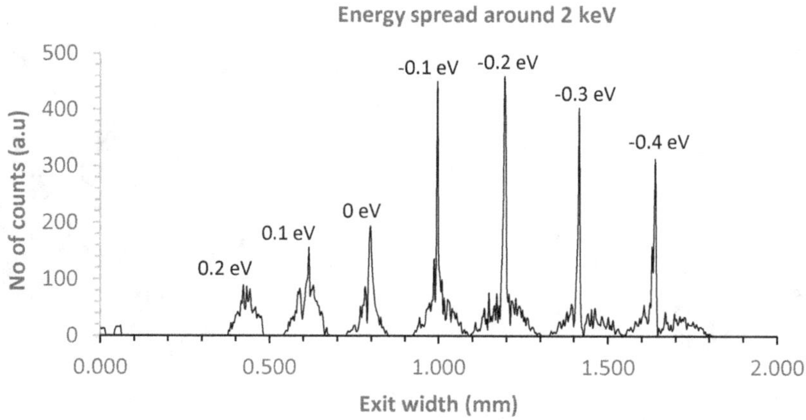

Fig. 6-10. Histogram of simulated electrons on the detector plane for a bin width of 3.3 μm.

Table 6-1. Full-width-half-maximum (FWHM) data extracted from the histogram of simulated electrons at the detector plane for different bin widths.

Energy spread (eV)	**FWHM) (eV) for a 2 keV energy beam retarded to a 10 eV pass energy (deceleration factor of 200)**		
	(3.3 μm bin width)	(6.6 μm bin width)	(9.9 μm bin width)
-0.4	0.017	0.013	0.022
-0.3	0.0066	0.013	0.022
-0.2	0.011	0.013	0.022
-0.1	0.0079	0.013	0.022
0.0	0.033	0.022	0.028
0.1	0.026	0.035	0.046
0.2	0.053	0.079	0.075

Appendix 1

The Electric Radial Mirror Analyzer (RMA) Attachment Design

The electric radial mirror electron energy spectrometer was specifically designed to be a band-pass electron energy analyzer attachment for the SEM. The analyzer uses a rotationally symmetric electric field distribution to transport electrons/ions emitted from a central point source in a radial direction on to a ring-shaped collection/detection area, as shown in Fig. A1-1a. The name radial mirror analyzer (RMA), derives from an electric field being used to mirror and analyze the energies of electrons/ions traveling in a radial direction from a central point source. In order to do this, segmented electrodes above the entrance grid V_1, V_2, V_3 and the curved deflecting electrode V_d, are biased negatively. The outer spectrometer radius, R, is chosen to be 5 cm. Its minimum working distance (W) is typically of the same size as the objective lens lower pole-piece radius, which can in principle be as small as 5 mm, allowing for high spatial resolution. The analyzer is conveniently located above the specimen, much like the conventional backscattered detector. Two grounded electrostatic grids are used to cover the entrance and the exit of the spectrometer in order to avoid distortion of the electric field near these regions.

Figure A1-1b shows simulated electron paths through the RMA design at the pass energy for a polar angular spread of ±6° around the central angle of 32.6° using Lorentz 2D by Integrated Engineering Software.

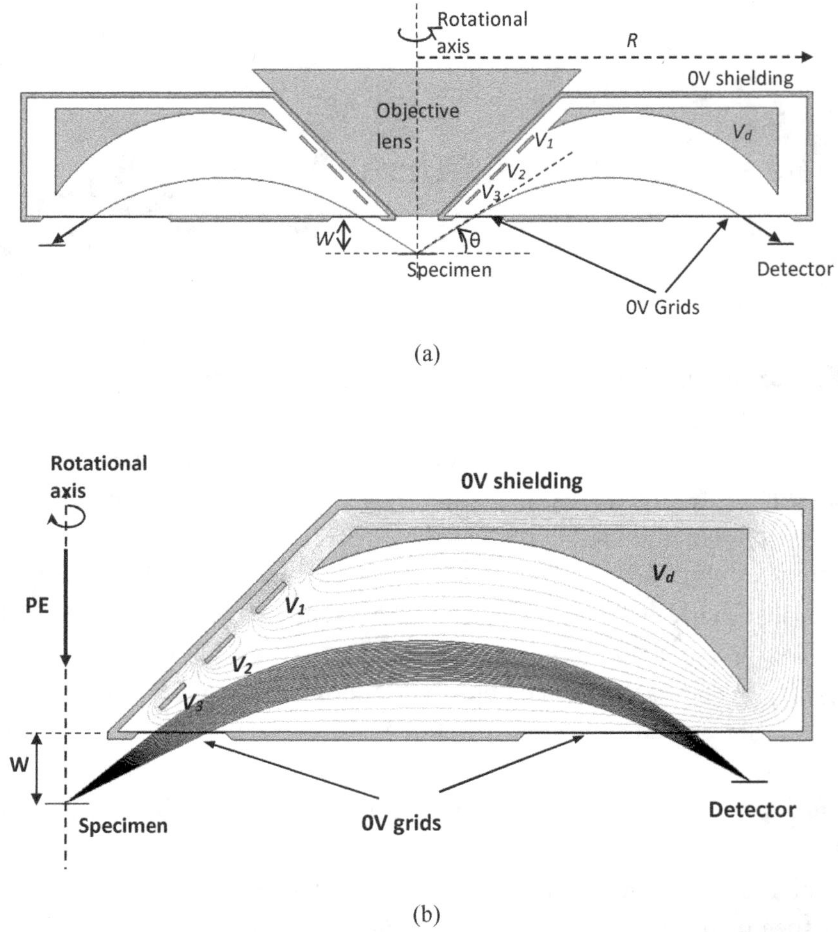

Fig. A1-1. The radial mirror analyzer (RMA) design. (a) Schematic diagram. The segmented electrodes are biased by V_1, V_2, V_3 and the curvature deflecting electrode is biased at V_d. Parameter W defines the working distance. In terms of the pass energy E_p, $V_d = 0.5714\ E_p$, while $V_1 = V_d$, $V_2 = 0.711\ V_d$, and $V_3 = 0.302\ V_d$. (b) Simulated ray paths of electrons through the spectrometer at the pass energy for a wide variety of entrance angles. The central ray enters in at 32.6° and 21 trajectories are plot over uniform steps for an input angular spread varying from −6° to 6°.

Simulation results for the RMA design predict that it has second-order focusing properties, and that the best energy resolution does not lie at the output Gaussian focal plane, but occurs a little distance before it (Hoang and Khursheed, 2011). The best simulated relative energy resolution was found to be 0.025% for an input angular spread of $\pm$ 6°, a factor of 4 times better than the Gaussian plane value. This high energy resolution performance is graphically demonstrated in Fig. A1-2, in which electrons for an input polar angular spread of $\pm$ 6° at the detection plane for three different energies are plot, the energy spread is $\pm$ 0.1% of the pass energy. Figure A1-2 shows that the electrons at the detection plane for these energies are well separated and confirms that the best simulated relative energy resolution is around 0.025%. This predicted energy resolution is around an order of magnitude better than that of the CMA (Sar-El, 1970) and the wide-angle toroidal energy analyzer (Khursheed and Hoang, 2008) for the same entrance angular spread, and is comparable to the HDA operating in its retardation mode (Benis and Zouros, 2008).

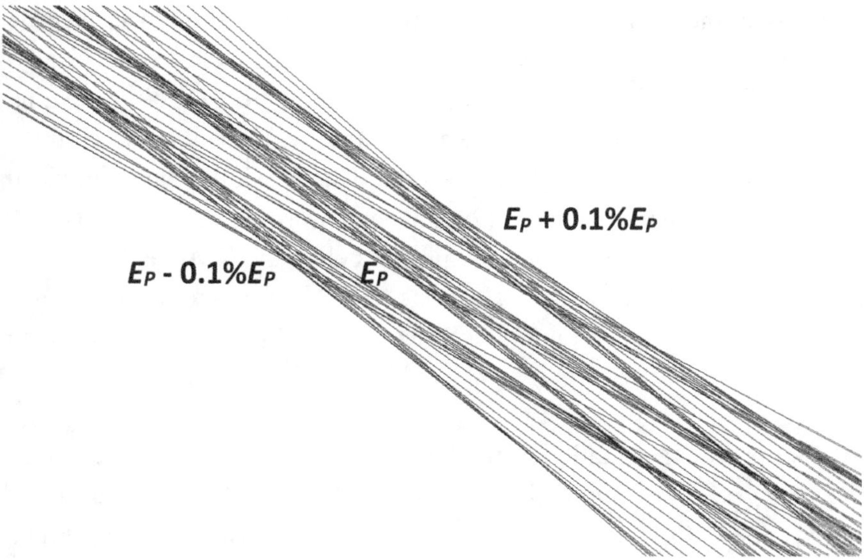

Fig. A1-2. Simulated trajectories around the output focal plane for 3 emission energies E_P – 0.1%E_P, E_P and E_P + 0.1%E_P, where E_p is the pass energy, and the input angles range from -6° to 6° around the central ray in uniform steps.

Since the RMA has a ring detection plane, it is predicted to have high transmittance (signal-to-noise), like the CMA or wide-angle toroidal energy analyzer, with the added advantage of significantly better predicted energy resolution. It is also better adapted to fit around the lower-pole piece of the final objective lens, unlike the wide-angle toroidal energy analyzer, which needs to fit below it. This means that for a given analyzer size, the RMA can achieve smaller working distances than the wide-angle toroidal energy analyzer.

Despite the RMA having several design advantages over the wide-angle toroidal energy analyzer, it has some additional practical limitations. One practical disadvantage is that the RMA requires a large ring detector, and electron detector geometries of this kind are not readily available, it needs to be custom made. Electrons exiting the analyzer do not naturally converge to a point, as in the wide-angle toroidal energy analyzer design. One way of solving that problem would be to use an array of electron detectors in the azimuthal angular direction, which may need to provide additional focusing for the higher energy scattered electrons. Ideally, some kind of post-analyzer arrangement is required to redirect exiting electrons towards the primary beam axis and focus them on to a single detector placed below the specimen stage. This possibility was simulated by Srinivasan and Khursheed (2015a), and is shown in Fig. A1-3. A toroidal electric sector analyzer is placed below the RMA exit aperture slit, deflecting all transmitted electrons back towards the primary beam axis, and focusing them on to single detector placed below the specimen stage.

Another practical limitation of the RMA design is created by its entrance and exit grids. The simulation used to predict its energy resolution of 0.025% for an input polar angular spread of ± 6° assumed the grids to be perfect equipotential surfaces. In practice, more realistic simulation which takes into account the non-uniformity of the electric potential around grid structures predicts that the presence of entrance/exit grids will lower the energy resolution performance. Srinivasan and Khursheed

(2015a) performed three-dimensional simulations of radial wire entrance/exit grids in the RMA and found that the steeply rising negative potential inside the analyzer above the entrance grid punches through the grid wire gaps, and significantly changes the trajectory paths of electrons through the analyzer. Their simulations predicted that for a 50% transparency factor, the energy resolution of electrons following different trajectory paths in the R-Z plane could be degraded by as much as a factor of two, while the energy resolution of some electrons experiencing non-radial trajectory deflections could be brought down by a factor of four. Srinivasan and Khursheed did not estimate how much the overall energy resolution would be degraded by such effects, and also stressed the need to model other possible grid structure geometries. No meaningful estimates can therefore be made at the moment on just how much the energy resolution will degrade for a given grid transparency, more simulation work is required.

Srinivasan and Khursheed (2015a) made a prototype RMA which they used to carry out preliminary experiments for a primary beam energy of 10 keV (beam current 150 pA) inside a Philips ESEM XL30 FEG SEM. The working distance of their prototype RMA was 7 mm. Although experimental SE energy spectral signals were recorded, the output signal was relatively low, caused by the fact that they used only a single PMT detector positioned relatively far from the exit aperture. Further experimental progress is dependent on making a better detection system.

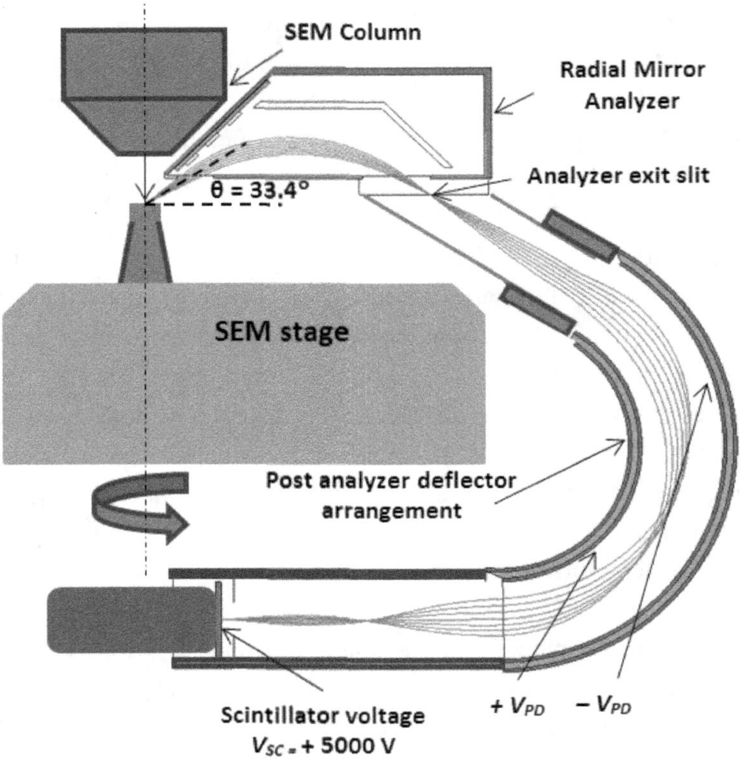

Fig. A1-3. Post analyzer toroidal electric sector arrangement. 7 rays are plot over a polar angular spread (Δθ) of ± 6° in uniform angular steps, shown here from the specimen to the scintillator of the PMT through the post analyzer deflector, at the central energy E_P. The magnitude of V_{PD} was experimentally calculated to be 0.436 E_P.

Appendix 2

The Parallel Radial Mirror Analyzer (PRMA) Attachment Design

In the category of parallel electron energy analyzer attachments proposed for AES in the SEM, the electric parallel radial mirror analyzer (PRMA) design (Khursheed *et al.*, 2012) has several significant advantages over its predecessors such as the hyperbolic field analyzer (HFA). The HFA was first proposed for AES applications by Jacka *et al.* (1999), and adapted for the SEM by Cubric *et al.* (2011). Although the detection energy range of the HFA can be relatively large, 75 eV to 2600 eV in the Jacka *et al.* design, its entrance angular spread is limited to around 2.2° (± 1.1°) in both polar and azimuthal angular directions. The PRMA in comparison, was designed to be fully rotationally symmetric (0 to 360° azimuthal angle collection) and has polar angular spreads of around ± 3°, giving it orders of magnitude better predicted output signal-to-noise ratios than the HFA for a comparable energy range of detection. In addition, the PRMA was simulated to have second-order focusing characteristics on its detection plane, superior to the first-order focusing characteristics of the HFA, giving it a predicted energy resolution of around one order of magnitude better than the HFA for the same input polar angular spread (Khursheed *et al.*, 2012). In principle, the PRMA can be designed to acquire the SE energy spectrum in parallel, and therefore has the potential to greatly speed up data acquisition times.

Figure A2-1 shows simulated electron trajectory paths for a modified version of the PRMA where its electrode potential values were determined by a Damped-Least Squares (DLS) module operating in combination with Lorentz 2D by Integrated Engineering Software for a distance of 7 mm between the specimen and analyzer entrance (Cheong *et al.*, 2015)).

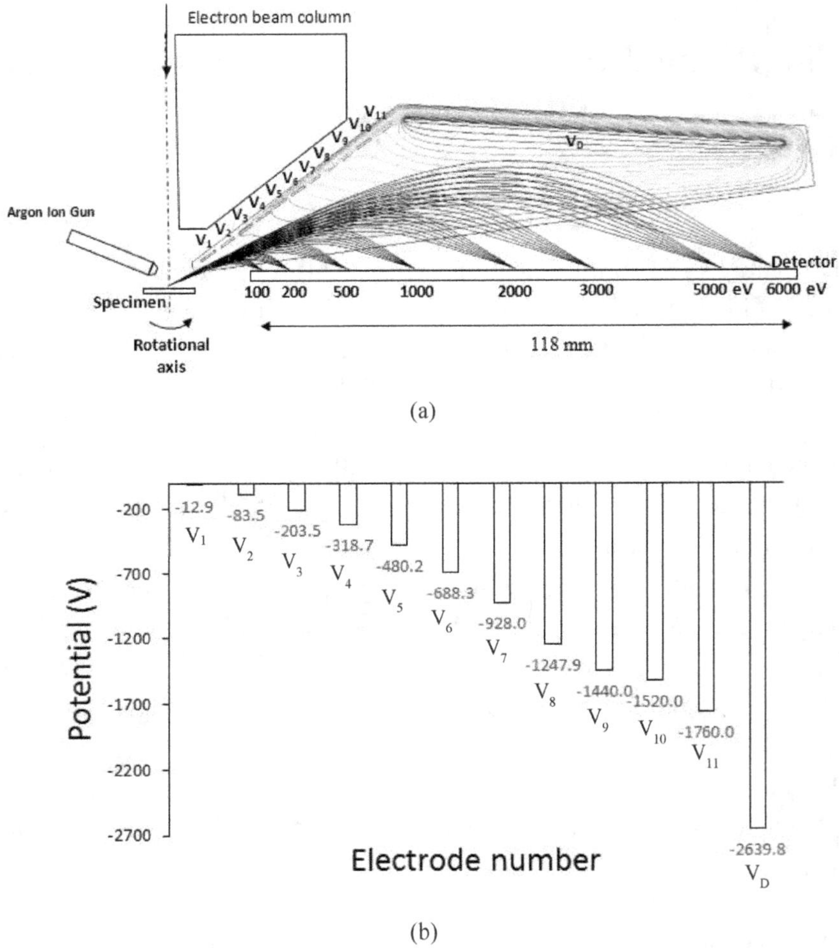

Fig. A2-1. PRMA design for an analyzer working distance of 7 mm (Cheong *et al.*, 2015). (a) Direct ray tracing of seven trajectories plotted evenly between −3° and 3° around a 24.2° polar entrance angle. Equipotential lines plot from −176 to -2464V in uniform steps of −176V are also indicated. (b) Electrode potentials.

Figure A2-2a shows ray tracing around the output focal plane at 1% above and below the selected energies of 600, 2000, 3500 and 5000 eV for an angular spread of ±3° in uniform steps of 1°. It indicates that the predicted energy resolution (corresponding to the half-width of the central ray) is around 0.2% for a wide range of energies. Figure A2-2b shows that the simulated focusing properties of the analyzer design are of second-order for three selected energies (third-order aberration dependence). Similar trace-width diagrams like these were obtained for other energies, confirming that second-order focusing is predicted for the entire electron energy range of detection.

Although the PRMA design has many potential benefits, not only for AES applications in the SEM, but also for SE energy spectroscopy, it has many practical challenges. Its twelve internal electrodes, most of which lie very close together, need to be biased to different negative voltages in a reliable way. Apart from the obvious challenging connection and power supply instability problems, there is also the issue of preventing electrical spark over for high pass energies. The other major practical difficulty comes from finding a suitable large ring/curvilinear position sensitive detector. In order for the PRMA to have a significant data acquisition time advantage over the wide-angle toroidal energy analyzer, it needs to have tens to hundreds of energy channels all working simultaneously, capturing different parts of the energy spectrum. This is because its operational polar angular spread is already around a factor of three lower than the one for the wide-angle toroidal energy analyzer, and therefore its higher level of averaging (required to reduce shot noise) can only be compensated by having a relatively high number of energy channels. A large ring/curvilinear position sensitive detector having hundreds of energy channels all working simultaneously requires the construction of a custom made detector system, and is not something that is easy for individual researchers to make by themselves. In the case of capturing the SE energy spectrum in parallel, there is the added complication that in order to be detected, the exiting electrons need to be accelerated to higher energies.

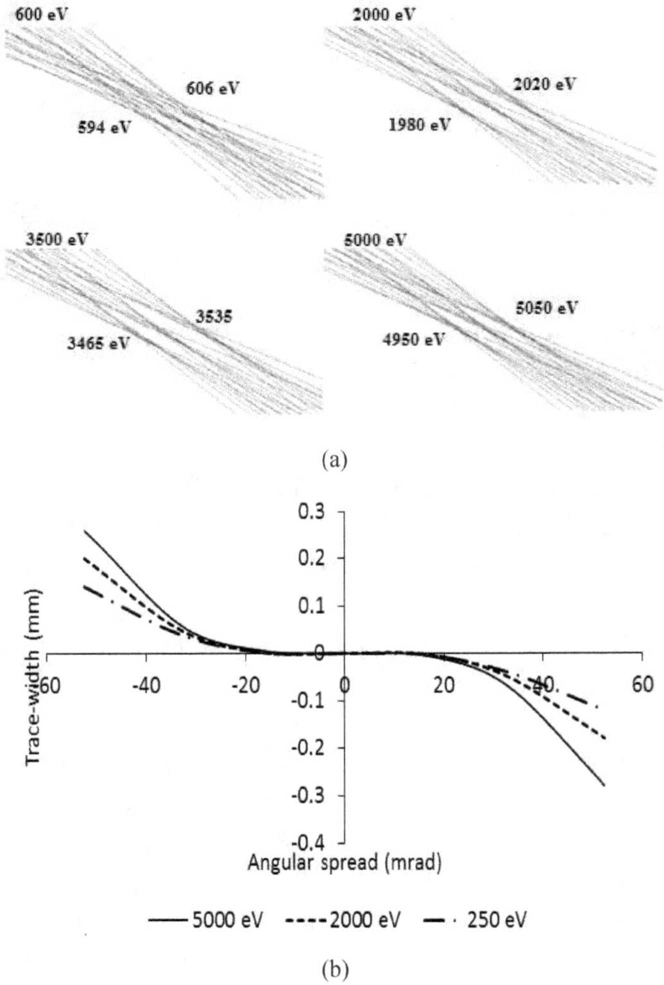

Fig. A2-2. Simulated characteristics on the focal plane at selected energies of the PRMA design, for an angular spread of ±3° in uniform steps of 1°. (a) Trajectory paths with energies at ±1% below and above the central energies of 600, 2000, 3500 and 5000 eV. (b) Trace-width as a function of polar angular spread on the Gaussian focal plane at selected energies. The polar angular spread ranges from −3° to +3°.

Appendix 3

TCAD Simulation of Test Semiconductor Samples with *p*-doped Stripes in *n*-substrate

Simulation of the *p*-stripe in *n*-type substrate semiconductor test sample was carried out by the TCAD Sentaurus software. The Sentaurus SPROCESS program (Sentaurus Process user guide, 2015) was used to simulate the ion implantation distribution of dopants into the silicon sample, and the Sentaurus SDEVICE program (Sentaurus Device user guide, 2015) was used to subsequently solve the semiconductor device field equations to find the final carrier, space-charge and electric potential distributions.

Ion implantation modeling parameters were first adjusted in order to fit a SRP experimentally supplied depth dopant profile for the P19 stripe. Figure A3-1a shows a schematic cross-section diagram for the *p*-stripe well geometry inside the *n*-type substrate, where the simulated mid-plane dopant depth profile was compared to the experimental SRP distribution. Some parameters in TCAD SPROCESS were obtained from the original device fabrication data, while others were adjusted to give the best fit. Figure A3-1b shows both simulated and experimental dopant depth profiles on the mid-plane, a fit with a RMSD error of around 1.1% was achieved. The simulation was done in 2D, assuming doped stripes to be infinitely long. The lateral profile direction range was adjusted to be around 40 microns, and the depth profile range was set to be around 20 microns. A step size of 5 nm was used in both directions. The dopant concentration of the phosphorus-doped silicon substrate was 10^{15} cm^{-3}. A 30 micron mask was specified at the top of the silicon, using Boron as the implanted species. The Ion implantation energy was 60 keV and the

Boron dose was given a value of 1.5×10^{15} cm^{-2}. A second step was simulated for the annealing process with a temperature of 1100 degrees for 40 minutes time in an inert gas environment. Figure A3-2 shows simulated dopant and electric potential distributions using the TCAD model.

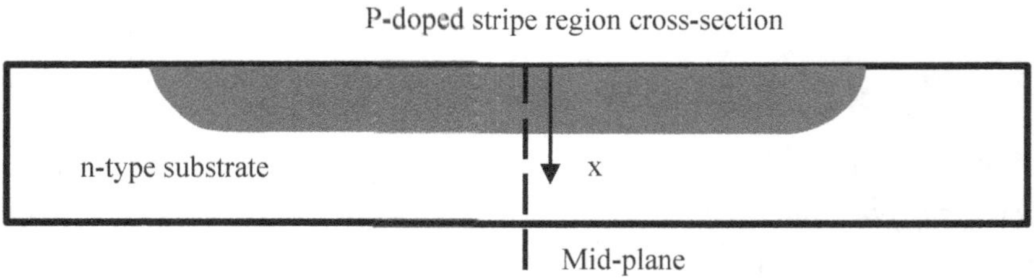

(a)

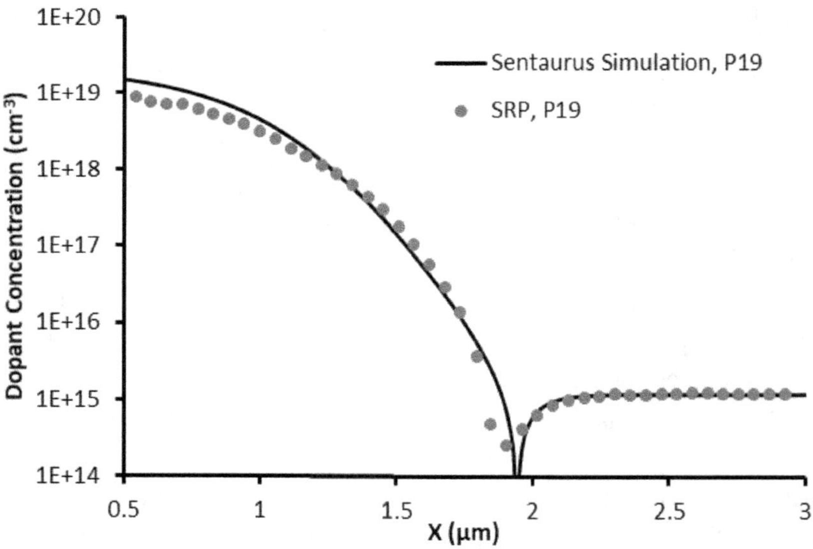

(b)

Fig. A3-1. Modeling and simulation for the P19 dopant depth profile on the test sample. (a) Schematic diagram of test sample cross-section for one *p*-doped stripe region. (b) TCAD Sentaurus SPROCESS simulation of P19 optimized to fit with SRP experimental depth profile. The simulation was based on implantation and diffusion parameters used in the sample fabrication.

(a)

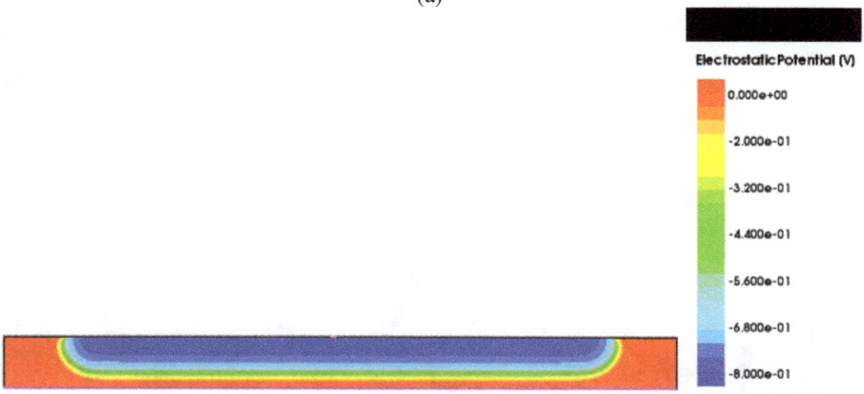

(b)

Fig. A3-2. Semiconductor device equation simulation of the P19 stripe on the test sample using TCAD Sentaurus. (a) Contour plot of the simulated dopant distribution obtained by the TCAD SPROCESS program. (b) The simulated electric potential distribution obtained through the use of the TCAD SDEVICE program.

Bibiliography

Agemura, T., Iwai, H., and Sekiguchi, T. (2017). Development of fountain detectors for spectroscopy of secondary electron in SEM, *Phys. Status Solidi C* 14, 1700057.

Agemura, T., Kimura, T., and Sekiguchi, T. (2018). Development of a fountain detector for spectroscopy of secondary electrons in scanning electron microscopy. *Japanese Journal of Applied Physics*, *57*(4), 046701.

Antolak, A. J., and Williamson, W. (1985). Electron backscattering from bulk materials. *Journal of Applied Physics*, *58*(1), 526–534.

Arat, K., T., Klimpel, T., and Hagen, C. W. (2019). Model improvements to simulate charging in scanning electron microscope, *J. Micro/Nanolith. MEMS MOEMS* **18**(4), 044003.

Archard, G. D. (1961). Back Scattering of Electrons. *Journal of Applied Physics*, *32*(8), 1505–1509.

Arnal, F., Verdier, P., and Vincensini, P. D. (1969). Coefficient de rétrodiffusion dans le cas d'électrons monocinétiques arrivant sur la cible sous une incidence oblique. *C. R. Acad. Sci.*, *268*, 1526–1529.

Assa'd, A., and Gomati, M. El. (1998). Backscattering Coefficients for Low Energy Electrons. *Scanning Microscopy*, *12*(1), 185.

Azzolini, M., Angelucci, M., Cimino, R., and Larciprete, R. (2019). Secondary electron emission and yield spectra of metals from Monte Carlo simulations and experiments, *Journal of Physics: Condensed Matter*, Volume 31, 055901.

Belhaj, M., Jbara, O., Odof, S., Msellak, K., Rau, E. I., Adrianov, M. V. (2000). An Anomalous Contrast in Scanning Electron Microscopy of Insulators: The Pseudo Mirror Effect., *SCANNING* VOL. 22, pp. 352–356.

Bindi, R., Lanteri, H., and Rostaing, P. (1979). Distributions energetiques experimentales des electrons secondaires emis par des cibles evaporees d'aluminium, de cuivre, d'or et d'argent. *Journal of Electron Spectroscopy and Related Phenomena*, *17*(4), 249–258.

Bock, R., Altermatt, P. P., and Schmidt, J. (2008). Accurate extraction of doping profiles from electrochemical capacitance voltage measurements. *23rd European Photovoltaic Solar Energy Conference*, (September), 4.

Broekman, L., A. Tadich, A., E. Huwald , E., Riley, J., Leckey, R., Seyller, T., Emtsev, K., and Leyb, L. (2005). First results from a second generation toroidal electron spectrometer, *Journal of Electron Spectroscopy and Related Phenomena* **144–147**, 1001–1004.

Burgess, S., James, H., Statham, P., and Xiaobing, L. (2013). Using Windowless EDS Analysis of 45-1000eV X-ray Lines to Extend the Boundaries of EDS Nanoanalysis in the SEM. *Microscopy and Microanalysis*, *19*(S2), 1142–1143.

Burgess, Simon, Sagar, J., Holland, J., Li, X., and Bauer, F. (2017). Ultra-Low kV EDS – A New Approach to Improved Spatial Resolution, Surface Sensitivity, and Light Element Compositional Imaging and Analysis in the SEM. *Microscopy Today*, *25*(2), 20–29.

Cazaux, J. (2011). Calculated effects of work function changes on the dispersion of secondary electron emission data: Application for Al and Si and related elements. *Journal of Applied Physics*, *110*(2).

Chang T. H. P. and Nixon W. C. (1967). Engineering Laboratory, Cambridge University, Cambridge, U.K., Solid-State Electronics Pergamon Press, Vol. 10, pp. 701–704.

Chee, A. K. W. (2016). Quantitative Dopant Profiling by Energy Filtering in the Scanning Electron Microscope. *IEEE Transactions on Device and Materials Reliability*, *16*(2), 138–148.

Chee, A. K. W. (2018). Enhancing doping contrast and optimising quantification in the scanning electron microscope by surface treatment and Fermi level pinning. *Scientific Reports*, (February), 2–11.

Chee, A. K. W., Broom, R. F., Humphreys, C. J., and Bosch, E. G. T. (2011). A quantitative model for doping contrast in the scanning electron microscope using calculated potential distributions and Monte Carlo simulations. *Journal of Applied Physics*, *109*(1), 1–9.

Cheong, K. H., Weiding Han, W., Khursheed, A, and Nelliyan, K. (2015). A Parallel Radial Mirror Energy Analyzer Attachment for the Scanning Electron Microscope, *Microscopy and Microanalysis*, Volume 21, Issue S4 (Proceedings Ninth International Conference on Charged Particle Optics), pp. 142–147.

Chung, M. S., and Everhart, T. E. (1974). Simple calculation of energy distribution of low energy secondary electrons emitted from metals under electron bombardment. *Journal of Applied Physics*, *45*(2), 707–709.

Cubric, D., De Fanis, A., Konishi, I., and Kumashiro, S. (2011). Parallel acquisition electrostatic electron energy analyzers for high throughput nano-analysis, *Nucl. Instrum. Meth. Phys. Res.* A **645**, pp. 227–233.

Da, B., Liu, J., Harada, Y., Cuong, N., T., Tsukagoshi, K., Hu, J., Yang, L., Ding, Z., Yoshikawa, H., and Tanuma, S. (2019). Observation of Plasmon Energy Gain for Emitted Secondary Electron in Vacuo, *J. Phys. Chem. Lett.* 2019, **10,** 5770–5775.

Dapor, M., Inkson, B. J., Rodenburg, C., and Rodenburg, J. M. (2008). A comprehensive Monte Carlo calculation of dopant contrast in secondary-electron imaging. *EPL (Europhysics Letters)*, *82*(3), 30006.

Dapor, M., Jepson, M. A. E., Inkson, B. J., and Rodenburg, C. (2009). The effect of oxide overlayers on secondary electron dopant mapping. *Microscopy and Microanalysis : The Official Journal of Microscopy Society of America, Microbeam Analysis Society, Microscopical Society of Canada*, *15*(3), 237–243.

Ding, Z. J., Li H. M., Shimizu, R., and Goto., K. (2008). On the Energy Distribution of Secondary Electrons Emitted from Metals, *Journal of Surface Analysis* Vol.15, No. 2, pp. 186–194.

Dinnis, A. R., Khursheed, A., and Smart, P. D. (1992). The time-of-flight spectrometer: first results, *Microelectronic Engineering* **16**, pp. 35–42.

Drescher, H., Reimer, L., and Seidel, H. (1970). Backscattering and secondary electron emission of 10-100 keV electrons in scanning electron microscopy. *Z. ANGEW PHYSIK*, *29*(6), 331–336.

Dubbledam, L. (1993). Electron Spectrometers and Voltage Measurements, in J. T. L. Thong, *Electron Beam testing Technology*, (Plenum Press, New York). pp. 211-239.

Dubbledam, L. and Kruit, P. (1987a). Signal-to-Noise ratio improvement in electron beam testing by using a dispersive analyser, *Scanning Microscopy* **1** (4), pp. 1647–1650.

Dubbledam, L. and Kruit, P. (1987b). First experimental results of an E-beam tester with dispersive secondary electron energy analyzer, *Microelectronic Engineering* **7**, Iss. 2-4, pp. 231–234.

El-Gomati, M. M., Wells, T. C. R., Müllerová, I., Frank, L., and Jayakody, H. (2004). Why is it that differently doped regions in semiconductors are visible in low voltage SEM? *IEEE Transactions on Electron Devices*, *51*(2), 288–292.

El-Gomati, M., Zaggout, F., Jayacody, H., Tear, S., and Wilson, K. (2005). Why is it possible to detect doped regions of semiconductors in low voltage SEM: A review and update. *Surface and Interface Analysis*, *37*(11), 901–911.

Elliott, S. L., Broom, R. F., and Humphreys, C. J. (2002). Dopant profiling with the scanning electron microscope - A study of Si. *Journal of Applied Physics*, *91*(11), 9116–9122.

Everhart T. E., Wells O. O., and Oatley C. W. (1959) Factors Affecting Contrast and Resolution in the Scanning Electron Microscope, *International Journal of Electronics,* 7:2, 97–111.

Everhart, T. E., Saeki, N., Shimizu, R., and Koshikawa, T. (1976). Measurement of structure in the energy distribution of slow secondary electrons from aluminum, *Journal of Applied Physics* **47**, p. 2941.

Farhang, H., Napchan, E., and Blott, B. H. (1993). Electron backscattering and secondary electron emission from carbon targets: comparison of experimental results with Monte Carlo simulations. *Journal of Physics D: Applied Physics*, *26*(12), 2266–2271.

Feuerbaum, H. P. (1979). VLSI testing using the electron probe, *Scanning Electron Microscopy* **I**, pp. 285–296.

Flemming, J. P. and Ward, E. W. (1970). A technique for accurate measurement and display of applied potential distributions using the scanning electron microscope, *Scanning Electron Microscopy* **I**, pp. 465–470.

Frosien, J. and Plies, E. (1987). High performance electron optical column for testing ICs with submicrometer design rules, *Microelectronic Engineering* 7, Iss. 2–4, pp. 163–172.

Girard, P, Charpenel, P. and Matin, H. (1992). Determination of the noncharging electron-beam energy on insulators, *Journal of Applied Physics* **71**, 2871.

Goldstein, J. I., Newbury, D. E., Michael, J. R., Ritchie, N. W. M., Scott, J. H. J., and Joy, D. C. (2017). *Scanning electron microscopy and X-ray microanalysis*. Springer.

Gopinath, A. (1977). Estimate of minimum measurable voltage in the SEM, *Journal of Physcis E: Scientific Instruments* Vol. **10**, pp. 911–913.

Gostev, A. V., Orlikovskii, N. A., Rau, E. I., and Trubitsyn, A. A. (2013). Updating of the toroidal electron spectrometer intended for a scanning electron microscope and its new applications in diagnostics of micro- and nanoelectronic structures. *Technical Physics*, *58*(3), 447–454.

Granneman, E. H. A. and Van Der Wiel, M. J. (1983). Transport, dispersion and detection of electrons, ions and neutrals, in E. E. Koch (ed.), *Handbook on Synchrotron Radiation*, Vol. 1, (North Holland publishing company).

Hachenberg, O., and Brauer, W. (1959). Secondary Electron Emission from Solids. *Advances in Electronics and Electron Physics*, *11*(C), 413–499.

Han, W., Srinivasan, A., Banerjee, A., Chew, M., and Khursheed, A. (2019). Beyond conventional secondary electron imaging using spectromicroscopy and its applications in dopant profiling, *Materials Today Advances* **2** 100012.

Hannah, J. M. (1974). Scanning electron microscope applications to integrated circuit testing. The University of Edinburgh.

Hashimoto, Y., Takeuchi, S., Sunaoshi, T., and Yamazawa, Y. (2020). Voltage contrast imaging with energy filtered signal in a field-emission scanning electron microscope, *Ultramiroscopy* **209**, 112889.

Henke, B. L., Liesegang, J., and Smith, S. D. (1979). Soft-x-ray-induced secondary-electron emission from semiconductors and insulators: Models and measurements. *Physical Review B*, *19*(6), 3004–3021.

Henrich, V. E. (1973). Role of Bulk and Surface Plasmons in the Emission of Slow Secondary Electrons: Polycrystalline Aluminum, *Physical Review* B 7(8) Volume 7, Number 8, pp. 5312–19.

Hoang, H. Q., and Khursheed, A. (2011). A radial mirror analyzer for scanning electron/ion microscopes. Nuclear Instruments and Methods in Physics Research, Section A: Accelerators, Spectrometers, Detectors and Associated Equipment, 635(1), 64–68.

Hoang, H. Q., Osterberg, M., and Khursheed, A. (2011a). A high signal-to-noise ratio toroidal electron spectrometer for the SEM. *Ultramicroscopy*, *111*(8), 1093–1100.

Hoang, H. Q., Osterberg, M., and Khursheed, A. (2011b). Experimental results from a second-order focusing toroidal energy spectrometer attachment for scanning electron microscopes. *Nuclear Instruments and Methods in Physics Research, Section A: Accelerators, Spectrometers, Detectors and Associated Equipment*, *645*(1), 241–244.

Hoffman, A., Nyberg, G. L., and Prawer, S. (1990). High-energy angle-resolved secondary-electron emission spectroscopy of highly oriented pyrolytic graphite, *J. Phys.: Condens. Matter* **2** 8099.

Insepov, Z., Ivanov, V., Jokela, S. J., Veryovkin, I., Zinovev, A., and Frisch, H. (2011). Comparison of secondary electron emission simulation to experiment. *Nuclear Instruments and Methods in Physics Research Section A: Accelerators, Spectrometers, Detectors and Associated Equipment*, *639*(1), 155–157.

Ishitani, T., and Tsuboi, H. (2006). Objective comparison of scanning ion and scanning electron microscope images. *Scanning*, *19*(7), 489–497.

Jackson, S. (1995). Determining hybridization differences for amorphous carbon from the XPS C 1s envelope. *Applied Surface Science*, *90*(2), 195–203.

Joy, D. C. and Joy, C. S. (1996). Low Voltage Scanning Electron Microscopy. *Micron* 27, pp. 247–263.

Joy, D. C., Khanna, David Braski, D. (1999). Secondary electron spectroscopy for microanalysis and defect review, *Proc. SPIE* 3677, Metrology, Inspection, and Process Control for Microlithography XIII, (14 June 1999), pp. 621–8.

Joy, D. C., Prasad, M. S., and Meyer, H. M. (2004). Experimental secondary electron spectra under SEM conditions. *Journal of Microscopy*, *215*(1), 77–85.

Joy, D. and Griffin, B. (2011). Secondary Electron Imaging - Doing it Better. *Microscopy and Microanalysis*, *17*(SupplementS2), 876–877.

Joy, David C. (1987). A model for calculating secondary and backscattered electron yields. *Journal of Microscopy*, *147*(1), 51–64.

Kazemian, P., Mentink, S. A. M., Rodenburg, C., and Humphreys, C. J. (2006). High resolution quantitative two-dimensional dopant mapping using energy-filtered secondary electron imaging. *Journal of Applied Physics*, *100*(5).

Kazemian, P., Mentink, S. A. M., Rodenburg, C., and Humphreys, C. J. (2007). Quantitative secondary electron energy filtering in a scanning electron microscope and its applications. *Ultramicroscopy*, *107*(2–3), 140–150.

Kazemian, P., Rodenburg, C., and Humphreys, C. J. (2004). Effect of experimental parameters on doping contrast of Si p-n junctions in a FEG-SEM. *Microelectronic Engineering*, *73–74*, 948–953.

Kessel, van L., Hagen, C. W., and Kruit, P. (2019). Surface effects in simulations of scanning electron microscopy images, *J. Micro/Nanolith. MEMS MOEMS* **18**(4), 044002.

Khursheed, A. (1983a). The computer aided design of electron detectors for the scanning electron microscope, PhD thesis, University of Edinburgh, Edinburgh, Scotland, Patent application no. 8308329 Voltage contrast detector for SEM, pp. 142–5.

Khursheed, A. (1983b). The computer aided design of electron detectors for the scanning electron microscope, PhD thesis, University of Edinburgh, Edinburgh, Scotland, chapter 5 Surface Fields, pp. 78–101

Khursheed, A., and Karuppiah, N. (2001). An add-on secondary electron energy spectrometer for scanning electron microscopes. *Review of Scientific Instruments*, *72*(3), 1708.

Khursheed, A. (2015). Energy Analyzer Attachments for the Scanning Electron Microscope. *Microscopy and Microanalysis*, *21*(S4), 130–135.

Khursheed, Anjam. (2011a). In-lens Improvements. In *Scanning Electron Microscope Optics and Spectrometers* (pp. 135–183). World Scientific.

Khursheed, Anjam. (2011b). Spectrometer design principles. In *Scanning Electron Microscope Optics and Spectrometers* (pp. 71–134). World Scientific Publishing Co. Pte. Ltd.

Khursheed, Anjam. (2011c). The emission hemisphere and BSE collection. In *Scanning Electron Microscope Optics and Spectrometers* (pp. 51–3). World Scientific Publishing Co. Pte. Ltd.

Khursheed, A., and Dinnis, A. R. (1984). A comparison of voltage contrast detectors, *Scanning* **6**, pp. 85–95.

Khursheed, A. and Dinnis, A. R. (1990). A time-of-flight voltage contrast detector for measurements on VLSI circuits, *Meas. Sci. Technol.* **1**, pp. 581–591.

Khursheed, A. and Dinnis, A. R. (1992). Experimental results from a time-of-flight spectrometer for electron beam testing, *Microelectronic Engineering*, **17**, pp. 451–4.

Khursheed, A. (1992). Multi-Channel vs Conventional Retarding Field Spectrometers, *Microelectronic Engineering*, **16**, No. 1–4, pp. 43–50.

Khursheed, A., and Hoang., H. Q. (2008a). Redesign of the scanning electron microscope for parallel energy spectral acquisition, *Ultramicroscopy* **108,** pp. 151–157.

Khursheed, Anjam, and Hoang, H. Q. (2008b). A second-order focusing electrostatic toroidal electron spectrometer with 2π radian collection. *Ultramicroscopy*, *109*(1), 104–110.

Khursheed, A. and Osterberg, M. (2006). Developments in the design of a spectroscopic scanning electron microscope, *Nucl. Instrum. and.Meth. A* **556,** p. 437.

Khursheed, Anjam, Hoang, H. Q., and Srinivasan, A. (2012). A wide-range Parallel Radial Mirror Analyzer for scanning electron/ion microscopes. *Journal of Electron Spectroscopy and Related Phenomena*, *184*(11–12), 525–532.

Kienle, M., and Plies, E. (2004). An off-axis multi-channel analyzer for secondary electrons. In *Nuclear Instruments and Methods in Physics Research, Section A: Accelerators, Spectrometers, Detectors and Associated Equipment*.

Kollath, R. (1947). Zur Energieverteilung der Sekundärelektronen II. Meßergebnisse und Diskussion. *Annalen Der Physik*, *436*(7–8), 357–380.

Konvalina, I., Mika, F., Krátký, S., Mikmeková, E. M., Müllerová, I. (2019). In-Lens Band-Pass Filter for Secondary Electrons in Ultrahigh Resolution SEM, *Materials* **2019**, 12, 2307.

Kruit, P. and Read, F. H. (1983). Magnetic field paralleliser for 2π electron-spectrometer and image magnifier, *J. Phys. E: Sci. Instrum.* **16**, pp. 313–324.

Krupka, J. (2013). Contactless methods of conductivity and sheet resistance measurement for semiconductors, conductors and superconductors, *Meas. Sci. Technol.* **24** (2013) 062001 (13 pp).

Kuhr, J. C. and Fitting, H. J. (1979). Monte Carlo simulation of electron emission from solids, *Journal of Electron Spectroscopy and Related Phenomena* **105** (1999) 257–273.

Lanteri, H., Bindi, R., and Rostaing, P. (1977). Effects of the working conditions of retarding field spectrometers on the energy distribution of secondary electrons, *Journal of Electron Spectroscopy and Related Phenomena,* **12**, pp. 451-458.

Lascovich, J. C., Giorgi, R., and Scaglione, S. (1991). Evaluation of the sp2/sp3 ratio in amorphous carbon structure by XPS and XAES. *Applied Surface Science*, *47*(1), 17–21.

Lascovich, J. C., and Scaglione, S. (1994). Comparison among XAES, PELS and XPS techniques for evaluation of Sp2 percentage in a-C:H. *Applied Surface Science*, *78*(1), 17–23.

Lasdon, L. S., Waren, A. D., Jain, A., and Ratner, M. (1978). Design and Testing of a Generalized Reduced Gradient Code for Nonlinear Programming. *ACM Transactions on Mathematical Software*, *4*(1), 34–50.

Leckey R. C. G. and J. D. (1985). A toroidal angle-resolving electron spectrometer for surface studies, *Applications of Surface Science* **22/23,** 196–205.

Leckey R. C. G. and J. D. (1988). Angle resolved spectrometer, US patent number 4, 758, 722.

Lesiak, B., Kövér, L., Tóth, J., Zemek, J., Jiricek, P., Kromka, A., and Rangam, N. (2018). C sp2/sp3 hybridisations in carbon nanomaterials – XPS and (X)AES study. *Applied Surface Science*, *452*, 223–231.

Li-Lung, L. (2005). Methodology of optimum-kV BSE application in SEM and summary of characteristic for low energy SEM/EDS/FIB. In *Istfa 2005: Proceedings of the 31st International Symposium for Testing and Failure Analysis*. ASM International.

Lide, D. R. *et al.* (2009). CRC Handbook of Chemistry and Physics, 2009–2010, 90th ed. CRC Handbook of Chemistry and Physics, 2009–2010, 90th ed . Edited by David R. Lide , Editor-in-Chief, and W. M. "Mickey" Haynes , Associate Editor (National Institute of Standards and Technology, *Journal of the American Chemical Society*, *131*(35), 12862–12862.

Lin, Y. and Joy, D. C. (2005). A new examination of secondary electron yield data. Surf. Interface Anal. 2005; 37: 895–900.

Lv, B., Qian, T. and Ding, H. Angle-resolved photoemission spectroscopy and its application to topological materials. *Nature Reviews Physics* **1**, 609–626 (2019).

Lye, R. G., and Dekker, A. J. (1957). Theory of Secondary Emission. *Physical Review*, *107*(4), 977–981.

Mathew, S., Annadi, A., Chan, T.K., Sasmara, T.C., Zhan, D., Wang, X.R., Azimi, S., Shen, Z., Rusydi, A., Breese, M.B.H. and Venkatesan, T. (2013). Tuning the interface conductivity of LaAlO3/SrTiO3 using ion beams: Implications for patterning. *ACS Nano* **7**(12), 10572–10581.

McMullan, D. (2006). Scanning electron microscopy 1928–1965. *Scanning*, *17*(3), 175–185.

McVicar, M. J., and Graves, W. J. (1997). The forensic comparison of soils by automated scanning electron microscopy. *Canadian Society of Forensic Science Journal*, *30*(4), 241–261.

Mizuhara, Y., Kato, J., Nagatomi, T., Takai, Y., and Inoue, M. (2002). Quantitative measurement of surface potential and amount of charging on insulator surface under electron beam irradiation, *J. Appl. Phys*. **92**, p. 6128.

Mizuhara, Y., Kato, J., Nagatomi,T., Takai, Y., and Inoue, M. (2003). Analysis of Dopant Concentration in Semiconductor Using Secondary Electron Method, *Jpn. J. Appl. Phys.* Vol. **42**, pp. L 709–L 711.

Mika, F., and Frank, L. (2008). Two-dimensional dopant profiling with low-energy SEM. *Journal of Microscopy*, *230*(1), 76–83.

Mika, F., Hovorka, M., and Frank, L. (2010). Imaging of Dopants with Presence of Surface Ad-Layers. In Proceedings of the 12th International Seminar on Recent Trends in Charged Particle Optics and Surface Physics Instrumentation. (pp. 35–36).

Mika, Filip, Konvalina, I., Krátký, S., and Müllerová, I. (2016). Bandpass Filter For Secondary Electrons In SEM - Experiments. In Filip Mika (Ed.), *Proceedings of the 15th International Seminar on Recent Trends in Charged Particle Optics and Surface Physics Instrumentation.* (pp. 36–37). Institute of Scientific Instruments The Czech Academy of Sciences, v. v. i.

Modinos, A. (1984). Secondary Electron Emission Spectroscopy. In *Field, Thermionic, and Secondary Electron Emission Spectroscopy* (p. 331). Boston, MA: Springer US.

Nakamae, K., Fujioka, H., and Ura, K. (1981). Local field effects on voltage contrast in the scanning electron microscope, *J. Phys. D: Appl. Phys.*, **14** (1981) pp. 1939–60.

O'Connor, D. J., Sexton, B. A., and Smart, R. S. C. (Eds.). (2003). *Surface Analysis Methods in Materials Science* (Vol. 23). Berlin, Heidelberg: Springer Berlin Heidelberg.

Oatley, C. W., Nixon, W. C., and Pease, R. F. W. (1965). Scanning Electron Microscopy. In *Advances in Electronics and Electron Physics* (Vol. 21, pp. 181–247).

Osterberg, M., Hoang., H. Q. and Khursheed, A. (2009). Initial experimental results on a magnetic beam separator spectrometer for the SEM, *Ultramicroscopy* **109**, pp. 1310–1314.

Pawley, J. (1984). Low voltage scanning electron microscopy. *Journal of Microscopy*, *136*(1), 45–68.

Perovic, D. D., Castell, M. R., Howie, A., Lavoie, C., Tiedje, T., and Cole, J. S. W. (1995). Field-emission SEM imaging of compositional and doping layer semiconductor superlattices. *Ultramicroscopy*, *58*(1), 104–113.

Peters, K. R. (1982). Conditions required for high quality high magnification images in secondary electron-I scanning electron microscopy. *Scanning Electron Microscopy*, (Pt 4), 1359–1372.

Pierron, J., Inguimbert, C., Belhaj, M., Gineste, T., Puech, J., and Raine, M. (2017). Electron emission yield for low energy electrons: Monte Carlo simulation and experimental comparison for Al, Ag, and Si, *J. Appl. Phys.* **121**, 215107.

Pillon, J., D. Roptin, D., and M. Cailler, M. (1976). Secondary electron emission from aluminium, *Surface Science* **59** (1976) 741–748.

Postek, M. T. (1996). An approach to the reduction of hydrocarbon contamination in the scanning electron microscope. *Scanning*, *18*(4), 269–274.

Postek, M. T., Vladár, A., Archie, C., and Ming, B. (2011). Review of current progress in nanometrology with the helium ion microscope, *Meas. Sci. Technol.* **22**, 024004, 1–14.

Prutton, M. and El-Gomat, M. M. (2006). Instrumentation chap3. In *Scanning Auger Electron Microscopy,* John Wiley & Sons.

Rau, E. I., and Robinson, V. N. E. (1996). An annular toroidal backscattered electron energy analyser for use in scanning electron microscopy. *Scanning*, *18*(8), 556–561.

Rau, E. I., and Tagachenkov, A. M. (2013). Image contrast of impurity regions of semiconductor crystals in scanning electron microscopy. *Bulletin of the Russian Academy of Sciences: Physics*, *77*(8), 943–947.

Reed, S. J. B. (2005). *Electron Microprobe Analysis and Scanning Electron Microscopy in Geology*. Cambridge University Press.

Reimer, L., and Tollkamp, C. (1980). Measuring the backscattering coefficient and secondary electron yield inside a scanning electron microscope. *Scanning*, *3*(1), 35–39.

Reimer, L. (1998a). *Scanning Electron Microscope, Physics of Image Formation and Microanalysis*, (Springer, 2nd edition).

Reimer, L. (1998b). EBIC chapter 7. In *Scanning Electron Microscope, physics of image formation and microanalysis*, (Springer, 2nd edition).

Riccardi, P., Cupolillo, A., Pisarra, M., Sindona, A., and Caputi, L. S. (2010). Observation of excited states of graphene on Ni(111) by secondary electron spectroscopy, *Appl. Phys. Lett.* **97**, 221909.

Richardson, N. and Muray, A. (1988). An improved magnetic-collimating secondary electron energy filter for very large scale integrated diagnostics. *J. Vac. Sci. Technol.* **B6**, pp. 417–421.

Rodenburg, C., R. Masters, K. A., Dapor, M., Krátký, S., and Mika, F. (2018). Secondary Electron Hyper Spectral Imaging in Helios Nanolab – Mapping Materials Properties or Artefacts? In Filip Mika and Z. Pokorná (Eds.), *Proceedings of the 16th International Seminar on Recent Trends in Charged Particle Optics and Surface Physics Instrumentation.* (pp. 68–69). Institute of Scientific Instruments The Czech Academy of Sciences, v. v. i.

Sah, C. T. (1991). *Fundamentals of Solid-state Electronics*. World Scientific.

Sar-el, E. Z. (1967). Cylindrical capacitor as an analyzer I. Nonrelativistic part, *The Review of Scientific Instruments* **38**, No. 6, pp. 1210–1216.

Schönjahn, C., Humphreys, C. J., and Glick, M. (2002). Energy-filtered imaging in a field-emission scanning electron microscope for dopant mapping in semiconductors. *Journal of Applied Physics*, *92*(12), 7667–7671.

Seebauer, E. G., and Barlaz, D. E. (2016). SIMS for analysis of nanostructures. *Current Opinion in Chemical Engineering*, *12*, 8–13.

Sekiguchi, T., Kimura, T. and Iwai, H. (2016). SEM observation of p-n junction in semiconductors using, *Superlattices and Microstructures* **99**, 165–168.

Sentaurus™ Device User Guide, (2015). K-2015.06, Synopsys, Inc.

Sentaurus™ Process User Guide, (2015). K-2015.06, Synopsys, Inc.

Shachal, D. and Bernadicou, M. (2000). Secondary electron spectroscopy and system, International PCT patent application, WO 00/70646, 23rd November.

Schäfer, J., Schoppe, R., Hölzl, J., and Feder, R. (1981). Experimental and theoretical study of the angular resolved secondary electron spectroscopy (ARSES) for W (100) in the energy range $0 < E < 20$ eV, *Surface Science* **107** (1981) 290–304.

Shatalov, V. M., Panchenko, O. F., Artamonov, O. M., Vinogradov, A. G., and Terekhov, A. N. (1988). Fine structure of angle-resolved secondary electron emission spectra in silicon. *Solid State Communications*, *68*(8), 719–722.

Sise, O., Martínez, G., Madesis, I., Laoutaris, A., Dimitriou, A., Fernández-Martín, M., and Zouros, T. M. J. (2016). The voltage optimization of a four-element lens used on a hemispherical spectrograph with virtual entry for highest energy resolution, *Journal of Electron Spectroscopy and Related Phenomena* **211**, pp. 19–31.

Srinivasan A. (2014). *Material characterization by energy filtered secondary electron signals inside the scanning electron microscope*, PhD Thesis, National University of Singapore, ScholarBank@NUS Repository.

Srinivasan, A. and Khursheed, A. (2015a). First Experiments Using a Radial Mirror Analyzer Attachment Prototype for Scanning Electron Microscopes, Volume 21, Issue S4 (Proceedings Ninth International Conference on Charged Particle Optics), pp. 136–141.

Srinivasan, A., and Khursheed, A. (2015b). Voltage and Dopant Concentration Measurements of Semiconductors using a Band-Pass Toroidal Energy Analyzer Inside a Scanning Electron Microscope. *Microscopy and Microanalysis*, *21*(04), 910–918.

Srinivasan, A., Han, W., and Khursheed, A. (2018). Secondary Electron Energy Contrast of Localized Buried Charge in Metal-Insulator-Silicon Structures. *Microscopy and Microanalysis*, *24*(5), 453–460.

Srinivasan, A., and Khursheed, A. (2014). Probing and analyzing buried interfaces of multifunctional oxides using a secondary electron energy analyzer. *Microscopy and Microanalysis : The Official Journal of Microscopy Society of America, Microbeam Analysis Society, Microscopical Society of Canada*, *20*(5), 1494–1498.

Sternglass, E. J. (1954). Backscattering of Kilovolt Electrons from Solids. *Physical Review*, *95*(2), 345–358.

Suga, S., and Sekiyama, A. (2014). *Photoelectron Spectroscopy, Bulk and Surface Electronic Structures, Springer Series in Optical Sciences*, Volume **176**, Editor-in-Chief W. T. Rhodes, Springer Publisher.

Suzuki, T., Kudo, M., Sakai, Y., and Ichinokawa, T. (2008). Material Contrast of Scanning Electron and Ion Microscope Images of Metals. *Microscopy Today*, *16*(1), 6–11.

Sekiguchi, T., Kimura, T. and Iwai, H. (2016). SEM observation of p-n junction in semiconductors using, *Superlattices and Microstructures* **99**, 165–168.

Thong, J. T. L. (1993a). *Electron Beam Testing Technology*. (John T. L. Thong, Ed.). Boston, MA: Springer US, chapter 1.

Thong, J. T. L. (1993b). *Electron Beam Testing Technology*. (John T. L. Thong, Ed.). Boston, MA: Springer US, chapters 6 and 7.

Thornhill, J. W. and Mackintosh, I. M. (1965). Westinghouse Research Laboratories, Pittsburgh 35, Pennsylvania. Microelectronics and Reliability Pergamon Press 1965, Vol. 4, pp. 97–100.

Tsurumi, D., and Hamada, K. (2013). Dopant mapping of semiconductors with scanning electron microscopy. *SEI Technical Review*, *1*(77), 141–146.

Ul-Hamid, A. (2018). *A Beginners' Guide to Scanning Electron Microscopy*, Springer publisher.

Valdes, L. B. (1954). Resistivity Measurements on Germanium for Transistors. *Proceedings of the IRE*, *42*(2), 420–427.

Vladar, A. E., Postek, Jr., M. T., and Vane, R. (2001). Active monitoring and control of electron-beam-induced contamination. In N. T. Sullivan (Ed.) (p. 835).

Vladar, A. and Postek, M. (2005). Electron Beam-Induced Sample Contamination in the SEM. *Microscopy and Microanalysis*, *11*(S02), 2005.

Volotsenko, I., Molotskii, M., Barkay, Z., Marczewski, J., Grabiec, P., Jaroszewicz, B., Meshulam, G., Grunbaum, E., and Rosenwaks, Y. (2010). Secondary electron doping contrast: Theory based on scanning electron microscope and Kelvin probe force microscopy measurements. *Journal of Applied Physics*, **107**(1), 014510.

Walker, C. G. H., El-Gomati, M. M., Assa'd, A. M. D., and Zadražil, M. (2008). The secondary electron emission yield for 24 solid elements excited by primary electrons in the range 250–5000 ev: A theory/experiment comparison. *Scanning* 30(5), 365–380.

Wang, H., Maiyalagan, T., and Wang, X. (2012). Review on Recent Progress in Nitrogen-Doped Graphene: Synthesis, Characterization, and Its Potential Applications. *ACS Catalysis*, *2*(5), 781–794.

Wells, O. C. and Bremer, C. G. (1968). Voltage measurement in the scanning electron microscope. *Journal of Physics E: Scientific Instruments*, *1*(9), 302.

Wells, O. C. and Bremer, C. G. (1969). Improved energy analyser for the scanning electron microscope. *Journal of Physics E: Scientific Instruments*, *2*(12), 437.

Werner, S. M. W., Ruocco, A., Offi, F., Iacobucci, S., Smekal, W., Winter, H., and Giovanni Stefani, G. (2008). Role of surface and bulk plasmon decay in secondary electron emission, Physical Review B **78**, 233403.

Willis, R. F. and Christensen, N. E. (1978). Secondary-electron-emission spectroscopy of tungsten: Angular dependence and phenomenology, *Physical Review B* volume 18, number 10, pp. 5140–5161.

Wirtz, T., De Castro, O., Audinot, J.-N., Philipp, P. (2019). Imaging and Analytics on the Helium Ion Microscope, *Annu. Rev. Anal. Chem.,* **12**:523–43.

Wuhrer, R. and Moran, K. (2016). Low voltage imaging and X-ray microanalysis in the SEM: challenges and opportunities IOP Conf. Ser.: Mater. Sci. Eng. 109 012019.

Xu, L., Weber, K., Phang, S. P., Fell, A., Brink, F., Yan, D., Yang, X., Franklin, E., and Chen, H. (2013). Secondary electron microscopy dopant contrast image (SEMDCI) for laser doping. *IEEE Journal of Photovoltaics*, *3*(2), 762–768.

Yong, Y. C. and Thong, J. T. L. (2000). Determination of Secondary Electron Spectra from Insulators, *SCANNING* VOL. 22, 161–166.

Zouros, T. J. M., and Benis, E. P. (2002). The hemispherical deflector analyser revisited. I. Motion in the ideal 1/r potential, generalized entry conditions, Kepler orbits and spectrometer basic equation. *Journal of Electron Spectroscopy and Related Phenomena*, *125*(3), 221–248.

Index

add-on lens attachment, 39, 40
angle resolved photon emission spectroscopy (ARPES), 11, 267, 268, 274
angle resolved secondary electron emission spectroscopy (ARSEES), 9, 10, 11, 267, 274
Auger electron spectroscopy (AES), 8
  spectrometer design, 80
Auger electrons, 3
Auger electrons in SEM, 154

Backscattered Electrons (BSE), 1, 2
  atomic number, 139
  detector, 3
  spectra, 154
  yield, 108, 134, 159

charge mapping
  metal-insulator-semiconductor (MIS), 245
  multifunctional oxide interfaces, 258, 263
  small-angle toroidal analyzer, 242, 243, 244
contamination, 3, 55, 91
  background signal, 95, 129
  density of states, 177
  detection & mitigation in SEM, 154
  dopant contrast, 66, 205
  Energy-Dispersive X-ray Spectroscopy (EDS), 125
  lateral dopant profiling, 208
  metal-insulator-semiconductor (MIS), 249
  secondary electron energy spectra, 102
  solar cell, 223
cylindrical mirror analyzer (CMA), 8, 16, 79

Density of States (DOS), 9, 10, 164
  Density Field Theory (DFT), 172, 267, 270
dopant contrast, 15, 17
  dopant profiling, 179
  doped stripe test sample, 182, 197
  gallium-irradiated silicon, 234
  high-pass energy filtering, 66
  low-pass energy filtering, 56, 57, 62
  native oxide, 202
  SE spectra, 75
  solar cell, 221

electron beam testers, 23, 25, 26, 27, 38, 111
electron energy loss spectroscopy (EELS), vii
energy analyzer designs
  add-on lens spectrometer, 41, 42, 44, 46
  cylindrical deflection analyzer (CDA), 16
  cylindrical mirror analyzer (CMA), 8, 16, 62, 73, 75, 77, 79
  hemispherical deflector analyzer (HDA), 9, 12, 62, 73, 287

hyperbolic field analyzer (HFA), 299
parallel radial mirror analyzer (PRMA), 82, 299
radial mirror analyzer (RMA), 82, 293
retarding field analyzer, 9, 12, 21, 29, 30, 31, 32, 53, 72, 73, 79
small angle toroidal analyzer, 75, 76, 78
spherical deflector analyzer (SDA), 46
wide-angle toroidal analyzer, 77, 78, 79, 82
analyzer design, 77, 82
experimental setup, 83
shot noise, 93, 94
signal-to-noise, 87, 91, 101, 123
Energy-Dispersive X-ray Spectroscopy (EDS), 125
elemental mapping, 148
gallium-irradiated silicon, 234
extraction field, 21, 25, 27, 39, 112

immersion magnetic lens. *See* Add-on lens attachment
interaction volume, 1, 125, 158, 159, 262, 268

low-voltage scanning electron microscopes (LVSEM), 46, 53, 268, 269
charge mapping, 256
in-lens SEM, 274
material contrast, 125, 126, 138, 151, 177
spectrometer design, 80

magnetic beam separator spectrometer, 50, 279
metal-insulator-semiconductor (MIS). *See* charge mapping
Monte Carlo simulation
SE spectra, 6, 9, 12, 71, 75
secondary electron yield, 160
specimen-beam interaction, 93, 159, 229, 250, 263

plasmon
surface, 11, 12, 13, 281, 282

reflection electron energy loss spectroscopy (REELS), 11, 281
retarding energy analyzer. *See* Energy analyzer design
retarding field objective lens, 49, 80, 278, 279
root-mean-square deviation (RMSD), 102, 103

scanning Auger microprobe, 8, 79
scanning Auger microprobe (SAM). *See* Auger electron spectroscopy
secondary electrons (SE), 1
analytical energy distribution, 5
collection efficiency, 23
ion-induced SE (iSE), 160, 161, 162, 163
S-curve, 21
SE1, 159, 161
SE2, 138, 159, 161, 250, 251
Spectrometer energy resolution
parallel radial mirror analyzer, 299
radial mirror analyzer, 296
small-angle toroidal analyzer, 75
wide-angle toroidal analyzer, 76, 77, 78, 87, 91

through-the-lens column, 4, 63, 188, 275
time-of-flight spectrometer, 35, 274

vacuum, 8, 103, 107

high vacuum (HV), 3, 79, 164, 266, 268
ultra-high vacuum (UHV), 3, 5, 6, 9, 11, 13, 62, 64, 75, 80, 104, 110, 164, 169, 177, 267
voltage contrast, 14, 15, 19
retarding field voltage contrast. *See* energy analyzer designs, retarding field analyzer
signal-to-noise, 31, 78
surface field effects, 111, 112, 120, 123
Wien filter analyzer, 40
work function
Chung-Everhart distribution, 32, 36, 37, 88, 89
metals, 102
working distance
electron beam testers, 38
SEM, 4
SEM spectrometer, 72, 78, 82
through-the-lens SEM, 68
voltage contrast spectrometers, 19, 23, 25

X-ray Photoelectron Spectroscopy (XPS), 91, 172, 267, 270